Study on Flow and Hang-up Characteristics of Caved Ore and Rock

Hao Sun · Lishan Zhao · He Wang · Xuan Qin

Study on Flow and Hang-up Characteristics of Caved Ore and Rock

Springer

Hao Sun
University of Science and Technology Beijing
Beijing, China

Lishan Zhao
University of Science and Technology Beijing
Beijing, China

He Wang
BGRIMM Technology Group
Beijing, China

Xuan Qin
China Academy of Safety Science and Technology
Beijing, China

ISBN 978-981-92-0334-5 ISBN 978-981-92-0335-2 (eBook)
https://doi.org/10.1007/978-981-92-0335-2

This Springer imprint is published by the registered company Springer Nature Singapore Pte Ltd.
The registered company address is: 152 Beach Road, #21-01/04 Gateway East, Singapore 189721, Singapore

Preface

Underground metal mining is fundamentally reliant on a critical component known as the draw. In caving operations, millions of tonnes of caved ore and rock, which exhibit significant variability in size, strength, and shape, must be safely, efficiently, and systematically released from the stope to the surface. As mining depths and stope footprints expand, this seemingly straightforward task evolves into a complex, multi-scale, and multi-physics process characterized by pronounced non-linearity. For instance, a sudden alteration in particle-size distribution can lead to increased local ore loss and dilution, while the collapse of a stress arch may result in a hazardous hang-up. Furthermore, the interaction between coarse particle packing and the percolation of fines can instigate spontaneous hang-ups. Traditional continuum theories, such as ellipsoid and stochastic-medium models, are inadequate for capturing the granular-scale phenomena of rotation, crushing, and collision that precipitate these occurrences; they also fall short in predicting the dynamic evolution of the Isolated Extraction Zone (IEZ) and the Isolated Movement Zone (IMZ). Consequently, discrete-element modelling (DEM), which meticulously tracks the movement and forces acting on each particle, has emerged as the predominant research tool, facilitating a paradigm shift from empirical draw control to quantitative draw management.

This monograph synthesizes a decade of experimental, theoretical, and numerical research conducted by our group on the coupled flow-hang-up behavior of caved ore and rock. Chapter 1 reviews the historical development of draw theories and quantifies the discrete characteristics of ore and rock particles—such as shape, size distribution, moisture content, friction, and cohesion—that govern their bulk response. Chapter 2 begins at the micro-scale, employing synchrotron 3-D imaging and digital-image correlation (DIC) to map crack propagation and energy dissipation in individual iron and copper ore particles subjected to varying strain rates. These data calibrate a Breakable Voronoi Block Model (BVBM) that accurately reproduces particle strength and fragment-size distributions, thus providing a reliable "particle genome" for larger-scale simulations. Chapter 3 transitions to the meso-scale, presenting direct shear tests on uniform and single-coarse-particle systems. DIC quantifies how a single coarse fragment non-linearly amplifies shear strength, dilation, and the migration trajectories of surrounding fines, offering a micro-mechanical explanation for the "coarse skeleton-fine percolation" segregation commonly observed near drawpoints. Chapter 4 introduces the upside-down drop shape draw theory and derives an analytical solution based on gravitational potential and frictional work. It systematically compares the movement zones of drop-shaped and ellipsoidal shapes in terms of volume error, dilution curves, and applicable depth-stress boundaries, thereby providing a novel theoretical framework for deep, high-stress, coarse-fragment cave mines. Chapter 5 integrates near-field physical experiments with far-field DEM simulations to elucidate the transition of 'rigid arches' near the drawpoint to 'flexible arches' in the remote zone. It clarifies the significant influences of drawpoint spacing, draw sequence, non-uniform fragmentation, and coarse-particle positioning on the coalescence or separation of multiple draw zones, offering quantitative guidelines for optimizing multi-point draw schedules and minimizing ore loss and dilution. Chapter 6 addresses the industry's most challenging issue—hang-ups—by constructing a three-dimensional hang-up phase diagram that encompasses particle-size ratio, internal friction, drawpoint diameter, and other critical parameters. A complex-network approach identifies 'vital hang-up particles' and proposes a dual-index criterion for hang-up precursors based on local force-chain intensity and global network connectivity.

The academic monograph is structured as a closed loop encompassing particle properties, crushing mechanisms, shear behavior, drop-shaped draw theory, flow characteristics, and hang-up mechanisms. The outcome presents a transferable experimental-numerical research paradigm for scientists, along with a set of immediately applicable tools for engineers tasked with optimizing draw and mitigating hang-up in operational mines. If this work can facilitate the transition of metal-mine draw control from an "experience-driven" approach to a "data-driven" methodology, thereby making each draw more controllable, safer, and more economical, then the decade of industry-university collaboration reflected in these pages will have been worthwhile.

The editorial responsibility was led by Hao Sun, the editor-in-chief from the School of Resources and Safety Engineering at the University of Science and Technology Beijing. He was supported by associate editors Lishan Zhao, also from the School of Resources and Safety Engineering at the University of Science and Technology Beijing; He Wang from the Institute of Mining Engineering at BGRIMM Technology Group; and Xuan Qin from the China Academy of Safety Science and Technology. Chapter 1 was authored by He Wang and Xuan Qin, while Chaps. 2 and 3 were contributed by Lishan Zhao. Hao Sun oversaw Chaps. 4, 5, and 6. The final collation and unification of the monograph were conducted by Hao Sun. Acknowledgment is due to the graduate students Kunlin Tang, Zesong Yin, Muya Li, Lichang Wei, Meichen Liu, Shenggui Zhou, and Xueqian Wang, as well as the undergraduates Hao Qin, Yuxin Zhu, and Jialu Li from the University of Science and Technology Beijing, for their extensive efforts in data collection, code writing, translation, figure drafting, and proofreading.

We also extend our sincere appreciation to the authors of both domestic and international publications consulted during the preparation of this monograph.

While we strive for accuracy, omissions and errors may occur; therefore, we warmly invite readers to provide constructive criticisms and corrections.

Beijing, China
February 2026

Hao Sun
Lishan Zhao
He Wang
Xuan Qin

Competing Interests The authors have no competing interests to declare that are relevant to the content of this manuscript.

Synopsis

This academic monograph addresses the central challenge of 'flow-hang-up' in draw-controlled caving of metal mines. It presents a cohesive experimental-to-numerical pipeline that encompasses single-particle crushing, particle system shear, gravity flow theory, and the hang-up of caved ore and rock. The work is organized into six thematic chapters: Introduction; Single-Particle Compression and Crushing; Shear Behavior of Ore-Rock System; Drop-Shaped Draw Theory; Flow Characteristics of Caved Ore and Rock; and Hang-Up Characteristics of Caved Ore and Rocks. Together, these chapters provide an integrated framework for modeling each stage of the draw process.

This monograph is suitable for senior undergraduates, postgraduates, and scholars in the fields of mining, geotechnical engineering, and granular mechanics. Additionally, it provides university lecturers, design institutes, and mine technical teams with practical tools, calibrated parameters, and field-ready data to optimize draw, control loss and dilution, and manage hang-ups.

Contents

Chapter 1
Introduction

1.1 Overview of the Draw Theories

Caving methods represent a category of mining techniques that manage ground pressure by facilitating the collapse of surrounding rock. As the stope face progresses, the surrounding rock is induced—either intentionally or naturally—to cave and fill the goaf, thereby containing the rock pressure. This method is primarily employed in the extraction of thick ore zones and steeply inclined medium-thick to thick ore zones. Unlike traditional mining methods, no separate ore rooms and pillars are retained; rather, the entire orebody is extracted in a singular, continuous sequence. The resulting system is characterized by its simplicity, high capacity, low cost, and ease of management, establishing it as one of the most efficient and profitable underground mining techniques. Based on the vertical unit that is caved, caving methods can be categorized into five fundamental types: single-slice caving method, layer-by-layer caving method, sublevel caving method, forced block caving method, and (natural) block caving method [1].

The block caving method is widely utilized in metal mines both domestically and internationally. Prior to 2000, the proportion of ore extracted through the block caving method in underground mining of black metal mines in China exceeded 85%, while in non-ferrous metal mines, it was approximately 40%. Internationally, mining operations employing this method accounted for around 25%. The defining characteristics of the block caving mining method include the direct contact of caved ore with the overlying waste rock, which is subsequently released from the drawpoint beneath the surrounding waste rock. Consequently, the loss and dilution rates of the ore are relatively high. If the draw process is inadequately managed or the design parameters of the stope structure are unreasonable, the draw results may deteriorate further, leading to waste of mineral resources and a decline in economic benefits for enterprises. With the exception of a few mines, the loss rate in block caving mines in China ranges from 20 to 30%, while the dilution rate varies from 25 to 35%. Compared to advanced international standards, these rates are generally 10–15%

H. Sun et al., *Study on Flow and Hang-up Characteristics of Caved Ore and Rock*,
https://doi.org/10.1007/978-981-92-0335-2_1

higher. The primary reasons for this discrepancy can be summarized as follows: (1) Inadequate planning and unreasonable structural parameters. (2) Suboptimal draw control and insufficient construction quality. While some degree of ore loss and dilution during mining operations is unavoidable, exceeding certain thresholds can significantly diminish the economic viability of the mine.

In the process of block caving mining, the determination of stope structure parameters is constrained by the physical and mechanical properties of the rock mass, structural planes, and their associated fracture mechanisms and processes. The draw parameters are influenced by the flow characteristics of the caved ore and rock. Implementing appropriate structural parameters and draw control measures is essential for minimizing loss and dilution, making it a critical and highly technical task for mining operations. Given that the caved ore and rock behave as an imperfect granular medium, their key characteristics include bulking property and fluidity. Consequently, the flow characteristics of the caved ore and rock serve as a fundamental basis for selecting draw control methods, calculating draw indicators, and ultimately achieving the objective of minimizing loss and dilution.

Currently, the primary research methods addressing the flow characteristics of caved ore and rock, particularly in relation to draw, can be categorized into three types: theoretical analysis, experimental analysis, and numerical simulation. The limitations inherent in theoretical frameworks, the complexity of various orebody occurrence boundary conditions, and the inadequate applicability of existing software have hindered both traditional continuous medium draw theory and contemporary computer simulation draw technologies from fulfilling the demands of modern mining practices. Specifically, while existing draw theories are relatively comprehensive in their exploration of flow behavior under isolated draw conditions, significant advancements are still required in the analysis of multi-drawpoint scenarios and conditions involving complex boundary interactions. Furthermore, research in draw theory has yet to progress to a stage where it can effectively predict ore loss and dilution. In an effort to mitigate mining and cutting workloads, numerous mining operations have initiated experimental studies focusing on large-spacing structural parameter mining. Consequently, the development of new draw theories and the resolution of coordination issues related to multi-sublevel stope advancement require additional experimental investigation.

The research on the draw issue in the caving method has a long history. Established draw theories include the ellipsoid draw theory, the stochastic medium draw theory, and several other related theories.

1.1.1 Ellipsoid Draw Theory

During the isolated draw operation, the spatial configuration formed by the caved ore and rock in the initial ore packing of the stope, prior to the commencement of the draw operation, is referred to as the Isolated Extraction Zone (IEZ). The configuration formed by all the moving caved ore and rock is defined as the Isolated

Movement Zone (IMZ). Results from draw tests indicate that the IEZ approximates an ellipsoid; thus, the ellipsoid equation serves as the expression for the IEZ. Based on the fundamental properties of the IEZ, a series of equations that reflect the draw law have been derived to elucidate the gravitational flow behavior of the caved ore and rock, commonly referred to as the 'ellipsoid draw theory' [2]. This theory is established on a substantial number of physical experiments, abstractly assuming that both the IEZ and IMZ are ellipsoidal in shape, and it remains the most classical draw theory to date. The number of scholars engaged in research on this theory is significant, both domestically and internationally, and its application is also the most extensive.

Research on the draw issue can be traced back to 1938, when the former Soviet scholar Minayev proposed that the IEZ was a rotating ellipsoid; however, no comprehensive theory was developed at that time. In 1952, Soviet scholar, Malakhov, published the work titled 'Draw of Block Caving Method' [3], which established the ellipsoid draw theory. Malakhov posited that both the IEZ and isochronous zones can be modeled as ellipsoids, indicating a transitional relationship between the two. Building upon this transitional relationship of the isochronous zone, he formulated equations for velocity and movement trajectories, proposed a draw management system for the block caving method, and developed a methodology for determining reasonable structural parameters for the bottom-pillar caving method.

In 1965, Kurikov published 'Joint Mining and Secondary Mining of Metal Deposits', in which he proposed the draw cone equation and asserted that the surface particles of the IEZ remain relatively unchanged during the movement of the IEZ. This work significantly advanced the development of draw theory, particularly the isolated draw theory. Kurikov's draw theory is regarded as a relatively mature and comprehensive ellipsoid theory, successfully passing all tests except for the shape test.

With the widespread application of the caving mining method in China, scholars from various research institutions, including the Northeast Institute of Technology, the Maanshan Institute of Mining Research, the Beijing Institute of Iron and Steel Technology, and the Central South Institute of Mining and Metallurgy, have gradually begun to conduct theoretical research on draw mechanisms. Most of this research is based on Malakhov's ellipsoidal draw theory. Notably, Professor Liu from the Northeast Institute of Technology has systematically summarized and refined Malakhov's ellipsoidal draw theory, addressing several apparent contradictions within the theoretical framework and enhancing its systematic coherence [4, 5]. Liu posited that maintaining a constant eccentricity can resolve the contradiction between the moving trace line and the ellipsoid surface transition theory. Furthermore, he proposed the isoeccentric draw theory [6] based on the transition relationship of the IEZ, establishing corresponding mathematical equations. Simultaneously, he utilized coordinate transformation equations to address the challenges associated with multi-drawpoint scenarios [7], achieving pioneering results in multi-drawpoint condition, quantitative calculations, and research on draw boundary conditions. However, experimental findings indicate that the eccentricity of the ellipsoid varies with the height of draw, leading to discrepancies between the calculation results of the isoeccentric draw theory and the experimental results.

Professor Li from the Xi'an Institute of Metallurgical Architecture [8] derived the eccentricity equation $1 - \varepsilon^2 = KH^{-n}$ through experimental research. Based on this equation, he established a unified mathematical framework for the fundamental principles of draw, thus formulating the draw theory of variable eccentricity truncated ellipsoids. This advancement significantly reduced the discrepancy between theoretical calculations and experimental results, thereby enhancing the ellipsoid draw theory.

At present, the generally accepted viewpoints in the ellipsoid draw theory are as follows:

(1) The shapes of IEZ and IMZ are both approximately spherical ellipsoids;
(2) During the draw process, the points with equal vertical velocities within IMZ are connected to form an isochronous zone;
(3) The surface particles of IEZ can simultaneously reach the drawpoint, and there is a transitional relationship during the movement;
(4) Within the movement zone, each horizontal layer subsides in a funnel shape, forming a draw cone; once this cone has extended down to the drawpoint, it is referred to as the extracted cone (Fig. 1.1).

Due to the practicality of the ellipsoid draw theory, it has maintained a dominant position in the research of the block caving method. Its application scope has expanded from metal mines to coal mines [9]. However, as a relatively traditional

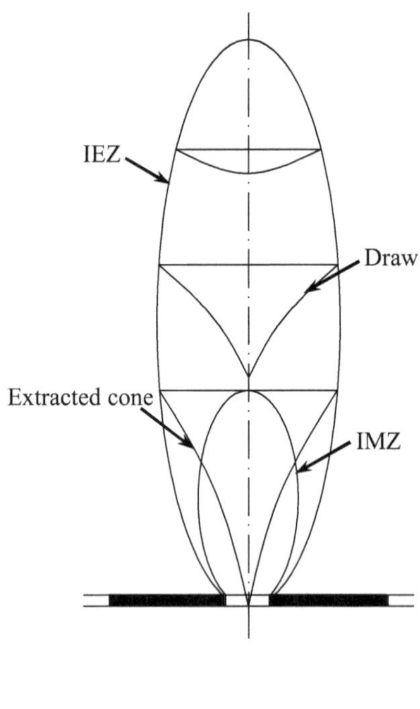

Fig. 1.1 IEZ, IMZ, draw cone and extracted cone schematic diagram

theory, the ellipsoid draw theory exhibits notable deficiencies, primarily manifested as [10]:

(1) The shape of the IEZ is central to the draw theory. Numerous experimental studies have demonstrated that it is not a standard ellipsoid but rather an approximate ellipsoid, with its shape varying according to the properties of the granular material and the draw conditions. Consequently, the discrepancy between the shape of the IEZ and the actual conditions undermines the foundation of the ellipsoidal draw theory.
(2) The ellipsoid draw theory presumes that the density field within the IMZ is both uniform and steady. However, the abrupt changes in the density of the granular material on the moving belt and at the boundary do not reflect the actual conditions.
(3) The velocity field described by the ellipsoid draw theory represents the velocity field of an ideal granular material, which leads to confusion regarding the distinction between ideal and actual granular materials. A significant issue arises at the boundary of the IMZ, where dual velocities are present; the velocity equation provides a velocity at each point, yet the velocity at each point on the moving boundary should be zero.
(4) The velocity formula derived from the ellipsoid draw theory, whether applied to ideal or real granular materials, fails to define the moving boundary. Consequently, a velocity equation that does not incorporate a moving boundary cannot be considered an accurate representation of the gravity flow law governing granular materials.

Despite these shortcomings, the ellipsoid draw theory remains valuable due to its simplicity and practicality, and it continues to play a significant role in the study of draw theory and its application in mining production.

1.1.2 Stochastic Medium Draw Theory

By simplifying granular materials into a continuously flowing stochastic medium and employing probability theory to examine the flow laws of these materials, a theoretical framework known as the stochastic medium draw theory has been established [11]. The application of probability theory to study the flow processes of granular materials dates back to the 1960s. Professor Litwiniszyn, a noted expert in surface movement from Poland [12], posited that the flow of granular materials is a stochastic process that can be analyzed using probabilistic methods. He developed a stochastic medium model, illustrated in Fig. 1.2, wherein each of the 3 square boxes A, B, and C contains a spherical ball (particle) influenced by gravity. When a particle in box A or box B shifts from its original position due to gravitational forces, particles in box C will random occupy that vacated position. Litwiniszyn did not investigate the draw process from the established model; instead, he derived inferences solely from the phenomena described within the model—specifically, the stochasticity of gravity

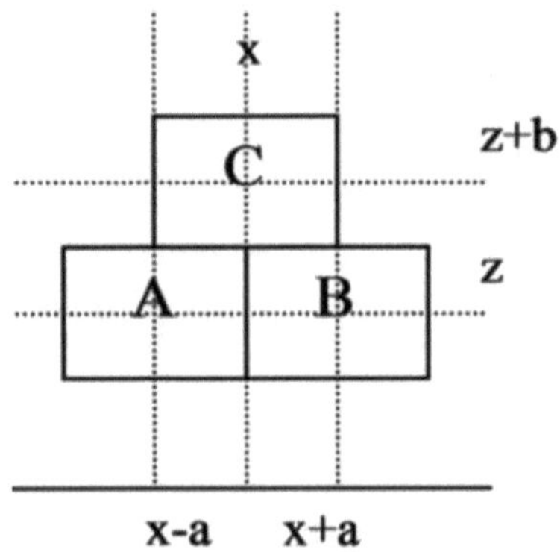

Fig. 1.2 Litwiniszyn's stochastic medium model

flow. That is, when a certain number of particles are extracted at the z level, particles at the original z + b level will descend with a specific probability. Litwiniszyn conceptualized granular materials as a continuously flowing stochastic medium and formulated the differential equation for the depth function *W* of the draw cone.

Wang [13] proposed a ball replacement model for the flow of granular materials (Fig. 1.3). This model is based on the equal probability of two adjacent balls filling the lower empty spaces, which led to the establishment of a probability field governing the flow of balls. Utilizing the central limit theorem, the author treated the ball medium continuously and introduced a statistical constant B for the granular material, which facilitated the formulation of the probability density equation for the flow of granular materials. By examining the relationship between the probability field and the vertical descent velocity field of the granular material, the flow speed of the granular material, the trace line equation, the extracted cone equation, and the IEZ equation were derived. This work established the stochastic medium draw theory, creating a relatively complete calculation system. The stochastic medium draw theory provides deeper insights into the essence of various phenomena related to the draw issue.

Professor Ren conducted a further investigation into the probability distribution of granular movement (Fig. 1.4). Through the analysis of experimental data, he derived the expression for variance. Additionally, based on both mathematical derivation and empirical analysis, he established the probability density equation for granular flow [14].

Ren adopted the granular material movement model, as illustrated in Fig. 1.4, employing a method that combines theoretical analysis with draw tests. To analyze the movement process of granular materials under various boundary conditions, a systematic theoretical equation was established. Two parameters, α and β, which reflect the characteristics of gravity flow and draw conditions, were introduced. This effectively addressed the issue of shape distortion described by the equation and enhanced the calculation accuracy.

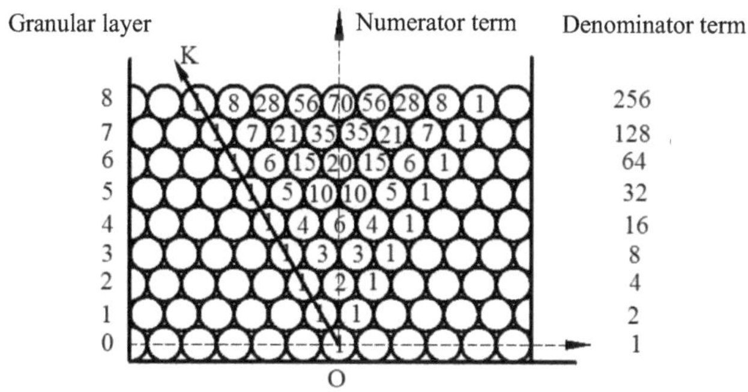

Fig. 1.3 Wang's stochastic medium model

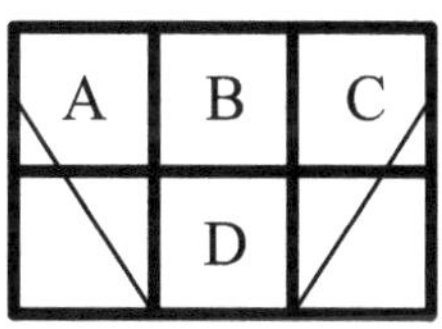

(a) Granular movement zone

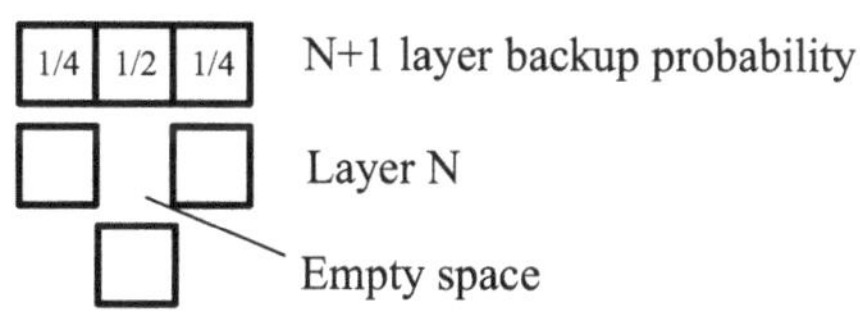

(b) Granular movement conceptual model

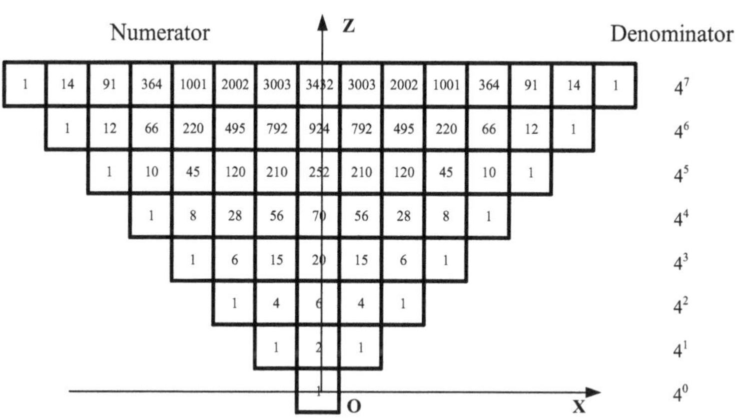

(c) Granular movement probability

Fig. 1.4 Ren's granular movement model

The existing draw theory primarily considers the drawpoint as a point source, often neglecting the impact of drawpoint size on the draw process. Tao [15] conducted an in-depth study on the influence of drawpoint size on the flow dynamics of granular materials and the stress field associated with granular material flow. He proposed a stochastic medium model for end draw, which he validated through empirical tests. Additionally, he introduced a flow function based on the percolation model, enabling the derivation of the velocity and stress fields for two-dimensional granular flow. Building on this foundation, Qiao [16] explored the relationship between granular flow and the distribution of flow probability, proposing similarity and unified conditions for investigating the flow dynamics of granular materials through the lens of stochastic medium theory. These research contributions have significantly advanced the stochastic medium draw theory and established a solid groundwork for further exploration of the flow dynamics of actual caved ore and rock utilizing this theoretical framework.

1.1.3 Other Draw Theories

In addition to the aforementioned classic draw theories, scholars both domestically and internationally have proposed several alternative theories, including the expected zone draw theory and the ellipsoid-like draw theory. Each of these theories presents distinct advantages and disadvantages, garnering varying levels of recognition and engineering application. This diversity contributes to the advancement and refinement of draw theories and technologies, while also enhancing our understanding of the gravity flow mechanisms associated with caved ore and rock.

1.1.3.1 Expected Zone Theory

The zeroing volume is a significant concept introduced in the study of draw, which has transformed the traditional experimental methods used to determine the shape of the IEZ [17]. Wei [18] processed and analyzed the experimental data, subsequently employing the zeroing volume curve to ascertain the IEZ, thereby introducing the notion of the expected zone. Building on this foundation, Gao [19] derived the spatial representation of the expected zone and elucidated the relationship between the total mass drawn and the spatial position of any point within the IEZ.

Starting from the expected zone, Yongtao Gao proposed and proved the principle of equal thickness transition of the IEZ [20]. He analyzed the fundamental laws governing granular materials, derived the draw cone equation, the extracted cone equation, the velocity equation, and the expression for the isochronous zone [21]. Utilizing the derived expression for the expected zone, he formulated the calculation for ore loss and dilution under isolated draw conditions [22]. Furthermore, he investigated several fundamental phenomena in the computer simulation of the draw process, discussed similarity issues based on the conclusions regarding the expected zone, and proposed a novel similarity method [23]. Through the research contributions of scholars such as Shanli Wei, Zhihong Su, and Yongtao Gao, a relatively complete theory of the expected zone has been established. Compared to the ellipsoid, the expected zone more accurately reflects the actual IEZ in terms of shape, size, and the trend of shape changes. For mines employing the block caving method, the draw parameters determined using the expected zone theory align more closely with real-world conditions, suggesting that the expected zone theory represents an advancement and continuation of the ellipsoid draw theory.

1.1.3.2 The Ellipsoid-Like Draw Theory

Li proposed the ellipsoid-like draw theory based on draw tests [24, 25], establishing both the ideal and actual equations of the theory, which differentiate between ideal and actual granular materials. He also formulated mathematical equations to describe the laws of gravity flow, demonstrating that the equation for ideal granular materials is a special case of the equation for actual granular materials. Furthermore, he introduced the IEZ surface equation, the movement trace equation, and the movement transition equation, among others.

The ellipsoid-like draw theory has significantly advanced the ellipsoid draw theory in terms of both depth and breadth. Specifically: (1) The IEZ surface equation is adaptable to variations in the shape of the IEZ; (2) The velocity field aligns with the fundamental assumptions of the ellipsoid-like theory; (3) It has initiated valuable discussions regarding the moving density field of granular materials. Furthermore, Rongfu Li and Shenhe Zhang proposed five test criteria for the draw theory: the continuous flow test, moving boundary test, IEZ shape test, granular material field test, and mass flux test. Through these tests, it was demonstrated that both the ellipsoid draw theory and the stochastic medium draw theory can only pass a subset of the tests, whereas the ellipsoid-like draw theory successfully passes all tests. Consequently, the ellipsoid-like draw theory is deemed superior to the aforementioned two types of draw theories [26].

However, the ellipsoid-like draw theory has not provided a scientific explanation for the formation of 'ellipsoid-like zones'. Furthermore, both the expected zone theory and the ellipsoid-like draw theory have primarily focused on the flow laws of isolated draw under infinite boundary conditions, neglecting a thorough examination of draw theories under semi-infinite and other complex boundary conditions. In practice, the boundaries of stopes are often intricate due to the constraints imposed by mining technology, making these boundary conditions significant challenges in the study of ore discharge theories. Therefore, the expected zone theory and the ellipsoid-like draw theory require further research and refinement to address the various boundary conditions encountered in actual mining operations.

1.1.3.3 The Draw Theory Based on Kinematic Model

The kinematic model was initially employed to predict the steady-state flow behavior of granular materials in silos or bins. Subsequently, it was utilized to analyze the non-steady flow patterns resulting from the expansion of granular materials [27]. In 1995, Nedderman [28] derived an equation capable of predicting the movement range of granular materials and accurately forecasted the velocity field distribution based on the experimental findings of Watson [29]. In 2003, Ferjani [30] conducted a comprehensive study on the movement range of granular materials in silos, utilizing an enhanced kinematic model.

In the field of draw theory, Melo, Vivanco, and other scholars from the University of Santiago, Chile, established a comprehensive theoretical framework based on a kinematic model. This model is derived from the Bergmark-Roos equation [31], and its fundamental assumptions include the following [32]: (1) The caved ore and rock are considered as a continuous medium, with the movement of granular materials from their initial position to the discharge point being a continuous process; (2) All granular materials are released from the coordinate origin, with the drawpoint acting as a point-source drawpoint, allowing for the simultaneous discharge of all particles that arrive at the same time; (3) The movement of granular materials is a stochastic process; (4) The vertical descent velocity of granular materials is directly proportional to their probability of downward movement; (5) Caved ore and rock exhibit homogeneity and isotropy in all directions. Building on research related to isolated draw, Melo, Vivanco, and their colleagues derived and validated the IMZ equations under various discharge conditions, including simultaneous and sequential draw. They also investigated the impact of different draw modes and drawpoint spacing on the shape, loss, and dilution of ore [33].

Although the kinematic model cannot fully represent the actual drawing conditions and states, it still provides valuable insights for guiding mine production. It effectively reproduces the results of physical draw experiments and clearly illustrates the shape change laws of the IEZ and the IMZ through analytical formulas. This is particularly evident in the flow characteristics of caved ore and rock under multi-drawpoint conditions. Therefore, the kinematic model serves as an effective tool for exploring the flow laws of caved ore and rock under varying draw conditions, thereby enriching and advancing draw theory.

1.2 Overview of the Physical and Mechanical Properties of Caved Ore and Rock

1.2.1 The Basic Concepts, Classification and Characteristics of Caved Ore and Rock

Granular material is a collective composed of mutually connected solid particles. Each individual solid particle in the granular medium is significantly smaller than the entire collective under study. The solid particles that make up the granular medium are referred to as granular particles or granular blocks. The accumulation of numerous granular particles forms what is known as a granular system. The packing system of caved ore and rock is classified as a granular system. In this monograph, the term 'granular medium' primarily refers to caved ore and rock.

Granular media are typically classified into two categories: ideal granular media and non-ideal granular media. Ideal granular media are defined as those that do not contain water or adhesive substances and exhibit no cohesion between the particles. In contrast, non-ideal granular media contain water and adhesive substances, resulting in some degree of cohesion among the particles. Caved ore and rock generally exhibit certain cohesion, categorizing them as non-ideal granular media. However, the cohesive strength between caved ore and rock is significantly lower than their intrinsic strength, often rendering it very weak. Consequently, these materials can easily undergo displacement and exhibit a tendency to flow.

Each particle within granular media exhibits solid characteristics, contributing to the formation of the granular media's skeletal structure. However, when considering the entire flow zone, these particles also display flow properties, thus exhibiting some fluid characteristics. This dual nature of granular media positions them uniquely, as they do not exclusively belong to either solids or fluids, but rather represent a medium possessing both solid and fluid properties; some scholars have referred to this particle system as a 'complex fluid'. This dual characteristic significantly influences the mechanical behavior during the gravity flow of caved ore and rock.

A scientific understanding and mastery of the physical and mechanical properties of caved ore and rock is essential for studying draw. The following sections will introduce the physical and mechanical properties of caved ore and rock that are relevant to draw research.

1.2.2 The Bulk Density of Caved Ore and Rock

The weight of a unit volume of granular material is called the bulk density of the granular material, and it can be calculated using Eq. (1.1):

$$\gamma_{\mathrm{k}} = \frac{W_{\mathrm{s}}}{Q_{\mathrm{s}}} \tag{1.1}$$

Here, γ_k is the bulk density of caved ore and rock, W_s is the mass of caved ore and rock, and Q_s is the volume of caved ore and rock.

The bulk density of caved ore and rock is a crucial physical parameter in the draw process. Accurate measurement of the bulk density of the particle system is essential for conducting draw calculations and research. This parameter can be categorized into various cases based on different conditions and measurement methods.

1.2.2.1 The Measurement of Bulk Density of Caved Ore and Rock

The steps for measuring the bulk density of caved ore and rock are as follows:

First, select an appropriately sized graduated cylinder as the measuring container based on the particle size of the caved ore and rock, as illustrated in Table 1.1.

Secondly, to correct the volume of the graduated cylinder, begin by weighing the empty graduated cylinder. Next, carefully pour the distilled water, which has been pre-measured for its specific gravity and temperature, into the graduated cylinder until it reaches the meniscus. Afterward, weigh the graduated cylinder containing the water. The volume of the graduated cylinder can be determined by taking the difference between the two weights and dividing it by the specific gravity of water at the measured temperature.

$$Q_{ch} = \frac{W_{ts} - W_{ch}}{\gamma_w}, \quad \text{L} \tag{1.2}$$

Here, Q_{ch} is the volume of the graduated cylinder/L, W_{ts} is the weight of the graduated cylinder with water/kg, W_{ch} is the weight of the graduated cylinder/kg, γ_w is the specific gravity of water at the measured temperature.

Then, the bulk density of caved ore and rock can be accurately determined using the following measurement method: First, the caved ore and rock are dried at a temperature of 105–110 °C until a constant weight is achieved. Subsequently, they are cooled to room temperature, and the particles are placed into a graduated cylinder. When utilizing a small graduated cylinder, the particles should be poured slowly and evenly through a sampling funnel. In contrast, when using a large graduated cylinder,

Table 1.1 Relationships between particle sizes and selected measuring cylinder specifications

Maximum particle diameter/mm	Capacity of the graduated cylinder/L	Specifications of the graduated cylinder/mm		Note
		Diameter	Height	
< 5	0.5	86.1	86.1	All are cylinders
5–10	5	185.5	185.5	
10–40	10	234	234	
40–80	20	294	294	
> 80	50	400	400	

it is essential to fill a small shovel and maintain a height of approximately 50 mm between the filling surface and the top of the cylinder while pouring evenly to prevent uneven distribution of the particles. After filling, the contents must be leveled off, and the weight should be measured again.

To determine the compacted bulk density of caved ore and rock, the measurement method differs slightly from the previously described technique. During the measurement process, when the particles are filled to one-third of the height of the graduated cylinder, they should be flattened, tamped, or pressed 25 times. When filled to two-thirds of the height, the particles must be tamped or pressed 25 additional times. Finally, when the cylinder is filled to its full height, tamping or pressing should occur 25 more times, followed by leveling and weighing. It is important to note that this method is only applicable to caved ore and rock with a maximum particle size of ≤ 40 mm.

The bulk density of caved ore and rock is calculated according to Eq. (1.3):

$$\gamma_{\mathrm{k}} = \frac{W_{\mathrm{ty}} - W_{\mathrm{ch}}}{Q_{\mathrm{ch}}},\ \mathrm{kg/L} \tag{1.3}$$

Here, W_{ty} is the weight of the graduated cylinder containing caved ore and rock.

Three parallel determinations should be conducted under identical operating conditions. The permissible error for the two parallel determinations is set at 20 kg/ m^3. For each determination, a new set of particles must be selected, and the arithmetic mean of the three determinations will be considered the final result.

1.2.2.2 The Filling Density of Caved Ore and Rock in the Model

The filling density of caved ore and rock is calculated by dividing the volume occupied by the loose or compacted particles in the model by the total weight of the sample. This relationship is expressed mathematically in Eq. (1.4):

$$\gamma_{\mathrm{m}} = \frac{W_{\mathrm{m}}}{Q_{\mathrm{m}}},\ \mathrm{kg/L} \tag{1.4}$$

Here, γ_{m} represents the filling density of caved ore and rock in the model, W_{m} represents the weight of caved ore and rock placed in the model, and Q_{m} represents the volume occupied by caved ore and rock in the model.

1.2.2.3 The Loading Density of Caved Ore and Rock

A standard mine car was employed as the measuring container at the site. Initially, the volume of the mine car was accurately calibrated. Subsequently, representative samples of caved ore and rock were evenly poured into the mine car until it was full and leveled. The weight of the caved ore and rock was then measured. The loading density of caved ore and rock was calculated using Eq. (1.5):

$$\gamma_{\mathrm{r}} = \frac{W_{\mathrm{r}}}{Q_{\mathrm{r}}}, \quad \text{ton/m}^3 \tag{1.5}$$

Here, γ_r represents the loading density of caved ore and rock, W_r is the weight of caved ore and rock loaded into the mine car, and Q_r is the loading volume of the mine car.

The loading density of caved ore and rock is usually used for on-site draw tests and for calculating the loading weight of transportation containers.

The bulk density of caved ore and rock serves as an indicator of the bulking property and compaction degree within the particle packing system. An increase in the bulk density of caved ore and rock correlates with a decrease in bulking property and an increase in compaction degree, resulting in a reduction in the size of the IEZ. Consequently, bulk density is a significant factor influencing the volume change of the IEZ. It is important to note that density is often associated with both bulking property and compaction degree, as elaborated in the following sections.

1.2.3 *The Bulking Property of Caved Ore and Rock*

The property of a rock mass that, after being fragmented by blasting, exhibits an increase in volume compared to its original state is referred to as the bulking property of caved ore and rock. This bulking property is typically quantified by the ratio of the volume of loose caved ore and rock to the volume of the intact rock mass, known as the bulking coefficient. The bulking coefficient can be calculated using Eq. (1.6):

$$K_{\mathrm{s}} = \frac{Q_{\mathrm{s}}}{Q_{\mathrm{p}}} \tag{1.6}$$

Here, K_s represents the bulking coefficient, and Q_s represents the volume of loose caved ore and rock, while Q_p represents the volume of the intact rock mass.

There exists a relationship as demonstrate in Eq. (1.7) between the bulking coefficient, the bulk density of caved ore and rock, and the solid density of the rock mass itself:

$$K_{\mathrm{s}} = \frac{\gamma_{\mathrm{z}}}{\gamma_{\mathrm{k}}} \tag{1.7}$$

Here, γ_z represents the solid density the entire rock mass, and γ_k represents the bulk density of caved ore and rock.

Depending on the external conditions at the time of measurement and the intended use, the bulking factor is subdivided into the loading bulking factor and the primary bulking factor.

1.2.3.1 Loading Bulking Coefficient

The value typically cited in geological reports is the loading bulking factor, which is determined after the rock mass has been blasted and allowed to loosen freely without any external confinement. In mine design, this factor is utilized to calculate the loaded weight for haulage and hoisting, as well as to quantify the looseness of the blast in open-pit operations.

1.2.3.2 Primary Bulking Factor

In the context of draw practice, the overall bulking that occurs when the in-situ rock mass is fragmented by the initial blast is referred to as primary bulking. The ratio of the volume of this newly loosened caved ore and rock to the original intact rock mass volume is known as the primary bulking factor, or primary bulking coefficient.

Although underground stope blasting generates a significant dynamic load, the confinement of the compensation space restricts the broken rock mass from loosening freely. Consequently, the primary bulking factor is typically smaller than the loading bulking factor. Typical values for the primary bulking factor under various formation conditions are as follows: deep-hole blasting with free compensation space ranges from 1.25 to 1.32, squeezed deep-hole blasting (vertical holes) ranges from 1.10 to 1.20, and squeezed chamber blasting ranges from 1.12 to 1.14.

1.2.4 The Void Rate and Compaction Degree of Caved Ore and Rock

1.2.4.1 The Void Rate

The void rate of caved ore and rock refers to the ratio of the volume of voids among the particles in a loose state to the total volume of the granular medium. It is calculated as follows:

$$n = \frac{Q_s - Q_z}{Q_s} \times 100, \quad \% \tag{1.8}$$

Here, n represents the void rate of caved ore and rock, and Q_z represents the volume of the solid particles within a granular medium.

The void rate of caved ore and rock can also be expressed in terms of the void ratio. The void ratio is defined as the ratio of the volume of voids to the volume of solid particles within a granular medium in a loose state. It is calculated as follows:

$$K_{kx} = \frac{Q_s - Q_z}{Q_z} \times 100, \quad \% \tag{1.9}$$

Here, K_{kx} represents the void ratio of caved ore and rock.

The relationship between the void rate and the void ratio of the granular medium is as follows:

$$n = \frac{K_{kz}}{1 + K_{kz}} \quad (1.10)$$

$$K_{kx} = \frac{n}{1 - n} \quad (1.11)$$

One important characteristic of the particle system of caved ore and rock is the presence of numerous voids among the particles. The void rate of this system varies based on the structural characteristics of the particle arrangement. Generally, the void rate is higher in coarse particle systems compared to fine particle systems. Additionally, the void rate is greater in particle systems composed of angular and irregular shapes, while it is lower in those made of smooth and regular shapes. When the compaction degree of caved ore and rock is high, the gravitational force exerted by the overlying rock is strong, and the retention time is prolonged, the void rate tends to be smaller; conversely, a lower compaction degree results in a larger void rate.

The method for measuring the void rate of the particle system of caved ore and rock is as follows:

Using a glass graduated cylinder, slowly pour water into the cylinder containing the particle sample until it reaches full capacity. Record the volume of water poured and subsequently calculate the relevant values using Eq. (1.12):

$$n = \frac{Q_{xf}}{Q_s} \times 100, \quad \% \quad (1.12)$$

Here, Q_{xf} represents the volume of water, and Q_s refers to the total volume of the particle sample in the granular state.

1.2.4.2 The Compaction Degree of Caved Ore and Rock

In the investigation of the particle packing system of caved ore and rock, relying solely on void rate and void ratio to characterize structural density is inadequate. Although the void ratio may remain constant, the varying shapes and gradation of the caved ore and rock result in differing properties. Therefore, it is essential to assess the compaction degree of the particle system, referred to as relative structural density, to gain insights into the loose and compacted states, as well as the structural stability of this particle system, both in its natural condition and post-compaction.

The compaction degree of the particle system of caved ore and rock refers to the extent to which the particles are compacted under the influence of external forces. It is defined as the ratio of the volume of the particle system of caved ore and rock after compaction to the total volume in its original loose state, as expressed by Eq. (1.13):

$$K_{y^s} = \frac{Q_{y^z}}{Q_s} \tag{1.13}$$

Here, K_y represents the compaction degree of caved ore and rock, and Q_y represents the volume of the compacted caved ore and rock.

Or it can be expressed in terms of the void ratio of the particle system of caved ore and rock:

$$K_{y^s} = \frac{K_{\text{kxmax}} - K_{\text{kx}}}{K_{\text{kxmax}} - K_{\text{kzmin}}} \tag{1.14}$$

Here, K_{kxmax} represents the void ratio of particle system in its fully loose state, K_{kzmin} represents the void ratio of particle system in its fully compacted state, and K_{kx} represents the void ratio of particle system in its natural state or a certain compacted state.

The compaction degree of the particle system of caved ore and rock is related to its bulking coefficient as follows:

$$K_{y^s} = \frac{K_y}{K_s} \tag{1.15}$$

Here, K_y represents the bulking coefficient of the compacted particle system of caved ore and rock, while K_s represents the original bulking coefficient of the particle system of caved ore and rock.

The compaction degree of the particle system of caved ore and rock is influenced by several factors, including the magnitude of dynamic and static loads, storage time, humidity, and particle size distribution. When the blasting seismic effect is pronounced and frequent, coupled with significant gravitational forces from overlying rock, increased humidity, and prolonged storage time, these conditions lead to a rearrangement of the caved ore and rock. Consequently, the structural integrity is altered, voids within the system are reduced, and the degree of compaction increases. This heightened compaction degree results in an elevated internal friction coefficient and cohesion force, alongside a reduction in the size of the IEZ, which deteriorates draw indicators and complicates the drawing process. Excessive compaction can cause the caved ore and rock to lose their loose characteristics. Nevertheless, provided that the adhesion is not excessively strong, the loose nature of the caved ore and rock can gradually be restored during the draw process.

1.2.5 The Humidity of Caved Ore and Rock

The moisture content of granular media is defined as the percentage of water contained within a specific quantity of granular material. In the domain of mining engineering, it is typically expressed as the ratio of the weight of water within the particle system of caved ore and rock to the weight of the dry caved ore and rock.

$$M = \frac{W_s - W_g}{W_g} \times 100, \quad \% \tag{1.16}$$

Here, M represents the moisture content of the particle system, W_s represents the weight of the particle system in its natural moisture state, and W_g is the weight of the particle system in its dry state.

Moist granular media refers to a state in which water exists solely as a film on the surface of granular materials. In contrast, wet granular media is characterized by water occupying the voids between particles.

Granular media devoid of surface water is termed air-dried granular media, whereas materials that contain only chemically bound water are classified as dry granular media.

The humidity of caved ore and rock is a crucial physical parameter that influences the draw conditions. When the particle system of caved ore and rock has a humidity level of less than 4–7%, it exhibits favorable loose and flow properties, resulting in optimal draw conditions and an effective draw outcome. Conversely, fine powder ore containing clay tends to self-compact during the draw process, leading to the formation of cavities and tubular IEZ, which complicates the draw process and negatively impacts the draw effect.

The humidity of caved ore and rock is influenced not only by the physical properties of the particles but also by the hydrogeological conditions of the mine and the amount of rainfall. Particularly when the collapse zone within the mining area reaches the surface, the humidity of caved ore and rock tends to vary with the seasons. During the rainy season, the humidity can exceed 8–20%, whereas in the dry season, it typically ranges from 2 to 8%.

An increase in the humidity of caved ore and rock enhances their adhesiveness, provided that the humidity remains within a certain range. However, once the humidity exceeds this threshold and the caved ore and rock reach a saturated state, the internal friction coefficient and cohesion between them decrease, leading to increased fluidity. In cases where the surface loess layer is thick and the collapse zone has reached the surface, the rainy season may trigger an 'underground debris flow', which poses a significant threat to production and safety, as well as increases the loss and dilution of the ore. Therefore, for mines facing similar risks, it is crucial to implement surface flood prevention measures.

The method for measuring the humidity of caved ore and rock is as follows: First, take the particle sample sealed in a wax-sealed container, remove it, and spread it flat. Next, divide the sample into four parts using the quartering method and weigh each portion. Subsequently, place the weighed samples into an oven and dry them at a temperature of 105–110 °C until the weight stabilizes. Finally, measure the weight of the dried samples. The weight humidity is calculated according to Eq. (1.16). The average value of the humidity results obtained from three or more parallel determinations is considered the humidity of caved ore and rock.

1.2.6 *The Size of Caved Ore and Rock*

The size of the granular medium pertains to the dimensions of caved ore and rock, as well as the percentage of each size category present.

In the current caving conditions within underground mining environments, the caved ore and rock consist of particles of varying sizes. In practical applications, the dimensions of these caved materials must adhere to specific requirements dictated by the product and the mining process itself; thus, both upper and lower size limits are essential. For instance, a maximum allowable dimension should be defined for larger particles; any particle exceeding this designated size is classified as an unqualified coarse particle. The proportion of coarse particles within the caved ore and rock system is referred to as the coarse particle output rate. The process of reducing coarse particles into smaller fragments is known as secondary fragmentation. Given that the geometric characteristics, gradation, and shape of caved ore and rock significantly influence their flow properties, draw strength, and the ore loss and dilution, it is imperative to conduct research in these areas.

1.2.6.1 The Geometric Parameters of Individual Particles

The size and shape of caved ore and rock are typically utilized to represent the geometric parameters of individual particles.

(1) Size of caved ore and rock

The dimensions of caved ore and rock can be quantified using linear dimensions, area, and volume. In mining engineering, linear dimensions are predominantly employed. The size of caved ore and rock is determined by measuring the maximum dimensions perpendicular to each other across three axes. The longest dimension, denoted as 'a', represents the length; the intermediate dimension, 'b', indicates the width; and the shortest dimension, 'c', signifies the thickness, as illustrated in Fig. 1.5. These three dimensions correspond to the sides of a parallelepiped that circumscribes the caved ore and rock.

In practical applications, it is often essential to quantify the size of caved ore and rock. Consequently, the diameter is commonly employed as a metric to represent their dimensions. The diameter of caved ore and rock can be articulated as follows:

$$d_{\mathrm{dk}} = b \tag{1.17}$$

$$d_{\mathrm{dk}} = \frac{a + b + c}{3} \tag{1.18}$$

$$d_{\mathrm{dk}} = \sqrt[3]{a \cdot b \cdot c} \tag{1.19}$$

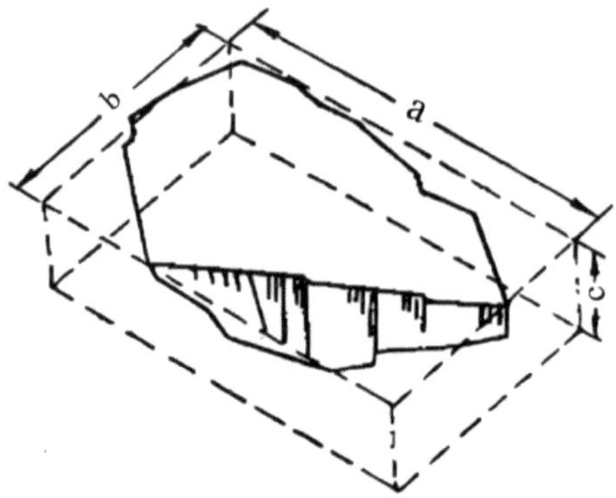

Fig. 1.5 Size measurement of caved ore and rock

Here, d_{dk} represents the average diameter of individual particles; a, b, and c are the length, width, and thickness of individual particles respectively.

When the length and width of caved ore and rock are relatively similar, their diameter can be expressed using Eq. (1.17). In cases where the length and width differ significantly, the diameter can be calculated according to Eq. (1.18). For enhanced accuracy, Eq. (1.19) should be utilized to determine the diameter of caved ore and rock. During the particle screening process, a specific particle can only pass through a designated sieve aperture size. Typically, the diameter of this sieve aperture is considered to represent the diameter of the particle.

(2) Shape of caved ore and rock

The shape characteristics of caved ore and rock can be represented by the ratios of their dimensions in three mutually perpendicular directions.

$$a : b : c \tag{1.20}$$

To facilitate the comparison of shapes among various caved ores and rocks, it is essential to utilize relative size ratios rather than absolute size ratios. These relative ratios should be expressed in terms of the width of the caved ore and rock as a unit. This approach provides a numerical characterization of the shapes of caved ore and rock.

$$\frac{a}{b} : 1 : \frac{c}{b} \tag{1.21}$$

Based on the aspect ratio of length and width, as well as the thickness and width of caved ore and rock, these materials can be classified into distinct shapes:

(1) Cube	$a = (1–1.3)b$	$c = (1–0.7)b$
(2) Cylindrical zone	$a > 1.3b$	$c = (1–0.7)b$
(3) Plate-like structure	$a = (1–1.3)b$	$c = (0.7–0.3)b$

(continued)

(continued)

(4) Long plate-like zone	$a > 1.3b$	$c = (0.7–0.3)b$
(5) Sheet-like structure	$a = (1–1.3)b$	$c < 0.3b$
(6) Long-shaped zone	$a > 1.3b$	$c < 0.3b$

In practical applications, the shape characteristics of caved ore and rock are often described using the difference coefficient of particle size K_x, which is defined as the ratio of the maximum dimension (length) to the minimum dimension (thickness):

$$K_x = \frac{a}{c} \tag{1.22}$$

For a regular cube, the difference coefficient is represented as $K_x = 1$. In contrast, the actual difference coefficient of caved ore and rock frequently $K_x > 1$, with the highest values observed in long plate-shaped and long sheet-shaped caved ore and rock.

1.2.6.2 Size Composition of Caved Ore and Rock

Based on the structure, physical, and mechanical properties of the rock mass, as well as the blasting parameters, the particle packing system generated by blasting consists of a collection of particles of varying sizes. In this collection, the proportion of the weight of particles of different size grades relative to the total weight is referred to as the size composition.

According to the requirements of the ore product, the mining process, the evaluation of the blasting effects, and scientific experimental research, caved ore and rock are typically classified into different size grades based on their dimensions. In production practice and the assessment of blasting efficacy, these materials can be categorized into three levels: large, medium, and small. Furthermore, in the experimental research related to draw, the size grades of caved ore and rock may be subdivided into three to seven levels depending on the specific test requirements.

The size of each level is typically characterized by the minimum and maximum particle sizes within the grade, or by the average diameter of the particles. Furthermore, all size grades of caved ore and rock can be represented by the average diameter of the particles.

The average diameter of caved ore and rock can be calculated using Eq. (1.23):

$$d_p = \frac{d_{max} - d_{min}}{2} \tag{1.23}$$

Here, d_p represents the average diameter of caved ore and rock within a certain level, d_{max} is the diameter of the largest or maximum-sized caved ore and rock within a certain level, d_{min} is the diameter of the smallest or minimum-sized caved ore and rock within a certain level.

1.2.6.3 Method for Determining the Particle Size Composition of Caved Ore and Rock

(1) Sieving method

Initially, it is essential to classify the size according to the specified requirements and select an appropriate sieve or grid sieve based on the number of grades. Subsequently, different sampling methods should be chosen in accordance with the mining and blasting methods employed. In the case of utilizing the sublevel caving method without a sill pillar, where vertical fan deep-hole stoping is applied, the size composition of the caved ore and rock will vary throughout the entire draw process. Therefore, samples should be collected from the same location at different stages of ore extraction; specifically, 1 to 3 tons should be sampled during the initial, middle, and final stages of the draw. Conversely, when employing the horizontal fan deep-hole stoping method with sublevel caving that includes a sill pillar, samples should be collected at various times from different hoppers, ensuring that 1–3 tons are sampled from each location. Furthermore, when using the vertical fan deep-hole blasting method alongside sublevel caving with a sill pillar, samples must be taken at different draw periods, ensuring their representativeness. Finally, the collected samples should be reduced using the quartering method until only 0.5–1 ton of caved ore and rock samples remain.

After measuring the total weight of the caved ore and rock samples, begin the sieving process using a sieve with a large aperture. To ensure the comparability of the sieving results, it is essential to adhere to the specified feeding system and oscillation duration. The weight of the residue collected from each sieve must be recorded. The percentage of the weight of each size fraction of caved ore and rock relative to the total weight is calculated using Eq. (1.24):

$$U = \frac{W_{\mathrm{j}}}{W_{\mathrm{z}}} \times 100, \quad \% \tag{1.24}$$

Here, U represents the percentage of weight at a certain size level, W_{j} indicates the residual weight of the sample at that level, and W_{z} is the total weight of the sieved sample.

To calculate the cumulative percentage of sieve residue for each type of sieve, one must determine the cumulative percentage for each size grade. This can be achieved by plotting the cumulative percentages on the vertical axis against the sieve hole sizes on the horizontal axis, thereby generating a screening curve. It is important to note that while the screening method is effective for assessing size composition, it is labor-intensive, time-consuming, challenging to implement in situ, and necessitates substantial equipment.

(2) Block-ruler area method

The block-ruler area method determines the percentage of the total survey area occupied by the horizontal projections of caved ore and rock in each size class on the surface of the particle pile. Through specialized size composition determination, it has been confirmed that the percentage of the area occupied by each particle size level on the surface of the particle packing system is generally consistent with the percentage of the volume occupied by each size level within the packing system. Consequently, size composition determination can be directly conducted on the leveled surface of the particle packing system by delineating a specific measurement area and laying out the block ruler. Transparent coordinate paper is then placed over the block ruler, where the contours of each level of caved ore and rock are depicted on the grid coordinate paper under adequate lighting conditions. Alternatively, the contours of the caved ore and rock shown in each grid of the block ruler can be copied onto the grid coordinate paper, as illustrated in Fig. 1.6. Subsequently, the total area of each level of caved ore and rock is calculated, along with their respective percentages of the entire measurement area, to obtain the size composition of caved ore and rock. The percentage of the area for each particle size level is calculated using Eq. (1.25):

$$U_{\mathrm{m}} = \frac{S_{\mathrm{j}}}{S_{\mathrm{c}}} \times 100, \quad \% \tag{1.25}$$

Here, U_{m} represents the percentage of area for a certain size grade, S_{j} represents the total area for a certain size grade, and S_{c} represents the total area of the measured range.

When determining the size composition of caved ore and rock using the block-ruler area method, it is essential to carefully select the measurement location. The chosen location should exhibit the caved ore and rock in a natural packing state and possess sufficient representativeness. The control range of the measurement area must be established based on the maximum size of the caved ore and rock within the measured particle system. Typically, the minimum side length of the controlled measurement area can be determined using Eq. (1.26):

$$l_{\mathrm{k}} = 3\sqrt{S_{\mathrm{z}}} \tag{1.26}$$

Here, l_{k} represents the minimum length of the controlled measurement area, and S_{z} is the horizontal projected area of the largest particle.

The structure of the block ruler is illustrated in Fig. 1.7. The overall dimensions of a block ruler are 1 m × 1 m. Each grid measures either 100 mm × 100 mm or 200 mm × 200 mm, and the ruler is designed to be foldable.

The method for determining particle size using the block-ruler area method is efficient, requires less labor intensity, and utilizes minimal equipment. Consequently, in practical production settings, it is often employed when the particle size is large and the occurrence of small pieces is infrequent. However, when numerous small pieces are present, the workload associated with manual description increases significantly. In such cases, employing the photographic method instead of manual description can greatly simplify the measurement process.

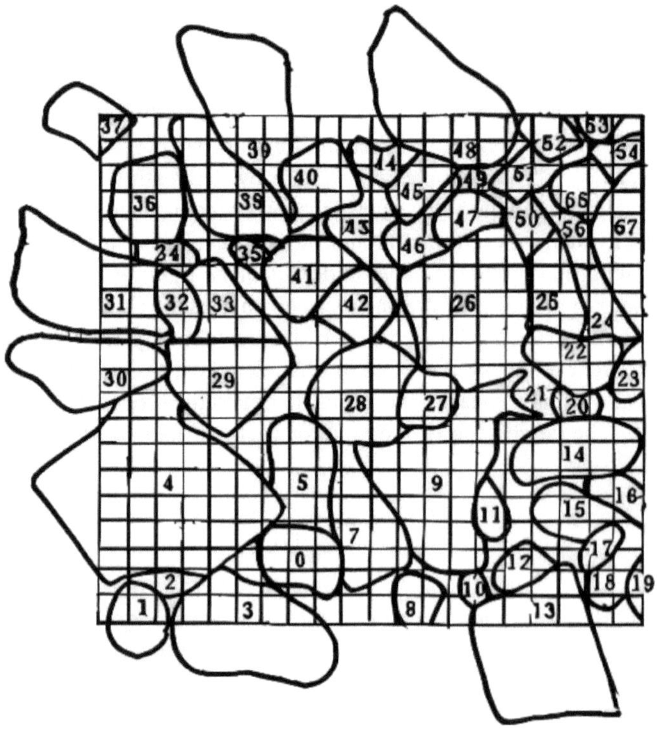

Fig. 1.6 Size composition of caved ore and rock based on the block-ruler area method

(3) Photography method

1) Photography area method. This method is based on the same principle as the block-ruler area method. A white block ruler, measuring 100 mm × 100 mm, is placed on the surface of the particle packing system composed of caved ore and rock. The total dimensions of the area under observation are 1 m × 2 m. The block ruler is utilized to capture images of the caved ore and rock within its boundaries. By calculating the percentage of the total area occupied by each particle size category within the entire ruler area depicted in the photograph, the particle size composition represented by each category can be accurately determined.

When measuring particle size, it is essential to select representative locations. During photography, the camera axis should generally be perpendicular to the surface being photographed to ensure frontal images. If there is an inclination angle between the camera axis and the surface, both the block ruler and the caved ore and rock in the photo may exhibit distortions. The scale between the object in the photo and the actual object can be set at 1:10 or 1:20. The particle size classification results of the caved ore and rock depicted in the photo are presented in Table 1.2.

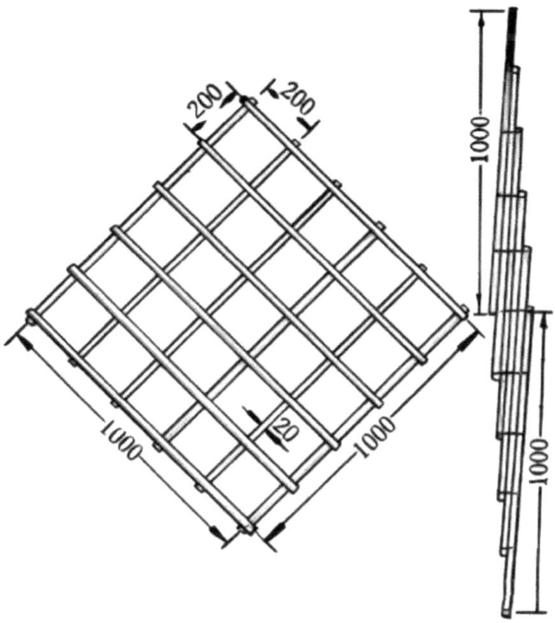

Fig. 1.7 Foldable block ruler

Table 1.2 Particle-size grading of caved ore and rock

Grading types	Grading level for particle size				
	1	2	3	4	5
On-site particle size grading/mm	< 20	20–100	100–300	300–500	> 500
Particle-size grading in the 1:20 photograph/mm	< 1	1–5	5–15	15–25	> 25
Area-based particle-size grading in the 1:20 photograph/mm	< 1	1–20	20–180	180–490	> 490
Particle-size grading based on the number of points in the photograph	1	1–20	20–180	180–490	> 490

The basis for the particle size grading illustrated in the photograph is that when caved ore and rock pass through a sieve with a specific diameter, the horizontal projection area of these materials will not exceed the area of the equivalent circle.

Ultimately, the caved ore and rock, which are circled and numbered in the photograph, are categorized to calculate their total area. Subsequently, the percentage of the total area for each category relative to the entire measured area is determined, thereby providing the size composition.

2) Photographic frequency method. This method also employs the photographic area technique, utilizing size classification and the numerical data presented in the photographs. For each grid within the size scale depicted in the images, the dominant size grade code is recorded. Once all grids have been annotated with the classification numbers, the frequency of each size grade can be computed.

Subsequently, the ratio of the frequency of each size grade to the total frequency, denoted as 'f', can be calculated to derive the relative frequency, which indicates the proportion of each size grade within the overall distribution. The frequency is typically determined using the election and vote-counting method, with a calculation example provided in Table 1.3.

The point frequency method can be effectively employed by overlaying a transparent glass sheet, which features a 2×2 mm^2 grid, onto the photograph. This grid, measuring 60×90 mm^2, contains 1350 evenly distributed points (30×45). By counting the number of points that intersect with caved ore and rock, the sizes of these materials are categorized into five distinct grades, as detailed in Table 1.3. The total points falling within each grade are tallied, and the percentage of points for each grade relative to the overall total is computed. This process also allows for the determination of the composition of sizes of caved ore and rock. Additionally, the calculation of points can be performed using a voting method.

The determination of the size composition of the particle packing system of caved ore and rock using the photographic method is efficient, requires minimal effort, and is both simple and reliable. This method is currently widely utilized in both production practice and scientific research.

(4) Polygonal area calculation method

This method represents the contour of caved ore and rock using highly detailed polygons, determining the coordinates of each vertex for every polygon. The area of each caved ore and rock is calculated based on these vertex coordinates. Subsequently, the total area for each grade of caved ore and rock, along with the percentage of the entire measured total area, is computed, allowing for the determination of size composition. Additionally, the projected area of the contour for each caved ore and rock can be calculated using Eq. (1.27):

Table 1.3 Calculation table for size-frequency distribution of caved ore and rock

Grading level	On-site particle size grading	Particle-size grading in the photograph	Frequency result calculated by the vote-counting method	Frequency f_i	Relative frequency f_e	Calculation formula $f_e = f_i/f$
1	< 20 mm	< 1 mm	30	30	0.15	$f_{e1} = f_1/f$
2	20–100 mm	1–5 mm	60	60	0.30	$f_{e2} = f_2/f$
3	100–300 mm	5–15 mm	70	70	0.35	$f_{e3} = f_3/f$
4	300–500 mm	15–25 mm	20	20	0.10	$f_{e4} = f_4/f$
5	> 500 mm	> 25 mm	20	20	0.10	$f_{e5} = f_5/f$
			In total	$f = 200$		

$$
\begin{aligned}
S_{\mathrm{d}} &= \frac{1}{2}\sum_{i=1}^{n}(x_{\mathrm{i}}y_{\mathrm{i}+1} - x_{\mathrm{i}+1}y_{\mathrm{i}}) \\
&= \frac{1}{2}[(x_1y_2 - x_2y_1) + (x_2y_3 - x_3y_2) + \cdots + (x_{\mathrm{n}}y_1 - x_1y_{\mathrm{n}})]
\end{aligned} \tag{1.27}
$$

Here, S_{d} represents the projected area of the shape of each particle, x_{i} and y_{i} represent the coordinates of the polygon vertices of the shape of each particle, and n is the number of polygon vertices of each particle.

The advantages of this method are as follows:

1) Using a video recorder and an image display device, the coordinates of the polygonal vertices of each caved ore and rock can be recorded and input into an electronic computer as data. The polygonal areas of each caved ore and rock can be calculated following a specific procedure, allowing for the computation of the total areas of each level of caved ore and rock. Furthermore, the percentage of the total area of each level of caved ore and rock relative to the entire measured total area can be determined, facilitating the analysis of particle size composition. Compared to the previously mentioned methods, video calculation offers automation, which reduces workload, increases measurement speed, and enhances convenience and reliability. Additionally, contemporary image identification technology can assist in the statistical analysis of particle gradation; this topic will be covered in a separate monograph and is therefore not addressed in detail here.
2) Due to the rapid measurement speed, the fragmentation quality of caved ore and rock can be evaluated swiftly, allowing for timely improvements to the blasting parameters of rock drilling. This, in turn, enhances the effectiveness and intensity of the draw process.
3) The results obtained from the automatic video electronic computer method for calculating polygonal areas are compared with those from the photographic area method, as illustrated in Figs. 1.8, 1.9 and Table 1.4. The percentage composition of particle sizes in caved ore and rock calculated by both methods is remarkably similar.

1.2.6.4 Uniformity Evaluation of the Size Composition of Caved Ore and Rock

The uniformity of rock mass fragmentation significantly influences the draw process. An inappropriate particle size, whether too large or too small, adversely affects gravity flow, draw intensity, and leads to ore loss and dilution. Therefore, it is essential to evaluate the uniformity of the size composition of caved ore and rock. Various indicators and diagrams can be employed for this evaluation, including average diameter, coarse particle rate, size heterogeneity coefficient, and size composition curve. In this context, only the evaluation method for the size heterogeneity coefficient will be discussed.

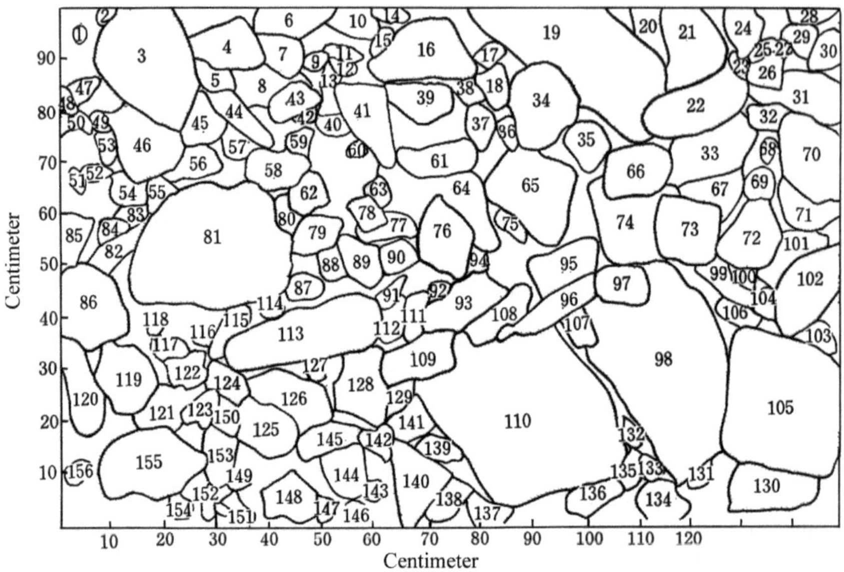

Fig. 1.8 Area measurement of the photo taken at a specific point based on the photographic area method

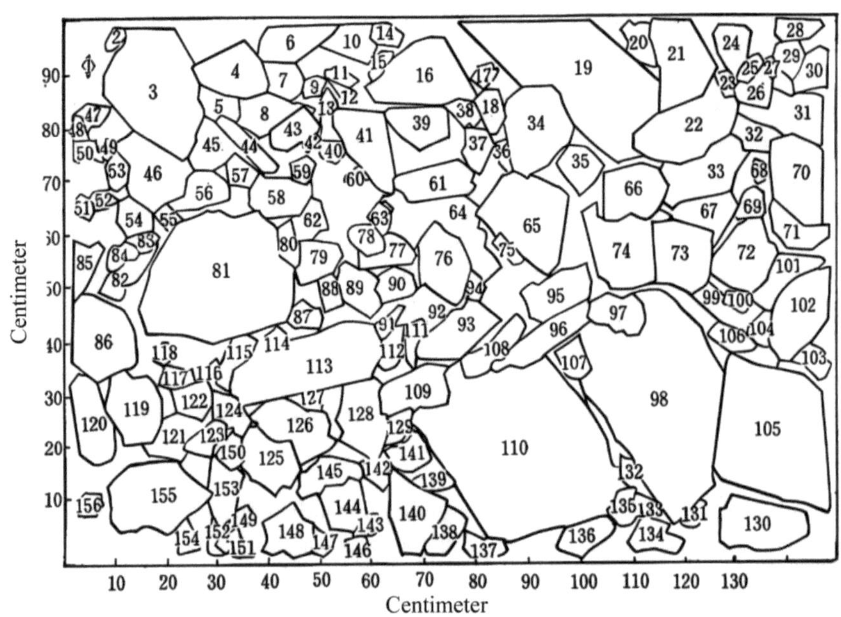

Fig. 1.9 Area measurement using the recorded coordinates of the measurement points based on the polygon area method

Table 1.4 Size composition of caved ore and rock measured by the two methods

Method name	Particle size grading/mm							
	0–25	25–50	50–100	100–150	150–200	200–250	250–300	300–350
Photographic area method	16.93	2.88	21.49	24.01	9.80	2.39	11.44	11.06
Automatic polygon area calculation method	18.91	3.72	21.12	22.03	9.76	2.34	11.09	11.06

The coefficient of size heterogeneity, commonly referred to in statistics as the coefficient of variation, is expressed by Eq. (1.28):

$$K_{\mathrm{b}} = \frac{\sigma}{d_{\mathrm{dj}}} \times 100 = \frac{\sqrt{\frac{\sum (d_{\mathrm{pi}} - d_{\mathrm{dj}})^2 u_{\mathrm{i}}}{\sum u_{\mathrm{i}}}}}{d_{\mathrm{dj}}} \times 100, \quad \% \tag{1.28}$$

Here, K_{b} represents the coefficient of size heterogeneity, σ is the standard deviation, d_{dj} is the average diameter of several different grades of particles, d_{pi} is the average diameter of a certain size level, and u_{i} is the percentage of the weight of the corresponding size level.

The criterion for assessing the uniformity of particle size in caved ore and rock is defined as follows: when $K_{\mathrm{b}} < 40\%$, the fragmentation is considered uniform; when K_{b} ranges from 40 to 60%, the fragmentation is regarded as moderately uniform; and when $K_{\mathrm{b}} > 60\%$, the fragmentation is classified as non-uniform.

The statistical results of the particle size of caved ore and rock post-blasting, obtained through on-site measurements, are presented in Table 1.5.

The coefficient of size heterogeneity is employed to evaluate the uniformity of the measured data concerning size composition. The calculation process is outlined as follows:

The initial step involves calculating the weighted average diameter. During large-scale blasting operations, the weighted average diameter for each particle size category is determined as follows:

Table 1.5 Statistical results of the particle sizes post-blasting obtained through on-site measurements for the two groups

Particle size/mm		0–100	100–200	200–300	300–400	400–600	600–1000
Average particle size/mm		50	150	250	350	500	800
Blasting method	Large-scale blasting (%)	18.0	10.0	3.0	6.0	15.0	48.0
	Shallow-hole blasting	12.0%	56.0%	28.0%	4.0%	–	–

$$d_{\mathrm{dj}} = \frac{\sum u_{\mathrm{i}} d_{\mathrm{pi}}}{u_{\mathrm{i}}} = \frac{u_1 d_{\mathrm{p1}} + u_2 d_{\mathrm{p2}} + \cdots + u_{\mathrm{n}} d_{\mathrm{pn}}}{u_1 + u_2 + \cdots + u_{\mathrm{n}}}$$
$$= \frac{18 \times 50 + 10 \times 150 + 3 \times 250 + 6 \times 350 + 15 \times 500 + 48 \times 800}{18 + 10 + 3 + 6 + 15 + 48} = 512\ \mathrm{mm}$$

During shallow-hole blasting, the weighted average particle diameter at all levels is:

$$d_{\mathrm{dj}} = \frac{12 \times 50 + 56 \times 150 + 28 \times 250 + 4 \times 350}{12 + 56 + 28 + 4} = 174\ \mathrm{mm}$$

The second step is to calculate the standard deviation. During a large-scale blasting, it is:

$$\sigma = \sqrt{\frac{(50 - 512)^2 \times 18 + (150 - 512)^2 \times 10 + (250 - 512)^2 \times 3}{18 + 10 + 3 + 6 + 15 + 48}} \rightarrow$$
$$\leftarrow \frac{+(350 - 512)^2 \times 6 + (500 - 512)^2 \times 15 + (800 - 512)^2 \times 48}{18 + 10 + 3 + 6 + 15 + 48}$$
$$= \sqrt{\frac{9504460}{100}} = \sqrt{95044.6} = \pm 308$$

For shallow-hole blasting, it is:

$$\sigma = \sqrt{\frac{(50 - 174)^2 \times 12 + (150 - 174)^2 \times 56 + (250 - 174)^2 \times 28 + (350 - 174)^2 \times 4}{12 + 56 + 28 + 4}}$$
$$= \sqrt{\frac{502400}{100}} = \sqrt{5024} = \pm 70.88$$

The third step is to calculate the coefficient of size heterogeneity. During large-scale blasting, it is:

$$K_{\mathrm{b}} = \frac{308}{512} \times 100 = 60.2\%$$

For shallow-hole blasting, it is:

$$K_{\mathrm{b}} = \frac{70.88}{174} \times 100 = 47.6\%$$

The calculation results indicate that during large-scale blasting, the value of K_{b} exceeded 60%. Although it only slightly surpassed this threshold, it suggests that the particle size distribution post-blasting was not sufficiently uniform. In contrast, during shallow-hole blasting, K_{b} values ranged between 40 and 60%, indicating a moderate level of uniformity in particle size distribution.

1.2.6.5 The Influence of Fine Ores and Clay on Draw

When the particle size of caved ore and rock is less than 5 mm, it is generally classified as fine ore. Fine ore can be further categorized based on particle size as follows: Caved ore and rock with a particle size ranging from 0.25 to 5 mm is referred to as coarse silt fines; those with a particle size between 0.05 and 0.25 mm are termed fine silt fines; particles sized from 0.005 to 0.05 mm are classified as dust particles; and those smaller than 0.005 mm are identified as clay particles.

Research has demonstrated that the clay content and humidity of caved ore and rock significantly influence the maximum width of the IEZ. When the humidity of caved ore and rock ranges from 4 to 7%, and the particle size is uniform without the presence of clayey substances, the flow properties are favorable, resulting in a larger maximum width of the IEZ. Conversely, when the clay content reaches 14%, self-compaction begins to occur; if the clay content increases to 25%, the adhesiveness markedly rises, leading to the formation of cavities or slender tube-like structures within the IEZ, thereby exacerbating the drawing conditions. In certain underground mines, the infiltration of fine waste rock into the ore particle layer through the ore-rock contact surface during the draw process can cause significant dilution, potentially becoming the primary source of dilution for the mine.

1.2.7 Repose Angle of Particle Packing System of Caved Ore and Rock

Granular materials, influenced by their own weight and occasionally by the additional effect of static pressure, exhibit the characteristic of forming a conical pile. The angle formed between the natural horizontal plane and the slope of the cone pile in its natural resting position is referred to as the angle of repose.

1.2.7.1 Factors Influencing the Angle of Repose

(1) Particle size

Studies have demonstrated that the repose angle of particle packing systems of caved ore and rock decreases as particle size increases, as illustrated in Fig. 1.10. It is evident from Fig. 1.10 that when the particle size increases from 1 to 80 mm, the repose angle diminishes, particularly in the smaller particle size range where the decrease is notably rapid. As the particle size approaches 80 mm, the repose angle begins to stabilize. This phenomenon occurs primarily because, with increasing particle size, the total specific surface area between the caved ore and rock decreases, leading to a corresponding reduction in friction between the particles. However, once the particle size exceeds a certain threshold, the potential for internal locking between

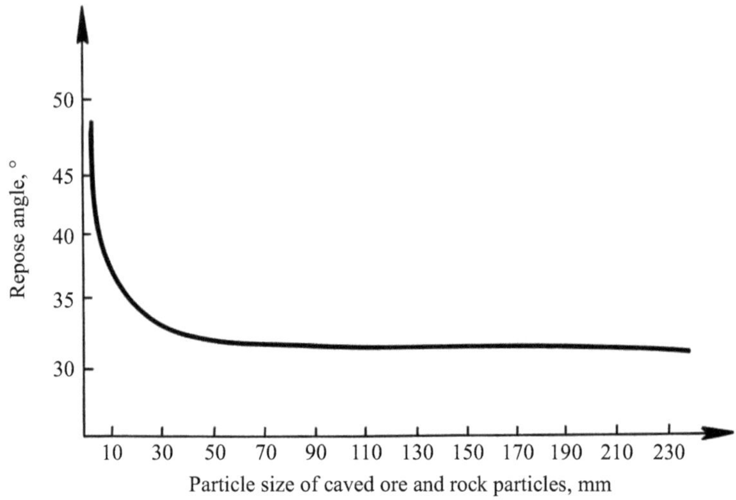

Fig. 1.10 Relationship between the particle size of caved ore and rock and their repose angle

Table 1.6 Relationship between the repose angle and humidity

Name of the granular media	Repose angle, °		
	Dry	Damp	Wet
Sand	28–35	30–40	22–27
Sandy clay	40–50	35–40	25–30
Various sizes of crushed ore	32–45	36–48	30–40

the particles increases, which counteracts the reduction in total specific surface area, resulting in the stabilization of the repose angle.

(2) Humidity

Generally, the repose angle increases with the humidity of the particle packing system of caved ore and rock. This phenomenon occurs because, as humidity rises, the cohesion among the caved ore and rock—particularly those with high clay content—increases. However, once the humidity reaches saturation, water occupies the voids between the caved ore and rock, leading to a decrease in frictional force and, consequently, a reduction in the repose angle, as illustrated in Table 1.6.

1.2.7.2 Determination Method for the Natural Angle of Repose

(1) Rotating-box free-flow method

The structure of the rotating box is illustrated in Fig. 1.11. The dimensions of the box are 400 mm × 200 mm × 200 mm. The bottom of the box is equipped with

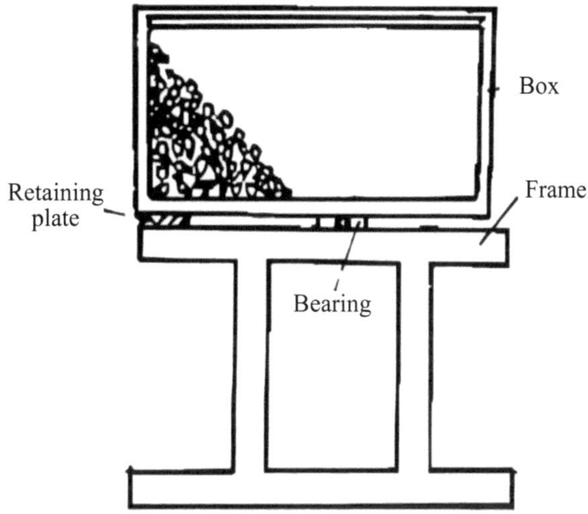

Fig. 1.11 Rotating box structure diagram

bearings, and transparent glass panels are installed on both sides for observation and angle measurement. During the measurement process, the sample of caved ore and rock to be analyzed is placed inside the box, which is then rotated around the central bearing to reach an upright position. The caved ore and rock sample is leveled within the box. Subsequently, the box is turned to a horizontal position, and a fixing plate is inserted to secure it. At this stage, the particle packing system of the caved ore and rock naturally forms a slope within the box. The angle between this slope and the horizontal plane represents the natural repose angle, which can be directly measured using a protractor.

(2) Grid-box free-flow method

The structure of the grid box is analogous to that of the rotating box, with the distinction that the aforementioned box is partitioned into two compartments of varying sizes by grids. During the measurement process, air-dried particle samples are uniformly distributed within the smaller compartment using a funnel, positioned approximately 30 mm below the box opening to ensure a level surface. Subsequently, the grid is gradually elevated, allowing the caved ore and rock to naturally descend into the adjacent compartment. Once the caved ore and rock cease rolling, the base length l and height h of the natural slope surface are measured. Use $\tan\varphi_{\text{z}} = \frac{h}{l}$ to calculate the natural repose angle φ_{z}.

(3) Open-bottom cylinder method

The specifications of the open-bottom cylinder, depicted in Fig. 1.12, are determined based on the particle size of the caved ore and rock to be measured. Generally, the diameter of the open-bottom cylinder should exceed five times the maximum particle size of the measured caved ore and rock. Specifically, for measuring the natural repose angle of caved ore and rock with a particle size of ≤ 15 mm, a cylinder with a diameter

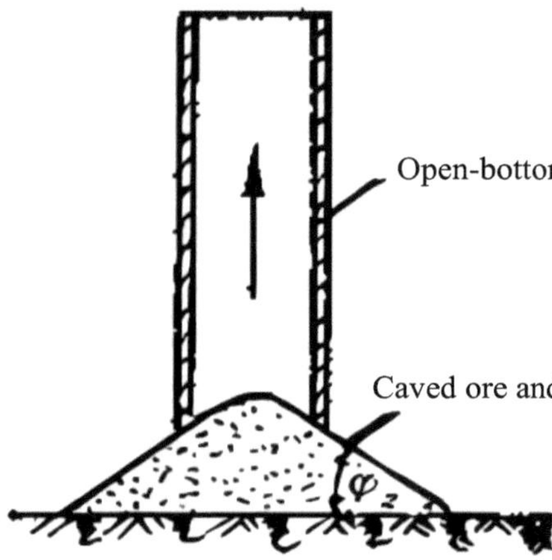

Fig. 1.12 Schematic of the natural angle-of-repose measurement using the bottomless-cylinder method

of 100 mm and a height of 300 mm is appropriate. Conversely, for particles with a size ranging from 25 to 100 mm, a cylinder with a diameter of 500 mm and a height of 750 mm is recommended.

During the measurement process, position the open-bottom cylinder on a flat surface and fill it with the caved ore and rock to be measured. Subsequently, utilize manual operation or a pulley system to gradually and smoothly lift the cylinder vertically, allowing the caved ore and rock to naturally form a conical shape. The angle formed between the cone's surface and the horizontal plane represents the natural repose angle of the caved ore and rock. Measure both the height of the cone and the diameter of its base, and apply Eq. (1.29) to calculate the natural repose angle:

$$\tan\varphi_z = \frac{2h_{zd}}{d_{zd}} \tag{1.29}$$

Here, φ_z represents the natural angle of repose, h_{zd} is the height of the cone, and d_{zd} is the diameter of the cone base.

The determination needs to be carried out multiple times and the arithmetic mean value should be taken. When measuring the diameter of the cone base, the average value of two mutually perpendicular axes should be used, and particles that are scattered far away should not be considered.

(4) Load-pressure simulation test box method

The structure of the load-pressure simulation test box method is illustrated in Fig. 1.13. This device is designed to measure the natural repose angle under simulated static pressure conditions typical of field environments. During the measurement process, the particle sample to be tested is placed in box (1) and leveled. A pressure uniform transmission plug (2) is then positioned on top of the sample. Subsequently, proportional weights (3) are added onto the transmission plug. Afterward,

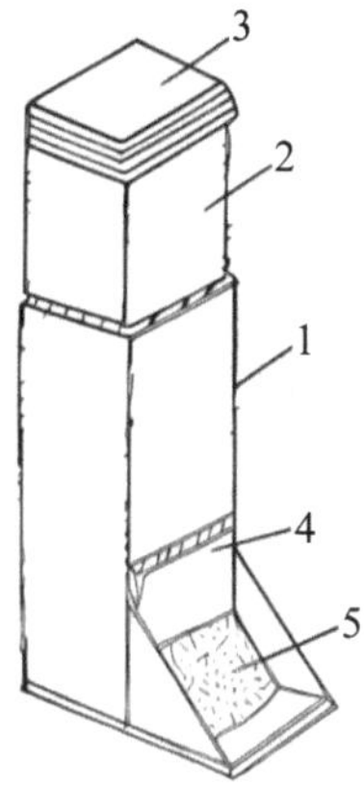

Fig. 1.13 Schematic of the load-pressure simulation test box

the rotating gate (4) located on the front wall of the box is opened. The particle sample (5) then flows naturally under the influence of gravity and static pressure, creating a natural slope surface. The angle between this slope surface and the bottom plate of the drawpoint is measured, which represents the natural repose angle φ_y under load. This angle can be calculated using Eq. (1.30):

$$\tan\varphi_y = \frac{h_m}{l_p} \tag{1.30}$$

Here, φ_y represents the natural angle of repose under pressure, h_m is the opening height of drawpoint, and l_p is the distance that the particle sample spreads on the bottom plate of the drawpoint.

The selection of the method for determining the natural angle of repose should be based on comprehensive considerations, including particle size, cohesion, and the measurement requirements of caved ore and rock. The natural angle of repose indicates the balance among frictional force, cohesive force, and gravity acting on caved ore and rock. However, it represents merely a surface condition of particle balance within the particle packing system of caved ore and rock, which is crucial for the design of the draw roadway and the draw device. In mining production practice, the natural angle of repose is closely related to the conditions of formation. It is frequently influenced by static pressure or dynamic impact during its formation, with scenarios where it remains unaffected by external forces being rare. Due to variations in measurement conditions and methods, the values of the natural angle of repose obtained can differ significantly. Therefore, when utilizing natural angle of repose data, it is essential to conduct a specific analysis; alternatively, separate and specialized measurements should be performed under specific conditions.

1.2.8 External Friction Angle and Internal Friction Angle of Caved Ore and Rock

The external friction angle of a granular medium is defined as the angle at which particles begin to slide down an inclined surface, transitioning from a stationary state to a moving state. This angle is measured between the slope and the horizontal plane. The tangent of the external friction angle is referred to as the external friction coefficient. In mining engineering, to facilitate the free sliding of caved ore and rock along a specific slope, the inclination angle α of the slope must exceed the external friction angle φ_w.

The external friction angle measuring device is illustrated in Fig. 1.14. During the measurement process, the particle packing system consisting of caved ore and rock is positioned within the horizontal rotating chute (2) on the device frame (1). The distance between the caved ore and rock and one end of the rotating shaft should be maintained at 8–10 cm. Subsequently, rope (3) is utilized to gradually and steadily elevate the rotating chute until the moment when the caved ore and rock begin to slide, at which point the lifting should be halted. At this moment, the angle formed between the inclined plane of the bottom plate of the rotating chute and the horizontal plane represents the external friction angle. This angle can either be directly measured using a protractor or calculated using Eq. (1.31):

$$\sin \varphi_w = \frac{h_{tg}}{l_{ch}} \tag{1.31}$$

Here, φ_w represents the external friction angle, h_{tg} is the height to which the rotating trough is lifted, and l_{ch} is the length of the rotating trough.

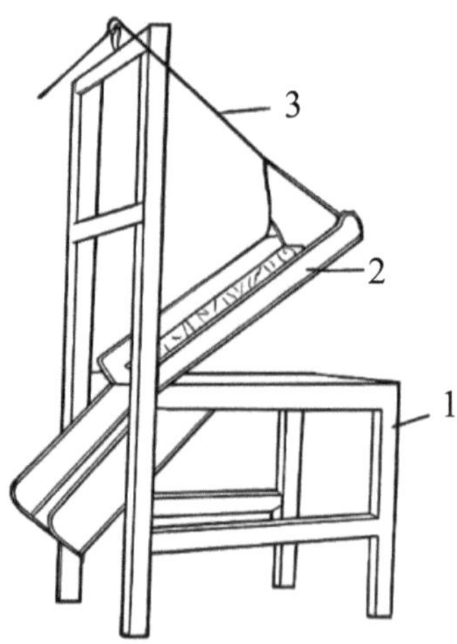

Fig. 1.14 Measuring device of external friction angle

The measurement results of the external friction angle for magnetite quartz rock, which has an overall weight of 3.4 tons/m^3, a bulk density of 2.2 tons/m^3, and a hardness coefficient ranging from 14 to 16, are presented in Table 1.7.

From Table 1.7, it is evident that several factors influence the external friction angle, including:

(1) The size and composition of caved ore and rock; smaller particle sizes of caved ore and rock result in a greater external friction angle.
(2) The moisture content of caved ore and rock; large, wet particles exhibit a higher external friction angle compared to smaller, naturally moist particles, while the external friction angle of fine, wet particles is also greater than that of naturally moist particles. However, when the moisture content exceeds a certain threshold, approaching the characteristics of a fluid-like slurry, the external friction angle decreases significantly. The moisture content at this point is referred to as the critical moisture of the caved ore and rock.
(3) The smoothness of the contact surface; rougher surfaces in contact with caved ore and rock lead to a higher external friction angle and an increased external friction coefficient. The external friction coefficients for various contact materials are presented in Table 1.8.

The internal friction angle of a granular medium refers to the angle formed between the normal stress and the resultant stress acting on the shear surface within the granular medium at the moment of shear failure, in the absence of cohesion.

Table 1.7 Measurement results of the external friction angle for magnetite quartz rock

Sample number	Grading of particle size/mm	External friction angle/°			
		Wooden base plate		Iron base plate	
		Natural humidity	Wet	Natural humidity	Wet
I	+ 160	28.0	–	23.0	–
II	– 160 + 80	32.0	–	26.0	–
III	– 80 + 40	32.5	27.0	26.5	24.5
IV	– 40 + 20	34.0	30.0	27.5	25.0
V	– 20 + 8	35.0	34.0	28.5	26.0
VI	– 8 + 5	37.0	46.0	30.0	40.0
VII	– 5	45.0	51.0	34.0	43.0

Table 1.8 External friction coefficient for static contact between different granular materials

Friction pair	Steel and iron ore	Steel and granite	Steel and sandstone	Granite slab and granite	Plank and stone	Concrete slab and stone material
Coefficient of friction	0.42	0.45	0.38	0.68	0.46–0.6	0.76

This angle can be determined using the shear strength test method for the granular medium. The resulting graph, which plots the normal stress σ against the shear stress τ based on the test results, illustrates the shear strength curve. The angle between this curve and the horizontal axis σ is defined as the internal friction angle φ. The tangent of the internal friction angle is referred to as the internal friction coefficient K. Consequently, the internal friction coefficient represents the ratio of the ultimate shear stress τ along the shear surface to the normal stress σ of the granular medium at the moment of failure:

$$K = \tan \varphi = \frac{\tau}{\sigma} \tag{1.32}$$

If the medium is imperfectly consolidated, it possesses a cohesion denoted as c. Consequently, the internal friction coefficient is defined as the difference between the shear stress τ and the cohesion c, divided by the normal stress σ:

$$K = \tan \varphi = \frac{\tau - c}{\sigma} \tag{1.33}$$

Within granular media, when particles in contact experience relative displacement, an internal frictional force invariably arises. This internal frictional force can be categorized into three types based on the contact conditions and states: (1) Sliding friction force, which hinders sliding when a portion of the particle packing system of caved ore and rock slides along another part on a plane or curved surface; (2) Static friction force, which occurs when the surfaces in contact within the particle packing system of caved ore and rock remain stationary despite a tendency for some parts to slide under applied force. This force that prevents the static state of caved ore and rock from transitioning into movement is termed static friction force; (3) Rolling friction force, which hinders rolling when particles roll on another surface. Each of these friction forces has associated friction coefficients and angles. In the study of the overall stress system and the sliding surface acting on the boundary of the flow zone during the draw process, the internal friction angle is a critical parameter. The internal friction angle commonly referenced corresponds to the static friction force.

1.2.8.1 Method for Measuring Internal Friction Angle

The internal friction angle φ and cohesion c, also referred to as the initial shear force, of the granular medium can be measured using direct shear tests and triaxial shear tests to evaluate the shear strength of the granular medium.

(1) Direct shear test

The direct shear test device is illustrated in Fig. 1.15. The testing procedure is as follows: To maintain a specific compacted state of the particle packing system of caved ore and rock, the particle packing system is initially compressed vertically using a constant pressure device. Subsequently, a vertical pressure, denoted as P, is applied, followed by the gradual application of a horizontal shear load, denoted as F. When the movable box (lower box) begins to move along CD, the value of the horizontal shear load F is recorded at that moment. The shear force acting on the shear surface at this time is referred to as the shear strength. Subsequently, by applying the same properties of the particle packing system of caved ore and rock, the same test is conducted under three or more different vertical pressures to obtain a series of corresponding P and F values. Then, using the cross-sectional area A of the particle packing system, the obtained P and F values can be removed to derive a series of corresponding normal stress σ and shear stress τ values. By taking the normal stress σ as the abscissa and the shear stress τ as the ordinate, one can plot the corresponding σ and τ values on the σ-τ coordinate system and connect the points to form a straight line. This straight line represents the shear strength curve of the particle packing system of caved ore and rock, as illustrated in Fig. 1.16.

To determine the internal friction angle φ, measure the angle between the shear strength curve and the horizontal axis σ. The desired cohesion c can be found at the intercept of the shear strength curve with the vertical axis τ.

The fundamental conditions for the direct shear test device stipulate that no moments act on the relative shear surface. The diameter or side length of the shear box must be sufficient to ensure that the caved ore and rock remain parallel and descend at the same velocity within a horizontal plane, thereby preventing the formation of an arch. To meet this requirement, the diameter of the shear box should be ten times the average particle size of the caved ore and rock. The gap between the upper and lower boxes should be minimized, yet it must be large enough to prevent blockage of the particles; thus, the typical clearance should exceed one-quarter of the maximum

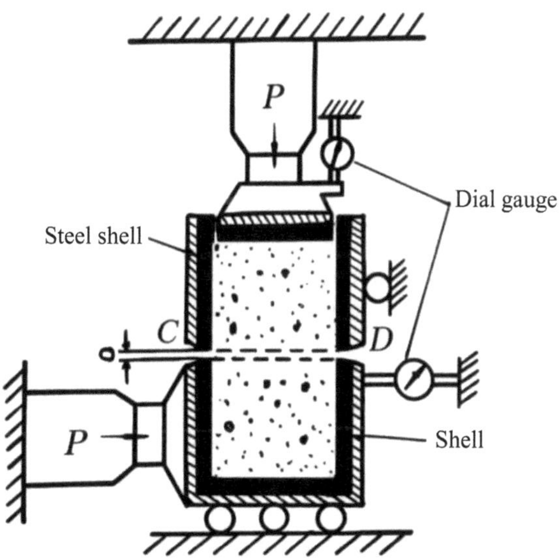

Fig. 1.15 Schematic of the direct shear test apparatus

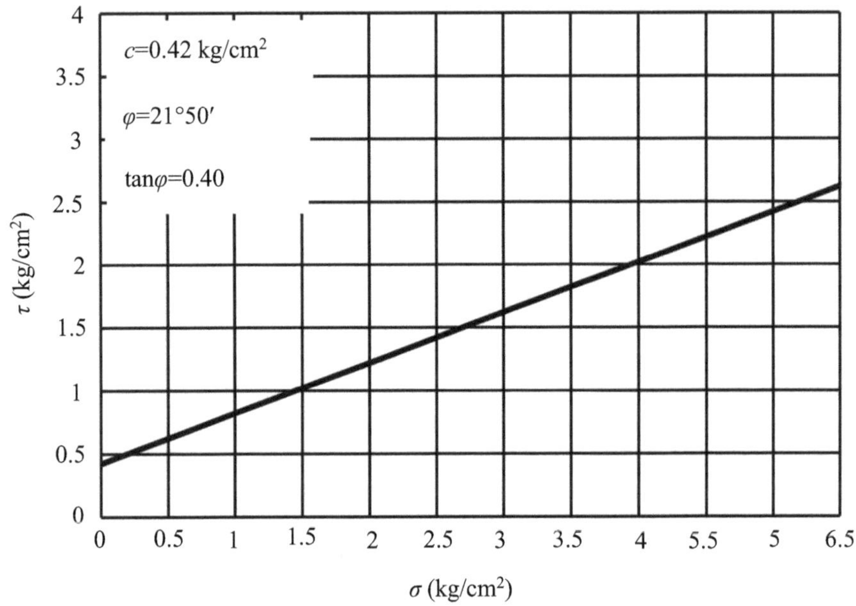

Fig. 1.16 Shear strength curve of particle packing system of caved ore and rock

particle diameter. For larger particle sizes of caved ore and rock, a toothed liner is often employed on the shear surface.

(2) Triaxial shear test

The primary component of the triaxial shear tester is the pressure cell, as illustrated in Fig. 1.17. This cell is crucial for determining the shear strength, internal friction angle, and cohesion of the particle packing system of caved ore and rock. It accomplishes this by utilizing principal stress, stress circles, and envelope lines as the particle packing system is progressively compromised under a specified lateral pressure.

For particles of varying sizes, it is essential to utilize caved ore and rock with corresponding diameters. Begin by placing the caved ore and rock within a rubber membrane, then introduce air to ensure that the internal pressure of the membrane reaches one atmosphere. Subsequently, secure the membrane onto the test seat, cover it with the test cap, and install the outer cover of the pressure cell. Next, introduce water with lateral pressure into the pressure cell until the desired lateral pressure is achieved. Finally, apply axial pressure to the end of the particle packing system of caved ore and rock, maintaining a specific lateral pressure until the particle packing system fails. The axial pressure exerted can be recorded using the dial on the force measuring ring.

Based on the values of the failure loads σ_1 and σ_2, first, draw a coordinate graph with total normal stress on the abscissa and shear stress on the ordinate. Then, with $(\sigma_1 + \sigma_2)/2$ as the center and $(\sigma_1 - \sigma_2)/2$ as the radius, draw the total normal

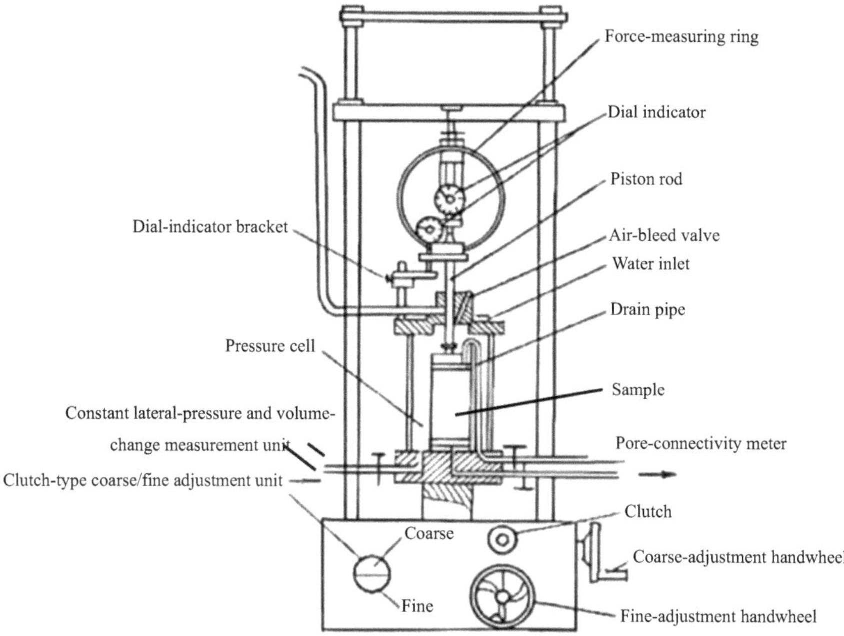

Fig. 1.17 Schematic of the pressure-cell assembly of the triaxial shear test apparatus

stress circle of damage on the abscissa. After drawing the total normal stress circles of damage under different lateral stresses, construct the common tangents of each circle to form the envelope line. This envelope line represents the total shear strength curve. The angle between this line and the abscissa σ is the internal friction angle φ, while the intercept with the ordinate τ represents the cohesion c.

1.2.8.2 Factors Affecting the Internal Friction Angle

The internal friction angle of caved ore and rock is influenced by several factors, including their granular characteristics, moisture content, size distribution, shape, surface roughness, and shear velocity.

(1) Bulking property. An increase in the bulking coefficient of caved ore and rock results in a decrease in the internal friction angle, internal friction coefficient, and cohesion, as illustrated in Table 1.9 and Fig. 1.18. Conversely, an increase in compaction leads to a rise in the internal friction angle, along with an increase in the shear force required to induce relative displacement of caved ore and rock. Therefore, a higher bulking coefficient and void ratio, coupled with a lower degree of compaction, are advantageous for gravity flow.
(2) Humidity. As the humidity of caved ore and rock increases, the cohesive force and shear strength generated by capillary action increase sharply. However,

Table 1.9 Relationship between the internal friction angle, cohesion and the granular coefficient

Bulking coefficient	1.30	1.35	1.45	1.55	1.65	1.75
Internal friction angle/°	46	45	43	40	38	36
Cohesion/kg/cm^2	0.5	0.26	0.20	0.05	0	0

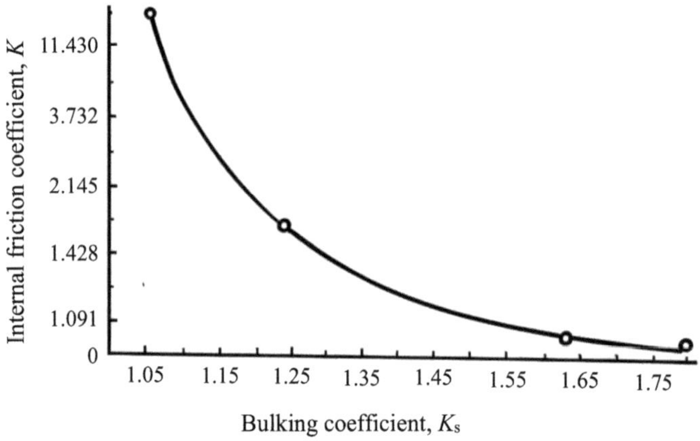

Fig. 1.18 Graph showing the relationship between internal friction coefficient and granular coefficient

when the humidity of the granular medium exceeds a certain threshold and becomes saturated with water, this cohesive force diminishes, leading to a significant reduction in the internal friction angle.

(3) Particle size composition. The presence of finer particles and clay particles, coupled with very low humidity, results in an increase in the internal friction angle of the granular medium.
(4) Particle shape and surface roughness. For particles of the same size, the internal friction angle of smooth caved ore and rock is lower. Conversely, particles that consist of multiple sizes with unequal axial apex angles exhibit a higher internal friction angle.
(5) Shear speed. An increase in shear speed leads to a decrease in the internal friction angle.

From the analysis of these factors, it is evident that the internal friction angle plays a significant role in influencing the conditions of gravity flow for caved ore and rock, as well as the shape and size of the IEZ and IMZ. Therefore, it is considered an essential physical and mechanical parameter.

1.2.9 Cohesion of Caved Ore and Rock

When cementing substances and water are present between the contact surfaces of caved ore and rock, the caved ore and rock exhibit a certain degree of stiffness and shear resistance, even in the absence of pressure. This initial shear resistance is referred to as the cohesion of caved ore and rock. Together with the internal friction force, it determines the shear strength of these materials.

The cohesion of caved ore and rock is influenced by several factors, including the quantity of clay particles present, moisture content, and the degree of compaction. Additionally, it is affected by the capillary action of water within the pores. As the concentration of cohesive particles and humidity increase, the application of pressure causes the caved ore and rock to consolidate, thereby enhancing cohesion. This cohesion plays a crucial role in the draw process and warrants careful consideration in draw research. Furthermore, the cohesion of caved ore and rock has been concurrently measured alongside the internal friction angle.

The property of caved ore and rock that leads to the loss of their loose nature and the formation of lumps is referred to as adhesiveness. When the dimensions of the drawpoint are such that large blocks cannot be freely discharged, it indicates a lack of adhesion. Conversely, the presence of cavities above the drawpoint suggests a higher degree of adhesiveness. Some researchers quantify this adhesion by measuring the area exposed by the cavities above the drawpoint. For instance, the measurement results for the degree of adhesion of caved ore and rock, with humidity levels ranging from 1.7% to 11% and clay particle content between 0 and 40%, are presented in Table 1.10.

From Table 1.10, it can be seen that as the clay content increases, as long as there is a certain humidity, the adhesiveness will also increase. With the increase in humidity, the adhesiveness also increases significantly. However, when the humidity reaches

Table 1.10 Determination results of the adhesiveness degree of caved ore and rock

Test serial number	Clay content/%	Number of experiments	Humidity at the maximum adhesion/%	Maximum exposed area of the upper void above the drawpoint/dm^2	Threshold humidity for adhesion onset/%
1	0	5	1.9–2.3	2.8	2.0–2.5
2	3	4	2.0–3.0	4	6.0
3	6	5	1.7–2.4	5	7.0
4	9	5	1.7–0.0	7	7.5
5	12	5	1.7–3.5	12	8.0
6	15	4	2.0–4.0	18	8.5
7	20	5	2.0–5.0	24	9.0
8	30	5	2.0–5.0	32	10.0
9	40	6	3.0–6.0	42	11.0

a certain level (depending on the clay content, the limit humidity can be 5–11%), it will instead reduce the adhesiveness.

The influence of adhesiveness on the mining process. The bonded caved ore and rocks are difficult to be released from the mining outlet, forming arches and cavities, which reduces the mining efficiency, poses a great threat to safety, and results in low recovery of pure ore and premature intrusion of waste rock into the cavities, causing significant losses and degradation.

1.3 Conclusions

This chapter separately discusses the principal theories of draw, as well as the key physical and mechanical properties of caved ore and rock that influence the draw process. These properties are crucial for understanding the laws of draw.

Caved ore and rock are classified as a non-ideal granular medium. In the learning proccss, it is essential to grasp the fundamental concepts, determination methods of various physical and mechanical properties, and their effects on gravity flow, as these form the foundation for draw research.

In future research, it is essential to focus on the rapid determination methods for particle size composition. Additionally, investigating the relationships among the bulking coefficient, void ratio, compaction degree, and their correlations with mechanical properties is crucial for understanding the shape and size of the IEZ and IMZ. Furthermore, it is important to study laboratory or field determination methods for the internal friction angle and cohesion of larger particle sizes in caved ore and rock, as this will serve as foundational work for draw simulation research and production practices.

References

1. Wang Q, Ren FY (2011) Mining engineering. Metallurgical Industry Press, Beijing
2. Wang CH (1982) Draw engineering. Metallurgical Industry Press, Beijing
3. Malakhov (1958) Draw from caved ore and rock. Metallurgical Industry Press, Beijing
4. Liu XG, Basic laws of ore movement during draw in caving methods (Part I): research on the ellipsoid theory. Nonferrous Metals 1979(4):38–45+26
5. Liu XG, Basic laws of ore movement during draw in caving methods (Part II): research on the ellipsoid theory. Nonferrous Metals 1979(5):40–44+33
6. Liu XG (1980) Movement of the ore-rock contact surface during draw in caving methods. Metal Mine 5:10–15
7. Wang YJ, Liu XG (1980) Investigation of the IEZ in multi-drawpoint draw. J Northeast Univ (Natural Science) 2:51–60
8. Li RF (1983) A unified mathematical equation for the fundamental draw laws. Nonferrous Metals (Mining Section) 1:1–8
9. Tian D, Shi HY, Fu EJ (2015) Study on relationship between coal caving step distance and top coal caving recovery rate based on ellipsoid theory. Coal Sci Technol 43(5):51–53, 143

10. Wu JJ (2009) Study on ore drawing law and structure parameter optimization in natural caving method. Central South University, Changsha
11. Ren FY (1994) Stochastic medium draw theory and its applications. Metallurgical Industry Press, Beijing
12. Litwiniszyn J (1956) Application of the equation of stochastic processes to mechanics of granular zones. Archiwum Mechaniki Stosowanej 8(4):393–411
13. Wang YJ (1962) New directions in draw theory: the stochastic medium approach. Donggong Loose-leaf Thesis Sel 8:535–538
14. Ren FY (1992) Stochastic medium theory for draw and its practical applications. Northeastern University, Shenyang
15. Tao GQ (2003) Investigation of flow property and their engineering applications. Northeastern University, Shenyang
16. Qiao DP (2003) Research on particle movement rule and ore drawing theory. Northeastern University, Shenyang
17. Liu XG, Wang YJ (1984) The zeroing volume and its applications. Nonferrous Metals (Mining Section) 5:17–21
18. Wei SL (1982) Two distinct properties of the two types of IEZs. University of Science and Technology Beijing, Beijing
19. Gao YT (1987) Theory of the expected draw zone. Metal Mine 11:20–27
20. Gao YT (1994) Proof of the transition relationship of the draw zone and its equal-thickness transition principle. Nonferrous Metals Eng 1:20–23
21. Gao YT (1993) The Velocity pattern and the constant velocity zone in the solo drawing hole. J Univ Sci Technol Beijing 4:327–331
22. Gao YT (1993) Estimating ore loss and dilution by expectated zone principle. Metal Mine 10:23–25
23. Gao YT (1994) Searching similarity of computer simulating drawing based on EVB draw theory. J Univ Sci Technol Beijing 3:201–206
24. Li RF (1994) Ideal equation of the ellipsoid-like draw theory. Nonferrous Metals (Mining Section) 5:38–44
25. Li RF (1994) Practical equation of the ellipsoid-like draw theory. Nonferrous Metals (Mining Section) 6:36–42
26. Zhang SH (2001) Draw theory and its verification. Xi'an University of Architecture and Technology, Xi'an
27. Pierce M (2010) A model for gravity flow of fragmented rock in block caving mines. University of Queensland, Brisbane
28. Nedderman RM (1995) The use of the kinematic model to predict the development of the stagnant zone boundary in the batch discharge of a bunker. Chem Eng Sci 50(6):959–965
29. Watson GR (1993) Flow patterns in flat-bottomed silos. University of Edinburgh, Edinburgh
30. Ferjani M (2003) An investigation of the plug/funnel-flow model. University of Minnesota, Minneapolis
31. Kuchta ME (2002) A revised form of the Bergmark-Roos equation for describing the gravity flow of broken rock. Miner Resour Eng 11(04):349–360
32. Melo F, Vivanco F, Fuentes C et al (2007) On drawzone shapes: from Bergmark-Roos to kinematic models. Int J Rock Mech Min Sci 44(1):77–86
33. Vivanco F, Watt T, Melo F (2011) The 3D shape of the loosening zone above multiple draw points in block caving through plasticity model with a dilation front. Int J Rock Mech Min Sci 48(3):406–411

Chapter 2
Compression Crushing Characteristics of Single Particle

Granular materials are discrete systems composed of a large number of macroscopic particles [1]. When the particle size of granular materials exceeds 1 m, they are referred to as granular media, where intermolecular forces weaken [2–4]. Granular materials are ubiquitous, existing in a spectrum between solid and fluid states, and exhibit complex mechanical properties. They play crucial roles in various industries, including agriculture, chemical engineering, construction, and civil engineering, and have been extensively studied in geotechnical engineering, geological engineering, and environmental science. Some granular materials occur naturally, such as snow, soil, and galactic clouds, while others are products of industrial processes, including grains, cement, concrete, and crushed stones. Additionally, some materials serve as intermediate products in manufacturing, such as ballast in railway engineering and soil-rock mixtures in slope engineering. During the mining process, caved ore and rock formed through natural or forced caving exhibit varying sizes, shapes, and complex arrangements, characteristic of a typical granular system. Mineral resources are essential for human advancement. Since the onset of the COVID-19 pandemic, the global mining market has experienced significant disruptions. In China, the consumption of mineral resources has been steadily increasing, yet domestic mineral production is insufficient to meet market demand [5, 6]. As China undergoes rapid development, addressing the imbalance between the supply and demand for mineral resources has become increasingly critical, necessitating the efficient exploitation and utilization of these resources.

The block caving method, recognized as one of the three principal mining techniques, is favored by numerous underground metal mines both domestically and internationally due to its advantages of low cost, high efficiency, and large-scale applicability. This method is extensively utilized in domestic metal mines. However, there remains a pressing need for continuous improvement in domestic mining technology. Currently, the loss and dilution rates in block caving mines in China exceed those observed abroad [8]. In the block caving method, the mined-out areas are filled with caved ore and rock through either forced or natural caving processes. Three

H. Sun et al., *Study on Flow and Hang-up Characteristics of Caved Ore and Rock*,
https://doi.org/10.1007/978-981-92-0335-2_2

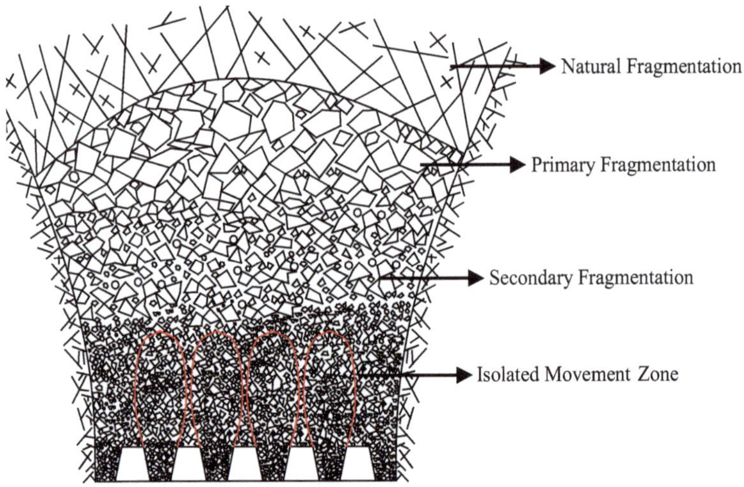

Fig. 2.1 Schematic of fragmentation processes of caved ore and rock in block caving method [7]

distinct types of caved ore and rock fragmentation phenomena can be identified: in-situ or natural fragmentation, primary fragmentation, and secondary fragmentation, as illustrated in Fig. 2.1 [7, 9, 10]. During the block caving mining process, natural fragmentation occurs prior to any artificial interference. As mining progresses, the stress state of the rock mass undergoes changes, resulting in primary fragmentation. Secondary fragmentation occurs during the subsequent draw process, driven by the continuous downward movement of caved ore and rock under the influence of gravity, as well as mechanical effects such as mutual compression and shearing between local caved ore and rock. The phenomenon of secondary fracturing in caved ore and rock introduces complexity into the ore drawing process, significantly impacting both the operational procedures and the final outcomes of ore extraction [7]. Brown [11] and Chitombo et al. [12] have posited that the occurrence of secondary fragmentation is a critical factor to consider prior to the implementation of the natural caving method in mining operations. This phenomenon directly influences the design and selection of various stope structural parameters, as well as the ultimate ore recovery rates.

In recent years, many mining professionals have primarily focused on the economic benefits of mines, leading to a deficiency in foundational theoretical research regarding the flow characteristics of caved ore and rock. This deficiency is particularly evident in the study of secondary fragmentation of these particles, as well as their transport and evolution mechanisms. Consequently, the development of domestic draw theory has lagged, which hinders the widespread application of the block caving method in China. Understanding the fragmentation mechanisms of caved ore and rock, along with their movement and evolution laws, is essential for minimizing the loss and dilution rates of ores. However, the current understanding of the secondary fragmentation mechanisms of caved ore and rock remains inadequate, which directly impacts the analysis and prediction of the particle size of these materials, resulting in unsatisfactory draw outcomes. In recent years, both domestic and international scholars have made progress in researching the secondary fragmentation of caved ore and rock. Currently, related research primarily focuses on two aspects:

(1) Factors influencing secondary fragmentation. The phenomenon of secondary fragmentation of caved ore and rock is complex and influenced by multiple factors. It is directly related to the physical and mechanical properties of the particles, including strength, size, shape, and the distribution and intensity of surface or internal fractures. Additionally, various external factors, such as the stress state of the mining area, structural parameters, and draw methods, also play a significant role [13–16]. Currently, research on the factors influencing the secondary fragmentation of caved ore and rock is relatively limited, with a lack of quantitative studies and no consensus within the field.
(2) Prediction of secondary fragmentation size. The size distribution of caved ore and rock is crucial in mining operations, significantly influencing production costs and profits. Given its close relationship with stope structural parameters and draw plan design, predicting the size of caved ore and rock remains a key focus in block caving research [17–23]. However, factors such as the invisibility of joints and fractures within the rock mass complicate this prediction. Currently, methods for predicting the secondary fragmentation size of caved ore and rock predominantly rely on specific data from individual mines, which limits their practicality and applicability in real-world scenarios. Furthermore, the prediction results often lack extensive engineering case studies for validation.

In addition, most physical draw tests do not account for the secondary fragmentation of caved ore and rock. This oversight arises because the raw materials used in these tests often inadequately represent the physical properties essential for accurate assessments, particularly in terms of mechanical properties. As a result, the effective fragmentation of caved ore and rock cannot be guaranteed during physical draw tests, rendering them insufficient for a precise examination of secondary fragmentation issues. In numerical draw studies, the spherical particle model or double particle model employed in the discrete element method is commonly used to simulate particle shapes. However, this approach significantly deviates from the actual irregular shapes of real caved ore and rock, failing to capture the interlocking forces between them. Consequently, this limitation leads to inevitable discrepancies in simulation outcomes, resulting in low reliability and practicality [7].

The geometric and physical mechanical properties of caved ore and rock significantly influence the macroscopic mechanical properties of the particle system. The single particle crushing test allows for precise control under loading conditions, making it a reliable method for investigating the compression and crushing characteristics of individual particles. This research can serve as a foundation for understanding the crushing characteristics and mechanical behavior of particle systems [24, 25]. In summary, exploring the crushing characteristics of caved ore and rock through single particle crushing tests, in conjunction with existing draw theories, can provide scientific guidance for engineering practices and help achieve optimal draw outcomes. The findings represent a valuable contribution to contemporary research on the secondary fragmentation of caved ore and rock. They establish a theoretical framework for optimizing blasting parameters and stope structure parameters, predicting dilution indices, and ensuring the safe and efficient extraction of mineral resources from underground mines. Furthermore, this research offers insights into the compression and fragmentation characteristics of various granular materials, including coal particles, coral particles, and rockfill particles.

2.1 Research on Single Particle Fragmentation

The common phenomenon of particle fragmentation refers to the shear or tensile failure of particles under external loads, wherein internal initial cracks or newly generated cracks in weak areas continue to expand, ultimately leading to penetration failure [26]. In experimental tests, particle fragmentation results in discrepancies between the particle system and its original structure. Research primarily integrates macroscopic physical phenomena with mesoscopic numerical tests to investigate the characteristics of particle fragmentation and its influence on the macroscopic and mesoscopic mechanical properties of the particle system.

In practical applications, both single-particle and multi-particle crushing occur simultaneously. Due to the ease of accurately controlling loading conditions in single-particle crushing tests, many scholars have investigated the macro and meso crushing mechanical responses and mechanisms of single particles through numerical simulations and laboratory tests. This research approach, which transitions from individual to collective analysis, enables a more systematic and comprehensive study of particle fragmentation, thereby enhancing the understanding of the damage characteristics and mechanical behavior of granular assemblies. In physical tests of single-particle crushing, novel technologies such as electronic computer tomography and laser three-dimensional scanning can be employed to investigate the characteristics of particle fragmentation at both macro and micro levels. However, limitations such as restricted test materials, high costs, and poor test repeatability persist. Currently, research on single-particle fragmentation primarily focuses on fragmentation modes, influencing factors, particle size distribution, and fragmentation strength.

2.1.1 Particle Crushing Mode

In the single particle crushing test, the results exhibit a certain degree of randomness, indicating diverse crushing modes. Guyon et al. [27] proposed a classification method for these single particle crushing modes. Based on the size differences of the fragments resulting from single particle crushing, three distinct types of crushing modes are identified, as illustrated in Fig. 2.2. Specifically, in Fig. 2.2a, the fragments produced are approximately equal in size to the original particle. In Fig. 2.2b, the particles are broken into the original particle size, with only the surface experiencing fragmentation; the resulting fragments consist of a slightly smaller particle along with several smaller particles. In Fig. 2.2c, the size of the original particles remains largely unchanged, while fine powdered particles are generated solely on the surface due to the grinding effect. Gorham et al. [28] conducted impact and compression crushing tests using uniform particle sizes. Their findings indicated that the classification of particle crushing modes varies under different loading conditions. In the impact tests, particle crushing modes were categorized into five distinct types, while in the compression tests, they were classified into four types. Additionally, the loading speed or corresponding load for the different crushing modes varied. Similarly, Antonyuk et al. [29] performed impact tests on needle-like crystal particles and static compression tests on cylindrical particle deposits. They observed that the particle fragmentation modes under these two loading conditions were distinct: needle-like crystal particles fractured into two sections upon impact, while cracks in the cylindrical particle accumulation propagated and penetrated along the loading direction, resulting in a similar fragmentation into two parts. Both Zhang et al. [30] and Zhang [24] proposed that particle fragmentation modes can be classified into three types: contact cracking, intermediate cracking, and random cracking. Furthermore, Ma et al. [31] indicated that the load–displacement curve in the single particle fragmentation test of coral is influenced by the particle failure mode, which can also be categorized into three typical types. Wang [32] highlighted the close relationship between particle crushing modes, particle shape, and the crushing load–displacement curve.

Particle crushing modes are primarily classified based on test results, which cannot be predicted prior to testing. Consequently, these modes can be categorized, and the proportion of each crushing mode across different testing methods can be analyzed using statistical methods, allowing for the proposal of a reasonable model. Furthermore, the classification of particle crushing modes often relies on manual statistics, which can introduce significant errors due to subjective factors. Therefore, the application of laser scanning and image identification technologies is recommended to enhance both the efficiency and accuracy of statistical analyses.

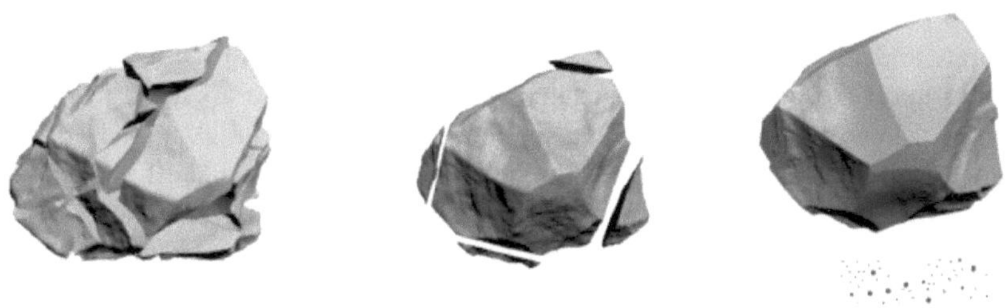

(a) Cracking (b) Fragmentation (c) Attrition

Fig. 2.2 Single particle crushing modes

2.1.2 Influencing Factors of Particle Fragmentation

McDowell et al. [33] demonstrated that the compressive crushing strength of a single particle can be modeled using the Weibull distribution. Nakata et al. [34] investigated various particle types, including sand and glass, confirming that the single particle crushing strength, when analyzed through the lens of the Weibull distribution, is significantly influenced by the particle's shape and material properties. Lim et al. [35] investigated various types and sizes of ballast particles, further substantiating the influence of intrinsic particle properties on the size effect observed in single particle crushing, as evidenced by the Weibull distribution of compressive crushing strength. Furthermore, Huang et al. [36] analyzed glass beads of differing sizes and observed that the characteristic crushing strength within the Weibull distribution for these particles also demonstrates a size effect. By employing image processing software alongside the discrete element method (DEM) for numerical simulation, Oskooei et al. [37] conducted a study on the compression of carbon black single particles, revealing that the crushing strength of particles with diverse shapes conformed to the Weibull distribution. The size effect and self-similarity were influenced by particle shape, loading direction (see Fig. 2.3), and other factors. Additionally, the evolution of the internal force chain during the crushing process of particles with different shapes was explored. Meng et al. [38] conducted single particle crushing tests on mudstone and sandstone with varying particle sizes. They employed Weibull statistical analysis to develop a model based on fractal dimensions for predicting single particle crushing strength and energy. Zhou et al. [39] conducted a single particle compression test on five groups of rockfill particles of varying sizes using the random granular discontinuous deformation analysis method. They observed that the compression curves for different particles exhibited a notable consistency. Their findings, supported by Weibull statistical analysis, suggest that the strength properties of the particles have a more significant influence on the characteristic strength than the particle size itself. Xu et al. [40] established a link between particle shape and the load-strain curve derived from single particle compression tests on marble, categorizing the curve into three distinct types. Ma et al. [31] performed a single particle crushing test involving approximately 300 coral particles under varying loading rates. Their results indicated that both particle crushing energy and fractal dimension increase with the loading strain rate and are linearly positively correlated with the logarithm of the strain rate. Building upon the mesoscopic mechanisms of single particle crushing in rockfill materials, Zhou et al. [41] proposed a single particle random crack model utilizing Particle Flow Code (PFC) and validated the model's rationality and effectiveness.

Domestic and international scholars have conducted extensive research on the crushing characteristics of single particles through physical tests and numerical simulations. Numerous influencing factors are at play, including particle type, shape, size, initial strength, loading rate, and loading direction, many of which are interdependent. Analyzing test results encompasses various aspects such as force–displacement curves, crushing strength, crushing modes, energy dissipation, fractal dimensions, crack propagation, and the evolution of internal force chains. Currently, there remains

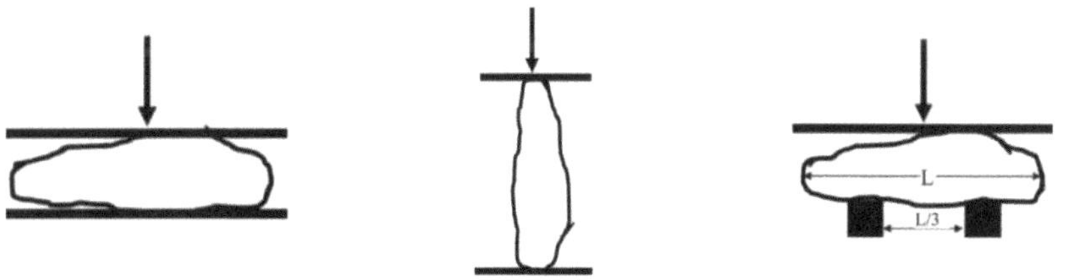

(a) Place the single particle horizontally along its longest dimension
(b) Place the single particle vertically along its longest dimension
(c) Position the single particle between two rigid supports along its longest dimension so that it acts as a beam

Fig. 2.3 Schematic of single-particle compression loading direction [37]

a need for further exploration in the study of single particle crushing characteristics. Firstly, in analyzing influencing factors such as particle shape, size, and loading direction, there is a lack of suitable physical quantities for quantitative descriptions, and the primary and secondary relationships among these factors have not been quantitatively defined. Secondly, no universal quantitative formula exists to relate size effects, fractal dimensions, energy dissipation, and crushing strength in the results of single particle crushing tests. Lastly, the mesoscopic aspects, including the internal structural characteristics and spatial arrangement of particles, require further investigation.

2.1.3 Size Distribution of Broken Fragments

Particle size distribution refers to the proportion of the number, mass, and surface area of particles across different size ranges relative to the total assembly of particles; it is also known as particle dispersion. In experiments, particle size distribution results are obtained using a laser particle size analyzer or through particle screening tests. Given the inherent variability in particle sizes within a system, an unavoidable size interval exists. The various fragments of a single particle collectively constitute a subpopulation of particles. Zhang [24] categorized individual particle fragments by mass instead of size, thereby obtaining a discrete particle fragment size distribution. To highlight regularities in the particle size distribution of these discrete fragments, the proportions of particles within each size range can be connected to form a continuous particle size distribution curve. This curve can subsequently be fitted using statistical distribution functions, which is referred to as the particle size distribution function. Table 2.1 presents a common particle size distribution function [32]. However, it is crucial to note that employing mathematical functions to quantitatively characterize the particle size distribution of fragments may not fully reflect actual results, as the correlation coefficient for the curve may vary depending on the selected parameters.

Table 2.1 Common particle size distribution functions

Distribution name	Distribution function
Normal distribution	$F(x) = \frac{1}{2}\left[1 + \text{erf}\left(\frac{x-\mu}{\sigma\sqrt{2}}\right)\right]$
Lognormal distribution	$F(x) = \frac{1}{2}\text{erf}\left(-\frac{\ln x-\mu}{\sqrt{2}\sigma}\right)$
Pareto distribution	$F(x) = \frac{k}{(k+x)^a}, x \geq 0$
Standard Pareto distribution	$F(x) = \frac{1}{(1+x)^a}, x \geq 0$
Rosin–Rammler distribution	$F(x) = \exp\left[-\left(\frac{x}{x_0}\right)^k\right]$
Weibull distribution	$F(x) = 1 - \exp\left[-\left(\frac{x}{x_0}\right)^k\right]$
Power-law distribution	$F(x) = \left(\frac{x}{x_0}\right)^k$

The caved ore and rock exhibit irregular shapes and uneven distributions, characterized by fractal properties in terms of shape, structure, strength, and deformation [42]. Consequently, the fragmentation distribution of rock particles cannot be adequately described using traditional geometric methods; rather, its dimensionality, specifically the fractal distribution, requires further exploration. The relationship between large-scale and small-scale characteristics can be quantified using the fractal dimension *D*. Building upon fractal theory, Mandelbrot [43] and Tyler et al. [44] proposed models for two-dimensional and three-dimensional particle size distributions, respectively. In China, Xie [45, 46] was the first to apply fractal theory to the study of rock mechanics and geotechnical mechanics. In terms of particle fragmentation, Xu et al. [47] utilized fractal theory, indicating that the fractal dimension *D* of particle fragmentation ranged from 2.0 to 3.0 based on their experimental analysis.

In the field of mining engineering, the prediction of particle size distribution of caved ore and rock has yielded significant research outcomes. Esterhuizen [17] introduced the empirical method known as Block Caving Fragmentation (BCF) for predicting the particle size of caved ore and rock, which was later refined [18]. Butcher et al. [19] developed a novel particle size prediction method based on numerical simulation, enhancing prediction accuracy. Pierce [20] improved the functionality of the Rapid Emulator Based on PFC3D (REBOP) software [21], thereby increasing the accuracy of REBOP's predictions for particle fragmentation by addressing issues related to fine particle flow and the secondary fragmentation of caved ore and rock. Additionally, Kojovic [22] and Rogers et al. [23] proposed corresponding prediction methods for rock fragmentation based on the particle fragmentation law and the discrete fracture network model (DFN), respectively. However, these methods heavily depend on extensive field-measured data to quantify input parameters, which limits their technical and economic feasibility. Furthermore, the predictive results lack validation through engineering examples. Therefore, further exploration into the prediction of particle size distribution of caved ore and rock is essential.

2.1.4 Particle Crushing Strength

The crushing strength of particles is typically quantified using parameters such as stress and energy. Among these parameters, the calculation of stress can be categorized into several types. Jaeger [48] noted that the rock crushing strength is determined using the Eq. (2.1):

$$\sigma_f = \frac{F_f}{d^2} \tag{2.1}$$

Hiramatsu et al. [49] indicated that particle crushing generally involves tensile failure, and its crushing strength can be calculated using the Eq. (2.2):

$$\sigma_f = 0.9\frac{F_f}{d^2} \tag{2.2}$$

Additionally, Nakata et al. [50] focused on sand particles and found that their crushing strength can be determined using the Eq. (2.3):

$$\sigma_f = \frac{F_f}{A} = \frac{4}{\pi}\frac{F_f}{d^2} \tag{2.3}$$

Here, σ_f is the stress upon particle crushing, also termed the characteristic stress, N; F_f is the axial force upon particle crushing, N; d is the characteristic particle size, mm.

Under identical test conditions, the crushing strength of brittle particles exhibits discrete characteristics, which can be attributed to the inherent heterogeneity of the rock [51]. To accurately characterize the strength distribution of individual particles, it is essential to identify an appropriate statistical distribution model. Commonly employed models include the Weibull function, Lognormal function, and Logistic function [52]. Among these, the Weibull function is the most frequently utilized for fitting the data. Studies conducted by McDowell [33], Huang et al. [36], Oskooei et al. [37], Meng et al. [38], Zhou et al. [39] and Xu et al. [40] have demonstrated that varying factors such as particle size, strength properties, loading rate, and loading direction can effectively model the crushing strength distribution of brittle single particles using the Weibull function. Furthermore, the variations in parameters within this function can reflect differences in strength distribution. Additionally, advancements in research technology have led to continuous improvements in methodologies for assessing particle crushing strength, thereby facilitating a more comprehensive understanding of mesoscopic crushing mechanisms through the application of discrete element software [53–55].

2.1.5 Research Status of Distinct DEM Simulation Approaches

Currently, scholars have conducted numerical simulation studies on the compression and fragmentation of individual particles of various materials using numerical simulation methods such as the DEM [37, 56–58], the Combined Finite Discrete Element Method (FDEM) [59–61], and the Extended Finite Element Method (XFEM) [62, 63]. The DEM method, characterized by its straightforward theoretical framework, clear physical interpretation of input parameters, and ease of implementation, has gained widespread recognition and application in simulating single particle fragmentation. Typically, particle fragmentation simulations based on DEM employ two primary methods: the Fragment Replacement Method (FRM) and the Particle Bonding Method (PBM) [57]. FRM emphasizes the criteria for particle fragmentation and the definition of the replacement mode for fragmented particles. In this method, the particle model disintegrates upon meeting the fragmentation criteria, with equivalent fragments substituting the original particle [64, 65]. Conversely, PBM, which is more commonly utilized, employs bonding keys to connect several sub-particles into a crushable cluster and simulates the fragmentation process through the fracture of these bonding keys [66, 67]. However, similar to conventional DEM simulations of single-axis compression of rocks, the current PBM-based single-particle crushing simulations exhibit drawbacks, including excessively ‘smooth’ spherical particles, excessive inter-particle voids, and limited applicability of the contact model, all of which diminish the accuracy and reliability of DEM simulations of single-particle fragmentation. Consequently, there is an urgent need to propose a new model for the compression and crushing of irregular single particles.

2.2 Experimental Investigation of Single-Particle Compression Crushing

To investigate the characteristics of single-particle compression crushing of caved ore and rock with varying shapes, three-dimensional image data of the particles were acquired using a three-dimensional scanner. The shape characteristics of the caved ore and rock were quantitatively analyzed. Subsequently, a single-particle compression test was conducted on the caved ore and rock.

2.2.1 Experimental Process of Single-Particle Compression Crushing

2.2.1.1 Introduction of Experimental Materials

As illustrated in Fig. 2.4, the single-particle compression test utilized magnetite particles sourced from the Shouyun Iron Mine, with the ore grade determined to be approximately 31% via X-ray diffraction (XRD). The particle strength of the Shouyun iron ore is classified as medium. Standard rock samples underwent uniaxial compressive strength and tensile strength tests, yielding fundamental physical and mechanical parameters, which are presented in Table 2.2. For this test, samples with a geometric average particle size ranging from 20 to 45 mm were selected as the subjects of research. To ensure single contact between the particles and the pressure plate, the selected particles were predominantly convex in shape. Furthermore, to minimize noticeable shaking of the particles during the compression process, there was only one contact point on the upper surface of the selected particles, while the lower contact surface was represented by an imaginary plane composed of at least three contact points. It is essential that the upper and lower contact surfaces are aligned vertically [68].

Fig. 2.4 Caved ore and rock used in single-particle compression crushing tests

Table 2.2 Basic physical and mechanical parameters of Shouyun iron ore samples

Sample	Compressive strength/MPa	Tensile strength/MPa	Elastic modulus/GPa	Poisson ratio	Density/(kg/m^{-3})
Iron ore	91.91	7.92	84.76	0.24	3500

2.2.1.2 Quantitative Characterization of Particle Shape

Before conducting single-particle compression crushing tests, it is essential to quantitatively characterize the shape of the selected irregular caved ore and rock. Currently, the quantification of particle shape predominantly relies on two-dimensional cross-sections, which do not accurately reflect their sphericity. Therefore, in this test, after selecting and serially numbering the particles, we will obtain the three-dimensional image information of the caved ore and rock through the following process:

(1) The EiScan-SE 3D white light scanner is employed for the automatic scanning of individual particles, effectively addressing occlusion issues. The outer surface shape feature information of each particle is captured through one or more scans, resulting in a comprehensive three-dimensional numerical representation of the individual particles, which is subsequently stored in the STL file format. The scanning process is depicted in step (a) of Fig. 2.5.
(2) The scanning accuracy of the scanner is 0.1 mm. To enhance the quality of the STL file, it is necessary to simplify the complex mesh using MeshLab. This process involves controlling the number of particle surfaces in the STL file to approximately 1000, as outlined in step (b) of Fig. 2.5.
(3) Based on the three-dimensional image information of a single particle, as illustrated in step (c) of Fig. 2.5, Rhino software is employed to determine the dimensions of the long axis L, short axis S, and middle long axis I of the single particle according to the minimum cuboid criterion. Additionally, basic shape

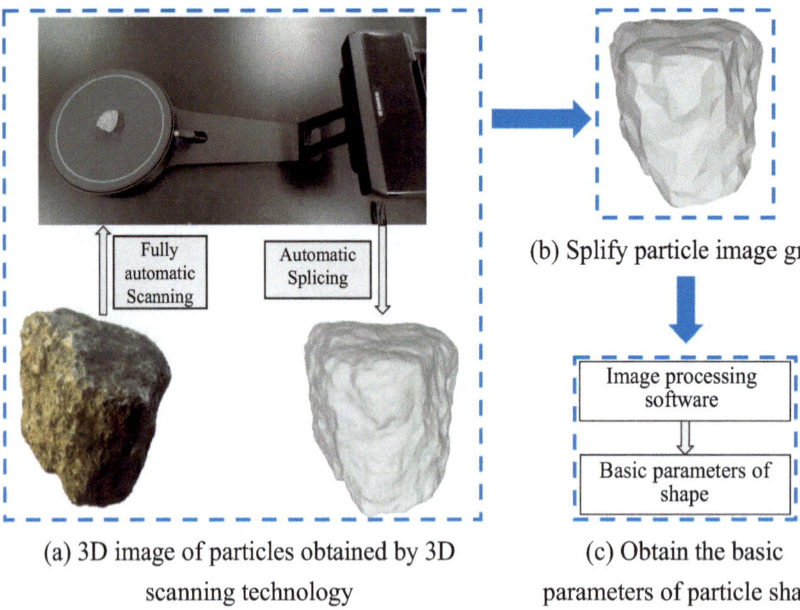

Fig. 2.5 Workflow for acquiring 3D image information of caved ore and rock

Table 2.3 Shape characterization parameters of caved ore and rock

Scale	Shape parameter	Definition	Formula
Macroscopy	Flatness f	The ratio of the short axis S to the middle and long axis I	$f = S/I$
	Elongation e	The ratio of the long axis L to the middle long axis S	$e = L/S$
	Sphericity ϕ	The ratio of equal volume sphere surface area S_p to particle surface area S_A indicates the degree of approximation between particles and spheres	$\phi = S_P/S_A$
Microscopic	Edge angle c	The ratio of the minimum convex shell volume V_{con} to the particle volume V represents the convexity of the particle contour	$c = V_{con}/V$

parameters such as surface area S_A and volume V are calculated, while the minimum convex hull volume V_{con} is derived using the Convexhull function in MATLAB software.

The characterization of irregular particle shapes can generally be categorized into three levels: contour shape, convexity, and surface texture. Contour shape and convexity describe the macroscopic and mesoscopic external characteristics of the particles, while surface texture pertains to the microscopic surface characteristics at a smaller particle size [69, 70]. For irregularly shaped particles, such as coral particles and caved ore and rock, the shape characteristics can be described using two levels of variables: contour shape and convexity [71]. Consequently, this study focuses exclusively on the macroscopic and mesoscopic scales of caved ore and rock, specifically examining contour shape and convexity. In this analysis, four commonly utilized shape characterizations—flatness, elongation, sphericity, and edge angle—were selected to delineate the differences in particle shape, as presented in Table 2.3. Based on the fundamental shape parameters of particles obtained through three-dimensional scanning, the range of macroscopic shape characterizations is depicted in Fig. 2.6. The range of particle flatness is between 0.347 and 0.975, while the range of particle sphericity spans from 0.671 to 0.880, and the range of particle elongation varies from 1.100 to 3.846. Furthermore, the edge angle of the particles is predominantly concentrated around 0.900, which can be attributed to the removal of prominent edges from the selected test particles prior to measurement due to factors such as extrusion and friction.

2.2.1.3 Introduction of Test Process

As shown in Fig. 2.7, the single-particle compression test of caved ore and rock was conducted using a microcomputer-controlled electro-hydraulic servo rock press, model YAW-600. The maximum test force was 600 kN, with a displacement measurement resolution of 0.002 mm. Prior to the test, the caved ore and rock samples were arranged horizontally on the lower platform, aligned along their longest dimension,

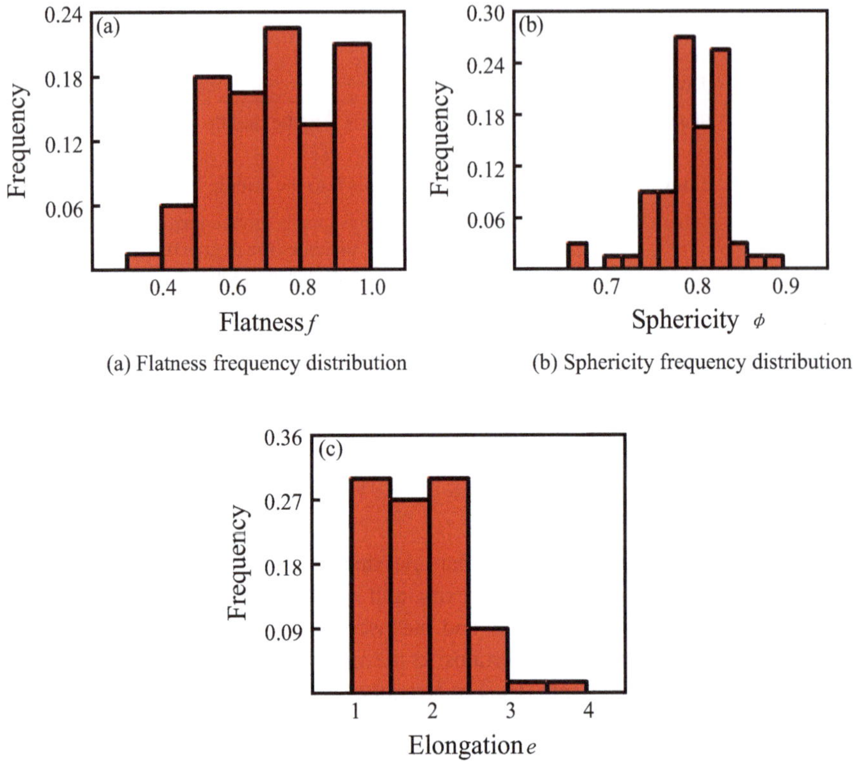

(a) Flatness frequency distribution

(b) Sphericity frequency distribution

(c) Elongation frequency distribution

Fig. 2.6 Distribution of shape-descriptor parameters that characterize the macro-scale contour shape of particles

and adjusted to ensure stability. The single-particle compression test procedure can be divided into three distinct steps:

(1) Preloading: Displacement served as the control parameter during the preloading phase. Initially, a high displacement loading rate was applied until the surface of the single particle was approached. This was followed by a continuous slow loading phase at a low displacement rate of 10%, until the upper pressure plate made contact with the upper surface of the single particle. At this juncture, the computer data recording panel displayed the pressure value, with the preload force set at 200 N.
(2) Compression: Single-particle compression tests were conducted at a constant displacement rate of 5 mm/min, with real-time recording of the load–displacement curve throughout the loading process.
(3) Termination Conditions: The local crushing induced during single-particle compression results in fluctuations in the load–displacement curve. The test can only be concluded when one of the following two conditions is met: (1)

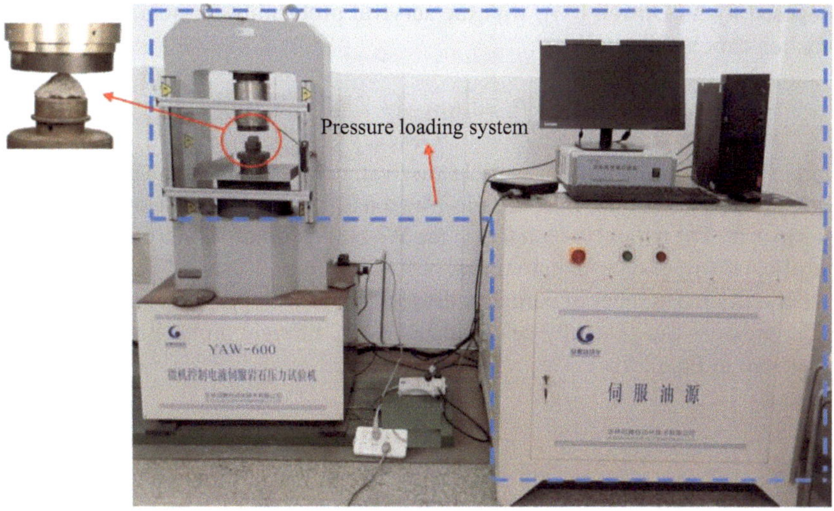

Fig. 2.7 Single-particle compression test device

the particle exhibits a macro penetrating crack; (2) the load–displacement curve demonstrates a significant decline [52]. To ensure the stability of the single-particle crushing strength, more than 30 tests are required. A total of 80 typical caved ore and rock samples were selected for single-particle compression tests. Based on the test results, 10 particles with obvious abnormalities were excluded, and the remaining 70 particles were analyzed for their crushing characteristics.

2.2.2 Analysis of Physical Test Results of Single-Particle Compression

Based on the results of single-particle compression tests conducted on caved ore and rock, this study analyzes the compressive crushing characteristics of irregular caved ore and rock from several perspectives. These perspectives include the distribution of particle crushing strength, the various modes of crushing, particle size distribution, and the morphological characteristics of fracture surfaces.

2.2.2.1 Distribution Characteristics of Crushing Strength

Given the irregular shape of caved ore and rock, it is essential to minimize measurement errors resulting from subjective human factors. Therefore, the particle crushing strength is represented by the peak pressure in the load–displacement curve. To model the distribution of particle crushing strength, we employ the Weibull function proposed by McDowell [33], with the survival probability formula of the Weibull function expressed as follows:

$$P_s = \exp(-[F/F_0]^m) \tag{2.4}$$

Here, P_s is the probability of particle survival, expressed as a percentage. F is the peak pressure at failure, measured in Newtons N. F_0 is the yield strength, also in Newtons N. The parameter m refers to the Weibull modulus, which characterizes the statistical distribution of failure times.

For a limited number of particles, the residual probability P_s of various particles under pressure can generally be calculated using different probability estimates, as illustrated in Eq. (2.5):

$$P_s = \frac{t}{n+1} \tag{2.5}$$

Here, n is the total particle number; t is the order of all the particle crushing strength values according to the ascending order.

According to the original strength distribution of particles and Eq. (2.5), the relationship curve between particle strength and residual probability $P_s \sim F$ can be drawn (Fig. 2.8a). The survival probability of particles subjected to peak pressure F is derived by taking the logarithm of both sides of the equality in Eq. (2.4):

$$\ln[\ln(1/P_s)] = m\ln(F/F_0) = m\ln(F) - m\ln(F_0) \tag{2.6}$$

It can be seen from Eq. (2.6) that $\ln[\ln(1/P_s)]$ has a linear relationship with $m \ln(F)$, and the slope obtained by fitting the straight line in the log–log plot is m. When F is equal to F_0, $P_s = 1/e$ or $\ln[\ln(1/P_s)] = 0$, F is the particle strength corresponding to 37% survival probability. Through Eq. (2.6), the double logarithmic points are calculated, and the correlation coefficient R^2 obtained by linear fitting

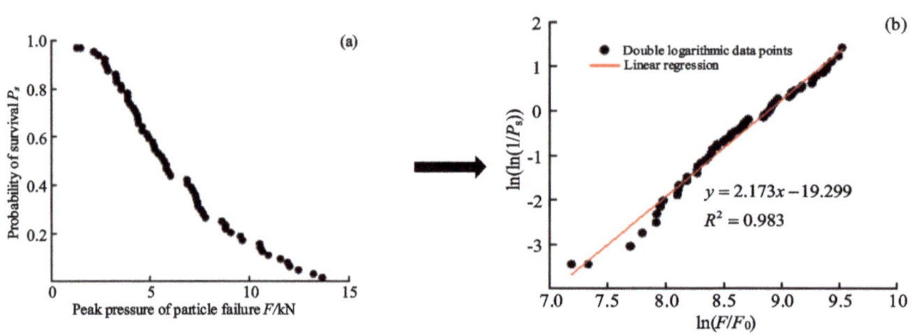

(a) Relationship curve between peak pressure and residual probability

(b) Weibull distribution fitting of particle crushing strength

Fig. 2.8 Weibull distribution fitting results of single-particle crushing strength of caved ore and rock

(Fig. 2.8b) is 0.983. According to the fitting Eq. (2.4) and the conversion Eq. (2.6), the corresponding Weibull parameter m is 2.173, and the characteristic strength F_0 is 7195.925 N. By comparing the original test data, the characteristic strength F_0 obtained is found to be closely aligned with the particle crushing strength, with an actual particle survival probability of 37%. The test results indicate that the crushing strength distribution of single particles of irregular caved ore and rock conforms to the Weibull function, similar to the behavior observed in brittle particle materials such as rockfill particles.

2.2.2.2 Crushing Modes

The crushing modes of particles under external load are primarily classified based on various criteria, including the size of the crushing fragments [27], the cracking mode [30], and the degree of particle crushing [72]. As illustrated in Fig. 2.9, the crushing modes of a single particle of caved ore and rock in this test be categorized into four typical types: mid-span crushing, edge attrition, through-crack failure, and random cracking.

(1) Mid-span crushing: Upon compression, particles are crushed from the center, resulting in a cross-section that is approximately perpendicular to the upper and lower contact surfaces.
(2) Edge attrition: Minor attrition occurs at a specific contact point between the particle surface and the platen.

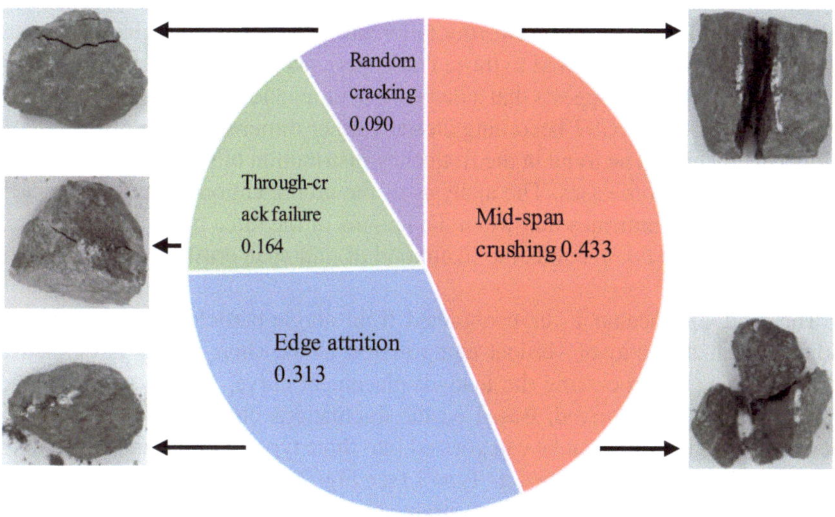

Fig. 2.9 Probabilistic statistics of single particle crushing modes

(3) Through-crack failure: The test ceases due to significant failure of the particle's main body, characterized by noticeable gaps, while the particles themselves remain intact.
(4) Random cracking: The location of particle damage is arbitrary, occurring independently of the contact point between the particle surface and the plate.

The results of the statistical analysis on particle crushing reveal that mid-span crushing particles constitute the highest proportion, accounting for 43.3%. In contrast, edge attrition particles represent 31.3%, while particles exhibiting through-crack failure account for 16.4%. The fewest particles are those that experienced random cracking, comprising only 9.0%.

During the testing process, it was observed that the shape of the particles is related to their mid-span crushing behavior. Further research was conducted to examine the effects of three shape characterization parameters—sphericity, elongation, and flatness—on the mid-span crushing modes of the particles (note that angularity, as a meso shape characterization parameter, is not included in this study). As illustrated in Fig. 2.10, a histogram was created, and a spline curve was employed to connect the midpoints (medians) of the histogram peaks for trend analysis. A comparison of the frequency distribution of shape characterization parameters for particles that underwent mid-span crushing with that of the original, untested particles revealed significant differences in terms of flatness. In the yellow zone of Fig. 2.10a, the frequency distribution of the original particles and those that underwent mid-span crushing reveals a significant difference in the region near 1.0. This observation indicates that as the flatness of the particles increases, the long axis I and the short axis S converge, resulting in a more square-shaped axial cross-section. Consequently, this geometric configuration leads to a reduced likelihood of mid-span crushing. Similarly, in the yellow zone of Fig. 2.10b, a noteworthy difference in frequency distribution is noted for original particles compared to those that experienced mid-span crushing, also in proximity to 1.0. This suggests that a decrease in elongation correlates with the long axis L and mid-long axis I becoming closer, further diminishing the probability of mid-span crushing. The trend in the frequency distribution of sphericity is generally consistent across both cases. The analysis of the aforementioned shape characterization parameters indicates that when the lengths of the three principal dimensions of the particles are comparable, the likelihood of mid-span crushing is significantly reduced.

The load–displacement curve obtained from single-particle compression tests indicates that the ratio of vertical compression to the short axis S is defined as vertical strain. By processing the load–displacement curve, various types of load-strain curves can be derived. Based on the fluctuations observed before the peak, these load-strain curves can be categorized into three types. For instance, the curves shown in Fig. 2.11 are classified as Type 1 (see Fig. 2.11a), Type 2 (see Fig. 2.11b), and Type 3 (see Fig. 2.11c). The load-strain curve of Type 1 approximates a straight line prior to the peak; the load-strain curve of Type 2 exhibits fluctuations before the peak, resulting in the formation of debris and small fragments; whereas the load-strain curve of Type 3 shows significant fluctuations before the peak, leading to the

Fig. 2.10 Frequency distribution of macroscopic shape characterization of original particles and mid-span crushing particles

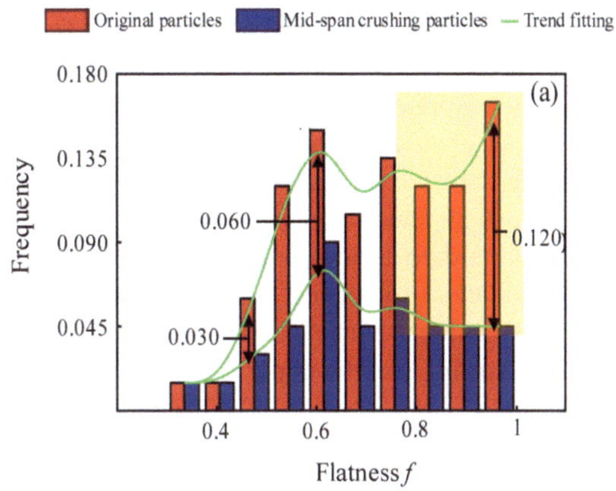

(a) Flatness frequency distribution

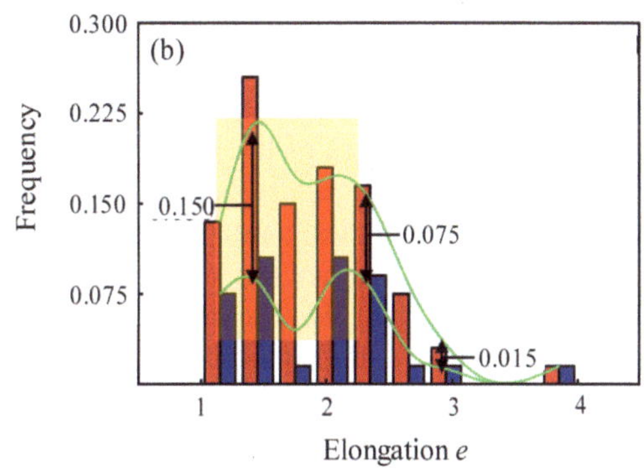

(b) Elongation frequency distribution

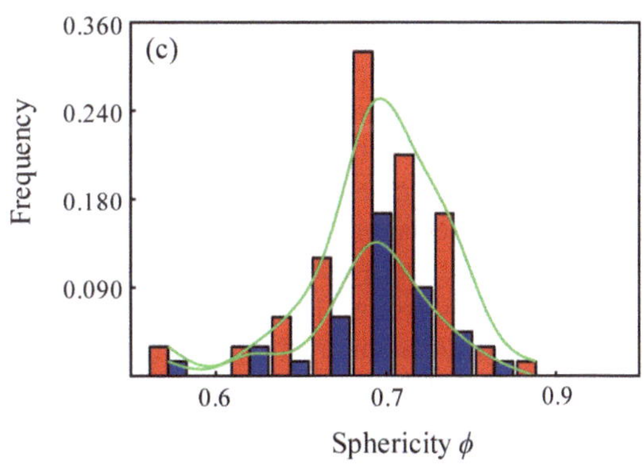

(c) Sphericity frequency distribution

generation of numerous fragments and the fragmentation of particles into multiple pieces. The fluctuations observed in the load-strain curves of Type 2 and Type 3, which occur prior to particle failure, are attributed to the presence of weak points on the surface of the caved ore and rock. These weak points lead to poor contact, resulting in grinding or local crushing before the main failure under external loading. Alternatively, the fluctuations may arise from the tilting of contact points after the surface has been crushed during compression, both of which are influenced by particle shape. Consequently, to achieve a load-strain curve that approximates a straight line before the peak in single-particle crushing tests, it is essential to select particles with fewer edges, ensuring that the upper and lower contact points are as vertically symmetrical as possible.

2.2.2.3 Distribution Characteristics of Fragments

Different crushing modes of single particles of caved ore and rock can produce fragments of varying sizes. After conducting compression tests on each particle, the resulting fragments are collected and weighed. The mass percentage of each fragment relative to the mass of the intact particle before fragmentation is calculated. The fragments are then ranked according to their mass percentage, with the fragment exhibiting the highest mass percentage referred to as the 'first size fragment', followed by the 'second size fragment', and so forth. Given that the sum of the mass percentages of the 'first size fragment' and 'second size fragment' produced by the fragmentation of any single particle in this test exceeds 90%, smaller fragments are not considered. The distribution of mass percentages for the 'first size fragments' of 70 particles is illustrated in Fig. 2.12a. A significant number of particles fall within the range of 0.9 to 1.0, which can be attributed to the partial detachment of fragments occurring solely on the surface of the particles due to through-crack failure and edge attrition. This results in fragments that are similar in size to the original particles. Comparative analysis of different crushing modes reveals that, as shown in Fig. 2.12b, the mass proportion distribution of the first size fragments for particles undergoing mid-span crushing is the most extensive, with a distribution range from 0.352 to 0.999. The mass proportions of the first size fragments for edge attrition particles are all above 0.748. Conversely, the mass proportion distribution of the first size fragments for randomly cracked particles is relatively random. For particles experiencing through-crack failure, the mass proportions of the first size fragments are consistently around 1.0. A significant number of particles fragmented into two parts during mid-span crushing, with the mass proportion distributions of the first and second size fragments illustrated in Fig. 2.13, both conforming to a normal distribution. The mean mass proportion of the first size fragments is approximately 0.65, while the mean mass proportion of the second size fragments is around 0.30.

Fig. 2.11 Load-strain curves obtained from single-particle compression test of different types of caved ore and rock

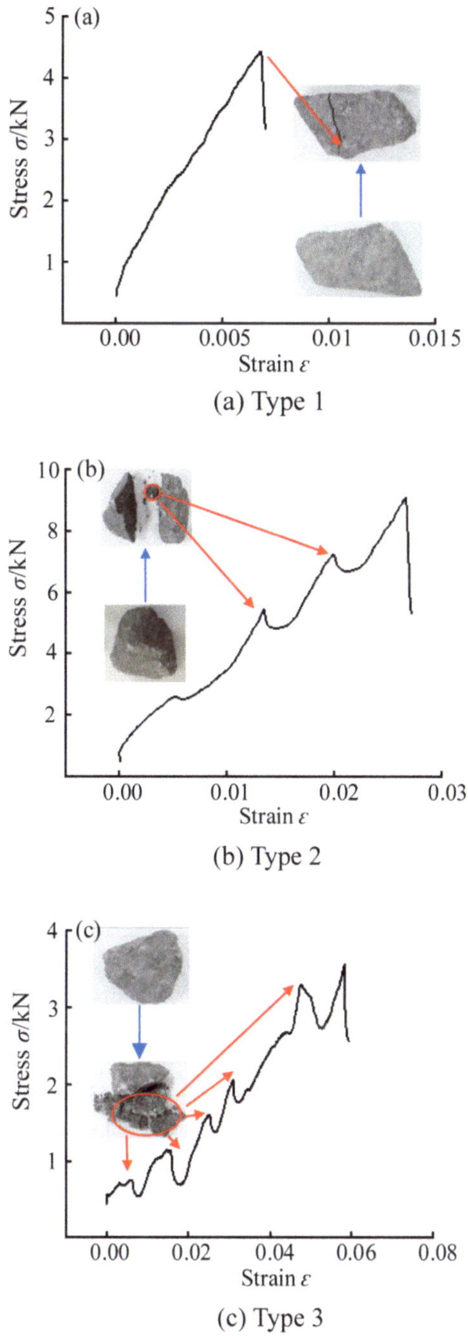

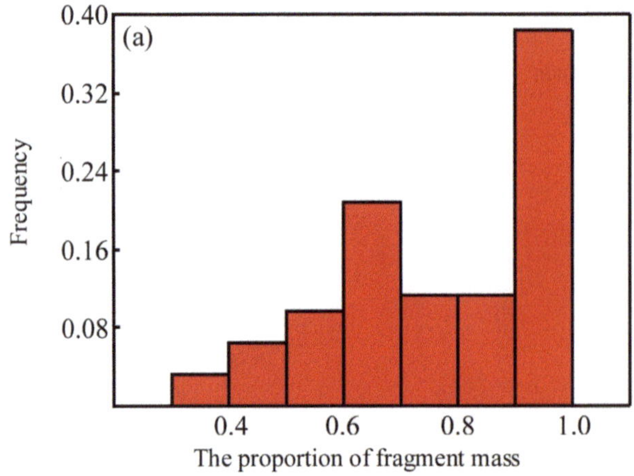

(a) Distribution of the mass fraction of all 'first size fragments'

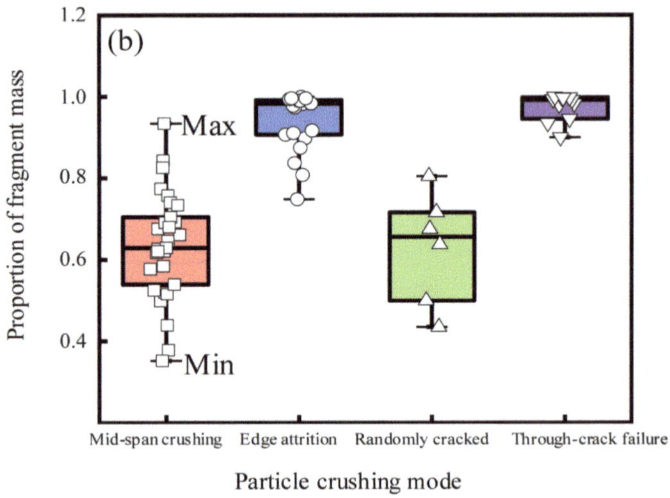

(b) Statistics of the mass fraction of 'first size fragments' in different particle crushing modes

Fig. 2.12 Mass-fraction distribution and statistics of fragments from single particle of caved ore and rock

2.2.3 3D Micro-morphological Characteristics of Fracture Surface of Caved Ore and Rock

To further investigate the crushing characteristics of caved ore and rock, this study not only examines macroscopic properties such as crushing strength, crushing modes, and particle size distribution, but also employs Scanning Electron Microscopy (SEM)

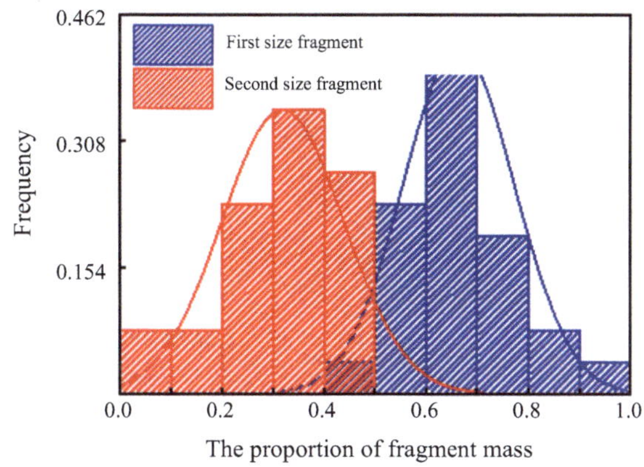

Fig. 2.13 Mass-fraction distribution of 'first size fragment' and 'second size fragment' of mid-span crushing particles

and 3D scanning technology to analyze the morphological characteristics of the fracture surfaces of caved ore and rock from a microscopic perspective.

2.2.3.1 Fracture Surface Morphology Analysis

To investigate the micro-morphological characteristics of fracture surfaces in caved ore and rock, we utilize the particle fracture surface illustrated in Fig. 2.14b as a case study, observing its micro-morphology through SEM. Figure 2.14a displays an SEM image of caved ore and rock slices at a magnification of 40, accompanied by energy dispersive spectroscopy (EDS) analysis to elucidate the mineral distribution. The white regions denote iron-containing metallic minerals, whereas the black regions signify gangue minerals. Figure 2.14c presents the SEM image depicting crack propagation on the fractured surfaces of the particles post-fragmentation. The analysis indicates that during crack propagation, the cracks navigate through gangue minerals, metallic minerals, and the interfaces between gangue and metallic minerals in a convoluted manner. This behavior is attributed to the presence of variably distributed pre-existing fissures or weak contact surfaces within the particles. The propagation of internal cracks facilitates the formation of micro-cracks or the opening of interfaces, ultimately leading to the development of external macroscopic cracks, which results in the fragmentation of the particles.

2.2.3.2 3D Construction of the Fracture Surface

This study focuses on the fracture surface of caved ore and rock, utilizing a 3D scanner to collect geometric morphology data of the cross-section. Three critical considerations must be addressed:

(a) SEM image of mineral distribution

(b) Example of fracture surface

(c) SEM image of crack propagation on the fracture surface

Fig. 2.14 SEM image of single particle mineral distribution and crack propagation

(1) Integrity of data: To ensure the integrity of the three-dimensional morphological data of the particle fracture surface, careful observation is essential. It is advisable to select fracture surfaces devoid of pronounced reflective characteristics. The reflective surfaces present on the particle fractures, which are minerals contained within the particles, can compromise the integrity of the scanned data and subsequently affect the reconstruction results of the fracture surface.
(2) Accuracy of data: To guarantee the accuracy of the three-dimensional morphological data, the selected particle fracture surface must remain stable on the rotating platform, avoiding significant shaking during rotation. This study employs a fully automated scanning method utilizing a rotary stage, wherein fractured particle pieces are positioned on the rotating platform. Any shaking during the scanning process may result in ghosting effects in the reconstructed particle fracture surface.
(3) Efficiency of data collection: To optimize the efficiency of the three-dimensional morphology data collection, continuous monitoring of the computer connected to the scanner is crucial throughout the scanning process. Parameters should be uniformly configured, and scanning data should be observed in real-time on the screen. Should any issues arise, it is recommended to halt and either restart or discard the scan to improve scanning efficiency.

After obtaining the point cloud data of the particle fracture surface from the scanning process, preprocessing is necessary. First, the range of the scanning data should be appropriately adjusted according to the research area. Second, the Rhino imaging software should be employed to reposition the center of the reconstructed fracture surface as close to the origin as possible. Subsequently, this data should be exported into a text file containing three-dimensional coordinates. Finally, the resulting point cloud data should be imported into MATLAB in DXF file format to construct a three-dimensional model of the fracture surface of caved ore and rock.

2.2.3.3 Quantitative Characterization Method of Fracture Surface Morphology

Fractals were first proposed by Mandelbrot [73] in 1967, leading to the development of fractal geometry. A fractal is defined as 'a rough or fragmented geometric shape that can be divided into parts, each of which is a reduced-scale copy of the whole' [74]. Self-similar structures are commonly found in nature, where local and global similarities exist. Fractal theory serves as an effective tool and provides strong support for the study of the roughness coefficient of rock mass structural surfaces. The two-dimensional fractal dimension [75] indicates that there are inevitable differences in the true profile surface of three-dimensional caved ore and rock. Quantifying the fractal characteristics of the three-dimensional morphology of structural surfaces is evidently superior. The indirect covering method replaces the direct covering method using two-dimensional Euclidean shapes such as circles and triangles as a single scale. Representative methods include the triangular prism surface area method [76], projection covering method [77], and the cube covering method [78]. The first two methods may encounter issues with approximate area calculations, leading to deviations in the resulting fractal dimensions, while the cube covering method provides more accurate results [79].

2.2.3.4 Fracture Surface Roughness Analysis

The three-dimensional reconstruction results of the fracture surface of the particles are illustrated in Fig. 2.15a, where the Z-axis depth reflects the degree of roughness of the fracture surface. The fractal dimension of the fracture surface is computed using the cube covering method [80], which quantitatively characterizes the roughness. The calculation principle is as follows: on a plane, there is a square grid with a side length of δ, and the height values at the four corners are $z(i, j)$, $z(i, j + 1)$, $z(i + 1, j)$ and $z(i + 1, j + 1)$ ($1 \le i, j \le m - 1$, where m is the number of measurement points along each edge). A cube with a side length of δ is used to cover the rough fracture surface starting from a certain uniform height, and the number of cubes within the covered area of $\delta \times \delta$ is calculated. Specifically, the number of cubes $N_{i,j}$ covering the fracture surface within the i, j grid is expressed as:

$$
\begin{aligned}
N_{i,j} = {} & \mathrm{INT}\left\{\delta^{-1}\left[\max(z(i,j), z(i,j+1), z(i+1,j), z(i+1,j+1))\right] + 1\right\} \\
& - \mathrm{INT}\left\{\delta^{-1}\left[\min(z(i,j), z(i,j+1), z(i+1,j), z(i+1,j+1))\right]\right\}
\end{aligned}
\tag{2.7}
$$

Here, INT is an up-integer function; $N_{i,j}$ is the number of cubes covering the rough surface; δ is the side length of the cube; i, j are lattice numbers.

Thus, the total number of cubes required to cover the entire fracture surface is:

$$
N(\delta) = \sum_{i,j=1}^{m-1} N_{i,j} \tag{2.8}
$$

Fig. 2.15 3D digital topography and quantitative roughness characterization of particle fracture surfaces

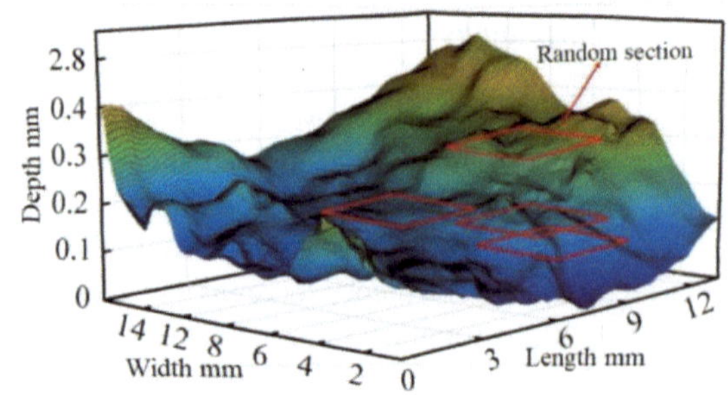

(a) 3D reconstruction of particle fracture surface morphology

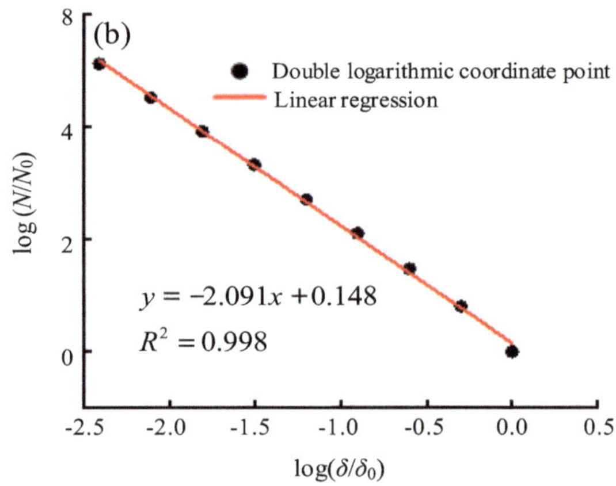

(b) Double logarithmic coordinate diagram of the total number of cubes $N(\delta)$ and the scale δ

The fracture surface is covered by repeatedly varying the side length δ of the cube, allowing for the determination of the total number of cubes $N(\delta)$ needed to cover the fracture surface, as expressed in Eqs. (2.7) and (2.8). This relationship between the scale δ and the fractal dimension D is established as follows:

$$N(\delta) \sim \delta^{-D} \tag{2.9}$$

Here, $N(\delta)$ is the total number of cubes; D is the fractal dimension.

Based on the principles outlined above, the scanned cross-sectional point cloud data is imported into MATLAB, where equidistant curved surfaces are extracted and interpolated to calculate the fractal dimension D of the cross-sections. Zhang et al. [81, 82] argue that when the scale δ exceeds 2 mm, the fractal dimensions of all cross-sections are equal to 2, indicating that low precision cannot adequately reflect fractal characteristics. Therefore, the selection of scale δ is as shown in the

Table 2.4 Relationship between the total number of cubes $N(\delta)$ and the scale δ

δ/mm	$N(\delta)$
0.005	1,472,761
0.01	368,114
0.02	92,061
0.04	22,915
0.08	5657
0.16	1410
0.32	331
0.64	70
1.28	11

range of δ in Table 2.4. The analysis results of randomly selected particle fracture surfaces are similar; for instance, the cross-section illustrated in Fig. 2.15a yields results as shown in Table 2.4. As the scale δ increases, the number of covered cubes decreases, leading to lower precision. A double logarithmic coordinate plot of the total number of cubes $N(\delta)$ against the scale δ is constructed (Fig. 2.15b), and a linear fit is performed, achieving a correlation coefficient R^2 of 0.998. This indicates that the degree of surface roughness of caved ore and rock fracture surfaces can be quantitatively characterized based on the fractal dimension, suggesting that the roughness of caved ore and rock fracture surfaces also conforms to fractal characteristics.

This study further explores the relationship between the roughness of fracture surfaces and the cross-sectional area *S*, specifically examining the size effect of the fractal characteristics of the fracture surface. We select three particle fracture surfaces, designated as P1, P2, and P3. For each fracture surface, we extract square regions of varying sizes from the same starting point, with side lengths ranging from 1 to 11 mm in 1 mm increments. The extracted regions are labeled as P1-1, P1-2, and so forth. Utilizing the cubic covering method previously described, we calculate the fractal dimension *D* for each square region. The fractal dimensions *D* exhibit variation across different cross-sectional areas, with all correlation coefficients R^2 exceeding 0.990. As illustrated in Fig. 2.16, the fractal dimensions derived from multiple cross-sections of P1, P2, and P3 display both maximum and minimum values, with their respective ratios being 1.077, 1.061, and 1.070. The relationship between the fractal dimension *D* of the cross-sections P1, P2, and P3 and the cross-sectional area S is consistent. A nonlinear fitting of all data points is conducted using an exponential function, as indicated by the yellow curve. Overall, *D* decreases with increasing cross-sectional area, gradually approaching equilibrium. When the cross-section is relatively small, *D* exhibits a broader range of fluctuations; however, it remains relatively stable when the cross-sectional area exceeds 36 mm^2. Given the smaller surface area of the particle fracture, we select a cross-sectional area of 36 mm^2 for the calculation of *D*, and we minimize the error by randomly selecting multiple cross-sections, as framed in Fig. 2.15a, to obtain the average fractal dimension $\overline{D}$.

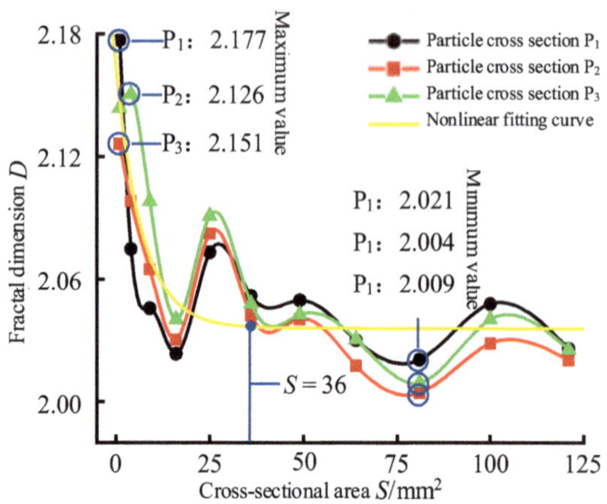

Fig. 2.16 Relationship between fractal dimension and cross-sectional area of particle fracture surface

2.2.3.5 Analysis of Influencing Factors on Fracture Surface

Generally, a greater crushing energy correlates with a larger fractal dimension of the particle fracture surface, which indicates a more severe degree of crushing [83, 84]. The crushing energy of brittle particles is typically defined as the elastic strain energy prior to complete failure, represented by the area enclosed by the load–displacement curve before particle failure and the horizontal axis. Therefore, the crushing energy E_b of the particles can be expressed as follows:

$$E_b = \int_0^{x_b} F_{MAX}\,dx \tag{2.10}$$

Here, F_{MAX} is the peak pressure of particle failure; x is the axial displacement during the test; x_b is the axial displacement at the maximum load point during the test.

The average fractal dimension of the fracture surfaces of irregular caved ore and rock is influenced by several factors beyond the energy required for particle fragmentation E_b. These factors include particle shape, size, and others. Particle size is represented by the equivalent particle diameter d, while particle shape is characterized by four shape parameters: sphericity ϕ, elongation e, flatness f, and angularity c. A multiple collinearity analysis was conducted using SPSS software on the selected seven parameters, and the variance inflation factors (VIF) [85] were all found to be less than 10, indicating that there is no serious multicollinearity among the four parameters. To quantitatively characterize the weights of the six influencing factors on the average fractal dimension and their statistical relationships, 12 typical results of particle fracture surfaces were selected, and normalization was performed on these 12 sets of data. The data concerning the influencing factors of the average fractal dimension of the fracture surfaces after dimensional unification is presented in Table 2.5.

Based on SPSS, a multiple linear regression analysis was conducted to explore the relationships among the aforementioned seven indicators. Equation (2.11) represents the established multiple linear regression model, with a correlation coefficient R^2 of 0.862, indicating a good fit based on 12 data sets and strong correlations among the parameters. The significance of the regression model is $p < 0.05$, demonstrating the statistical significance of this model. In the regression model, the significance test results for all influencing factors, except for the elongation rate *e* and the constant, have *p*-values less than 0.05, which indicates that these factors are statistically significant and should be retained. From the analysis of Eq. (2.11), it is observed that the fractal dimension is positively correlated with energy, flatness, and angularity, while it is negatively correlated with equivalent particle size and sphericity. Among the six factors, sphericity has the greatest influence weight, followed by flatness, energy, angularity, and equivalent particle size, all of which have approximately equal influence weights. The elongation rate does not show statistical differences in its impact. As the flatness of the particles increases, the short axis approaches the medium-long axis, resulting in a larger fractal dimension of the fracture surface after fragmentation. This indicates more severe damage and greater input energy, confirming the earlier conclusion that higher particle flatness reduces the likelihood of mid-span crushing.

$$\overline{D} = 0.877E_b - 0.844d - 1.415\phi + 0.908f + 0.875c \tag{2.11}$$

Table 2.5 Normalized data on factors influencing the mean fractal dimension of particle fracture surfaces

Serial number	Fractal dimension	Energy E_b/J	Sphericity ϕ	Elongation e	Flatness f	Edge angle c	Equivalent particle diameter d/mm
1	0.34	0.71	0.31	0.46	0.87	0.00	0.79
2	0.98	1.00	0.59	0.02	0.80	1.00	0.57
3	0.43	0.11	0.18	1.00	0.39	1.00	1.00
4	0.25	0.08	0.45	0.87	0.44	0.67	0.21
5	0.23	0.76	0.45	0.49	0.09	1.00	0.84
6	0.00	0.34	1.00	0.01	0.87	0.67	0.22
7	0.23	0.52	0.73	0.00	0.68	0.67	0.25
8	1.00	0.18	0.04	0.66	0.29	0.67	0.10
9	0.66	0.96	0.45	0.53	0.18	1.00	0.89
10	0.46	0.43	1.00	0.00	1.00	0.67	0.00
11	0.73	0.50	0.50	0.51	0.28	1.00	1.00
12	0.73	0.00	0.00	0.66	0.00	0.92	0.92

2.3 Numerical Study of Single-Particle Compression Crushing

Current research on the single-particle crushing characteristics of caved ore and rock remains limited. This section aims to address these gaps by employing 3D white light scanning and reconstruction technology. The proposed methodology utilizes the Voronoi block model within the PFC and a soft bond contact model to construct and analyze the macroscopic and mesoscopic characteristics of the Breakable Voronoi Block Model (BVBM) for irregular caved ore and rock. Furthermore, the study investigates the influence of various parameters, including ball diameter, loading speed, softening tensile strength factor, softening factor, and BVBM size, on single-particle crushing behavior. It also provides a recommended range of values for parameters that are suitable for single-particle modeling and research on crushing characteristics.

2.3.1 BVBM Construction

To construct the BVBM, the initial step involves utilizing 3D white light scanning and reconstruction technology to acquire actual three-dimensional imaging data of caved ore and rock, which is essential for forming irregular particle geometries. Subsequently, the 3D Voronoi subdivision method is applied to generate irregular Voronoi block models. Finally, the mesoscopic parameters of BVBM are established by comparing the Soft Bond Contact Model (SBM) with the physical test results of single particle compression. The specific modeling process for BVBM is detailed as follows:

2.3.1.1 Compression Test of Single Particle

(1) Test materials

The physical testing of single particle compression crushing was conducted using the particles detailed in Sect. 2.2.1.1. The macroscopic physical and mechanical properties of these ore particles are summarized in Table 2.2. To ensure the effectiveness of the selected granular material, it is essential that it fulfills two specific criteria [76]. First and foremost, the particles must exhibit a convex shape. This convexity is crucial as it guarantees that only a single point of contact occurs during the compression process, aligning with the requirements set forth by the PFC model. Additionally, it is important for the particles to possess one upper contact point along with a minimum of three lower contact points. This configuration helps create a lower contact surface characterized by multiple points of contact beneath the particle. It is particularly vital that the upper contact point is positioned so that its vertical projection is close to the center of the lower contact surface. This arrangement plays a significant role in

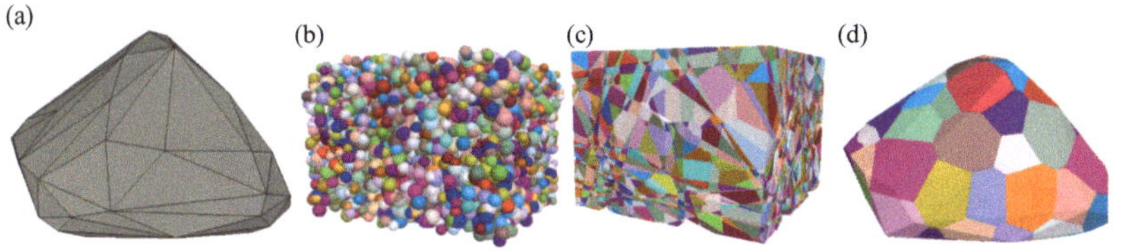

(a) Particle geometry creation (b) Ball generation (c) Voronoi subdivision (d) BVBM construction

Fig. 2.17 Workflow for constructing BVBM model

reducing any shaking of the particle that may occur during the compression process, thereby enhancing the accuracy and reliability of the test results.

The numerical compression test of a single particle is conducted using the discrete element software PFC. Prior to this, the BVBM of the irregular particle needs to be performed, and the specific process is illustrated in Fig. 2.17:

1) Particle geometry creation: The selected caved ore and rock samples were assigned numerical identifiers, and a EiScan-SE 3D white light scanner was employed to capture three-dimensional images of the particles. The point cloud data was exported in STL file format, and the dense mesh was locally simplified to approximately 1000 facets using Meshlab software. Care was taken to preserve the significant contact points on both the upper and lower contact surfaces. The simplified particle geometry was subsequently imported into PFC, where excess triangular faces were removed until the model comprised approximately 200 facets. It is noteworthy that only convex irregular blocks (Fig. 2.17a) were included in this modeling.
2) Ball generation: Balls with a void ratio of 0.36 and a size distribution ranging from 2.5 to 4.0 mm were produced within the minimum external cuboid that encompasses the actual particle geometry. The choice to utilize balls with a varied size distribution, rather than mono-sized balls, was intended to ensure that the irregular blocks generated through Voronoi partitioning closely resembled the irregular particles found in caved ore and rock. Subsequently, a static equilibrium calculation was conducted (Fig. 2.17b).
3) Voronoi subdivision: Voronoi subdivision and regularization are performed using the spherical coordinates of particle centers as random points, with the spherical radius defined as the Euclidean distance. The Voronoi subdivision polyhedron is generated within the minimum external cuboid encompassing the particle geometry, facilitating the creation of Voronoi subdivision cuboid models that accommodate various particle size distributions (Fig. 2.17c).
4) BVBM construction: The blocks outside the particle geometry were removed, and the rigid Voronoi blocks in contact with one another were bonded using SBM. This process yields the BVBM, which is shaped to resemble the actual caved ore and rock (Fig. 2.17d). The BVBM model is not merely a collection of individual spheres; instead, it constitutes a structure comprised of bonded Voronoi-divided polyhedral.

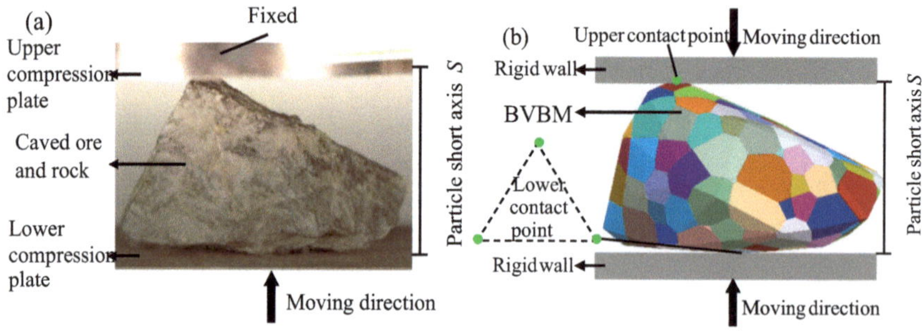

(a) Laboratory test (b) Numerical simulation

Fig. 2.18 Schematic of physical compression tests and numerical simulation for single particle of caved ore and rock

(2) Test process

In accordance with Sect. 2.2.1.3 regarding the physical test (as illustrated in Fig. 2.18a), a single particle compression test was simulated using the BVBM model. This simulation employed two rigid walls to replicate the upper and lower compression plates. The BVBM was stabilized on the bottom rigid wall in a manner consistent with the configuration utilized in the physical test. Additionally, a rigid wall was positioned at a specified distance above the highest vertex of the particle, ensuring that both the upper and lower sections of the particle experienced rigid constraints, akin to the physical test (refer to Fig. 2.18b). During the compression process, the two rigid walls moved towards each other at a constant rate. In contrast to the physical test, where only the bottom rigid wall was in motion, the simulation did not incorporate dynamic factors, resulting in the absence of discrepancies. Throughout the simulation, the forces exerted on the two rigid walls were recorded in real-time. The simulation was programmed to cease when the force on the rigid walls fell below 60% of the previously recorded maximum force. It is noteworthy that the same particle was simulated multiple times to derive an average value.

2.3.1.2 Brief Introduction to the SBM Principle

The SBM is well-suited for simulating rock materials with diverse attributes. The behavioral and rheological components of the soft-bond model are depicted in Fig. 2.19 [86, 87]. Here, γ denotes the softening tensile strength factor, which ranges from 0 to 1, while ζ represents the softening factor, with a value range extending from 0 to $+\infty$.

The mechanical response of the SBM before transitioning into a softening regime exhibits characteristics that are akin to the PBM. However, a notable distinction between the two models arises in their treatment of bond failure. In the case of SBM, the bond does not simply get removed upon failure, as illustrated in Fig. 2.20a [37]. Instead, the model allows for the possibility of entering a softening regime.

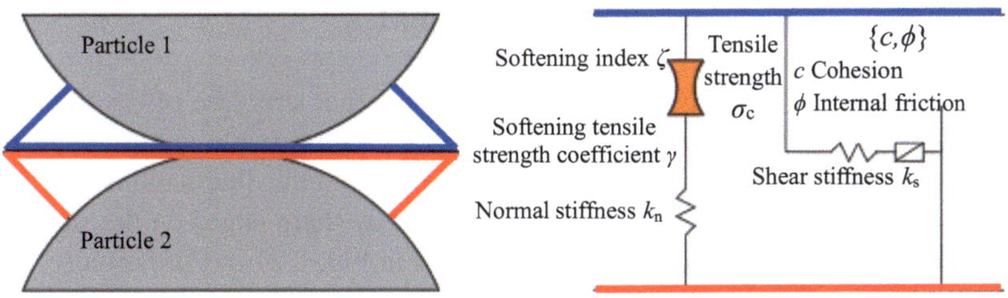

Fig. 2.19 Inter-particle bond state and rheological components of the SBM

In this regime, the bond experiences a reduction in bond stress until it reaches a predetermined threshold value. At this critical point, the bond is considered broken and is effectively removed from the model. As the process of softening unfolds, the tensile stress associated with the bond diminishes continuously. This reduction persists until the tensile stress falls below the product of two significant parameters: the softening tensile strength factor γ and the tensile strength σ_c, as depicted in Fig. 2.20b [88]. The rate at which this softening occurs is governed by the softening factor ζ. It is crucial to emphasize the implications of these parameters: when the softening factor ζ is set to 0, the bond behaves in a completely brittle manner, which renders the SBM functionally analogous to the PBM. Conversely, when the softening tensile strength factor γ is equal to 1.0, the bond's softening behavior is effectively inhibited. Hence, for the bond to demonstrate any form of tensile softening, it is essential that both conditions are met: ζ must not equal 0, and γ must not equal 1.0.

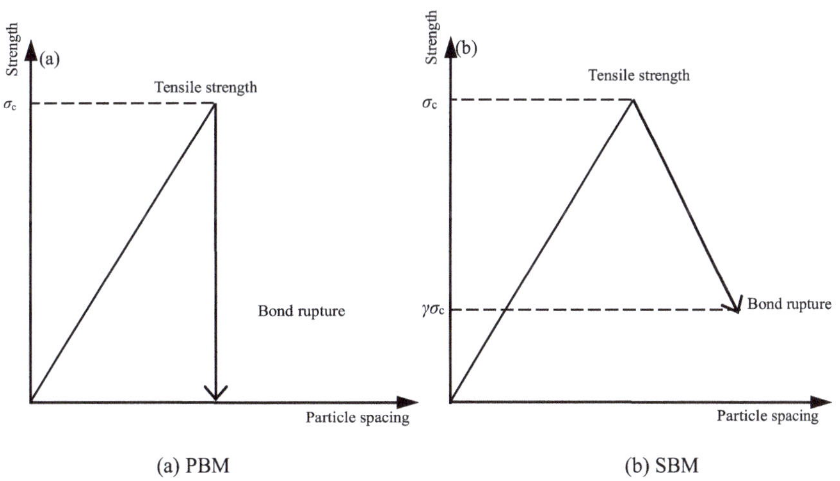

Fig. 2.20 Mechanical behavior of two bond types

2.3.1.3 Meso-parameter Calibration of BVBM

Due to the inherent variations in the shape and internal structure of caved ore and rock, the load–displacement curves observed in single particle compression tests exhibit considerable randomness. Prior to particle failure, fluctuations in certain load–displacement curves arise from the breaking of sharp edges on the surface or from vertical asymmetric contacts (as illustrated in Fig. 2.21a). Furthermore, some load–displacement curves display a nearly linear trend before reaching the peak, as indicated by the black curve in Fig. 2.21b. To enhance model fitting accuracy, this study selects rock particles characterized by quasi-linear load–displacement curves to determine the meso-parameters of the BVBM [37, 76, 88].

By adjusting the softening factor ζ, the softening tensile strength coefficient γ, the elastic modulus E_c, the damping coefficient *damp*, and other meso-parameters, the results of the BVBM match are shown in Fig. 2.21b. This demonstrates that the numerical simulation of BVBM employed in this test is consistent with the physical tests in terms of both the load–displacement curve and the particle crushing mode. This not only verifies the reliability of the BVBM construction method used but also indicates that SBM is suitable for numerical research on the compressive crushing of single particle of caved rock and ore. The matched meso-parameters of BVBM are provided in Table 2.6.

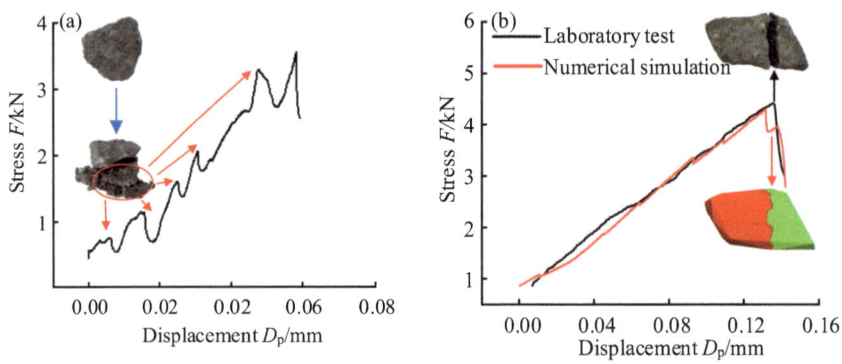

(a) Load-displacement curve with obvious fluctuations

(b) Quasi-linear load-displacement curve and matching results

Fig. 2.21 Load–displacement curves and BVBM matching results of different types of single caved ore and rock in physical tests

Table 2.6 BVBM macro-meso parameters

Macro-meso parameters	Values
Density/kg m^{-3}	2500
Minimum particle size/m	2.50e−3
The maximum and minimum particle size ratio	1.6
Local damping coefficient	0.7
BVBM short axis dimension/mm	20.624
Loading termination criterion	0.6
Loading rate/m m^{-1}	0.01
BVBM Size scaling factor	1.0
SBM	
Elastic modulus/Pa	1.95×10^9
Normal-shear stiffness ratio	2.5
Coefficient of friction	0.5
Cohesive strength and tensile strength/Pa	$50–1.15 \times 10^6$
Softening index	5
Softening tensile strength coefficient	0.8
Linear contact model	
Friction coefficient between blocks	0.5
Friction coefficient between block and wall	0.0
Normal stiffness of contact between block and wall/N m^{-1}	4.5×10^9
Normal-shear stiffness ratio k_n/k_s	1.0

2.3.2 *Crushing Characteristics of Single Particle Under the Influence of Multiple Parameters*

The force–displacement curve of individual ore particles, as illustrated in Fig. 2.21b, along with the BVBM parameters detailed in Table 2.6, serves as reference points for both macro and mesoscopic parameters. To investigate the impact of five distinct factors—namely ball diameter, loading speed, softening tensile strength factor, softening factor, and BVBM size—on the crushing characteristics of ore particles, a series of single particle numerical compression tests were conducted using the single variable method. The appropriate range of values for each factor was established. The mesoscopic parameters within the BVBM model were held constant, and a total of 10 distinct random seeds were employed to replicate the numerical tests pertaining to single particle compression. The peak force and displacement averages derived from these 10 numerical tests were utilized as indicators of the macroscopic crushing performance of single particle crushing for each set of modeling parameters.

2.3.2.1 Effect of the Number of Spherical Particles in the Model

The accuracy and computational efficiency of simulation outcomes are significantly influenced by the number of blocks in each BVBM. Consequently, selecting the right block size is crucial. In this section, we maintained the other mesoscopic parameters listed in Table 2.6 constant and developed BVBMs with varying numbers of blocks by modifying the average diameter $<d>$ of the balls. These BVBMs were created based on the actual particle profile of an identical geometry, as depicted in Fig. 2.17b. The benchmark value for the average ball diameter $<d>$ was established at 3.250 mm (Table 2.6). The diameter of the balls was scaled proportionally, with the scaling ratio ranging between 0.6 and 1.4, incrementing by 0.1. Given that the macroscopic short axis size S_{p} (Fig. 2.18) of an individual particle measured 20.624 mm, the ratio range of $S_{\text{p}}/<d>$ was approximately between 4 and 11.

Figure 2.22 illustrates the variations in peak pressure and displacement of single particles under different particle size influences. As shown in the light blue region of Fig. 2.22, when $S_{\text{p}}/<d>$ exceeds 4.9 (i.e., the average diameter $<d>$ of spherical particles is less than 4.225 mm), both the peak pressure and displacement of single particles are close to the reference values, and the standard deviation of the macroscopic results obtained from numerical experiments under different random seeds is relatively small. This indicates that the BVBM simulation results exhibit good stability. Conversely, as indicated in the light-yellow area of Fig. 2.22, when $S_{\text{p}}/<d>$ is less than 4.9, both the peak pressure and displacement of single mineral particles are significantly higher than the reference values, and the standard deviation of the macroscopic results from numerical tests under different random seeds is larger, indicating that the BVBM simulation results become unstable. Generally, when the block size is relatively large compared to the BVBM size, or when the number of blocks contained within the BVBM is small, inaccurate and unreasonable numerical simulation results for the compression of single particles may occur. This is due to the fact that larger blocks may not adequately simulate the true crystallinity of single particles. Therefore, considering both the accuracy of the simulation results and computational efficiency, it is recommended that the ratio of the macro short-axis size of the single particle in the BVBM to the average diameter of spherical particles, $S_{\text{p}}/<d>$, should be in the range of approximately 5–10. It is noteworthy that to clearly distinguish the significantly different simulation results under various parameters, light yellow and light blue background colors have been set in the subsequent parts of Fig. 2.22 in this chapter.

2.3.2.2 Influence of the Loading Speed

Existing research results from discrete element simulations [49, 89, 90] indicate that increasing the loading rate within quasi-static conditions during numerical compression tests can enhance calculation efficiency compared to physical compression tests.

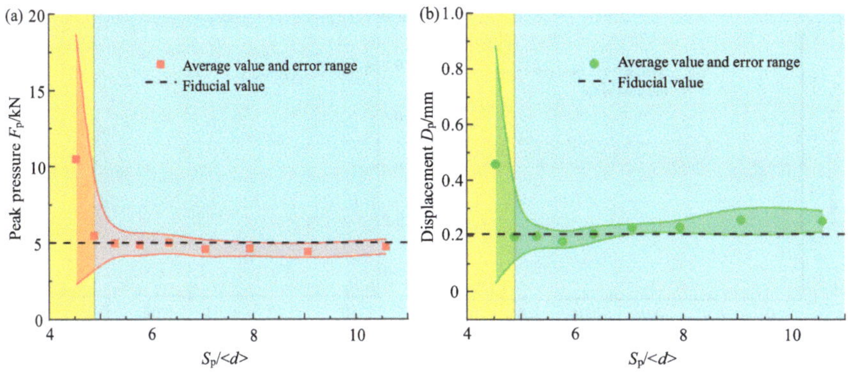

(a) Peak pressure (b) Displacement

Fig. 2.22 Evolutions of peak pressure and displacement of single particle under the influence of different spherical particle size

The displacement loading rate for single-particle numerical compression tests typically ranges from 0.001 to 0.1 m/s [91, 92], corresponding to a nominal strain rate range of 8.1×10^{-4} to 8.1×10^{-2} s^{-1}.

Figure 2.23 illustrates the variation in peak pressure and displacement of single particles under different loading rates. As shown in Fig. 2.23a, the peak pressures are all close to the baseline value and exhibit a slight increase with the increase in loading rate, while the standard deviation of peak pressure shows no significant change. This indicates that, under the condition of other parameters remaining constant, the loading rate has no significant effect on the peak pressure of single particles. In the light blue area of Fig. 2.23b, when the loading rate is less than 0.05 m/s, the displacements are all close to the baseline value with a small standard deviation, indicating that the BVBM simulation results exhibit good stability. Conversely, in the light-yellow area of Fig. 2.23b, when the loading rate v exceeds 0.05 m/s, the displacements are significantly higher than the baseline value, and the standard deviation increases markedly, suggesting that the BVBM simulation results become unstable and less reliable.

Figure 2.24 illustrates the load–displacement curves under the same random seed but with different loading rates. As depicted in Fig. 2.24, the pre-peak segments of the load–displacement curves overlap across various loading speeds, indicating that the loading speed does not influence the macroscopic behavior of single particle crushing in rock materials. With an increase in loading speed, there is a slight rise in peak pressure, while the post-peak curve transitions from a nearly vertical drop to a non-vertical oscillatory decline, accompanied by a gradual increase in displacement. This phenomenon occurs because, when a single particle fractures, an excessively high loading rate can prolong the contact duration between the rigid wall and the BVBM, resulting in the force on the pressure plate not immediately dropping to the predetermined stopping condition of 60% of the peak pressure. Consequently,

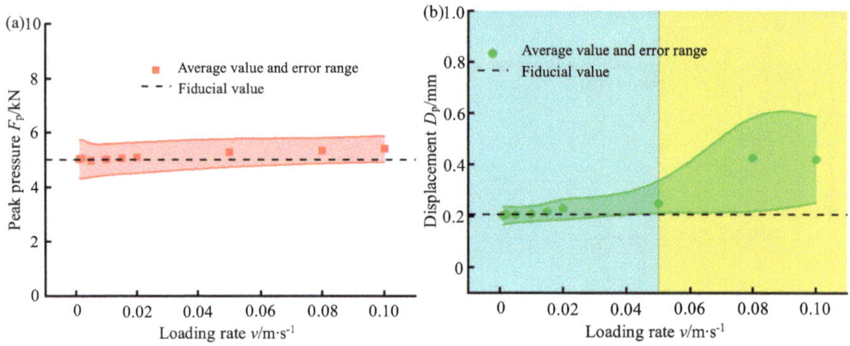

(a) Peak pressure (b) Displacement

Fig. 2.23 Evolutions of peak pressure and displacement of single particle under different loading speeds

Fig. 2.24 Load–displacement curves under different loading speeds with the same random seed

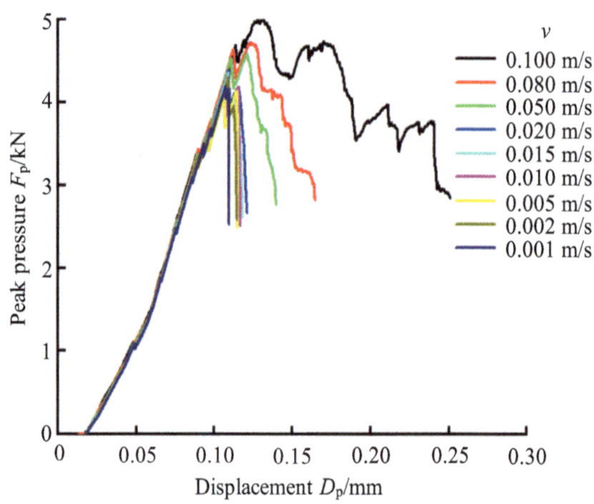

it is evident that the post-peak curve and displacement do not correspond with the actual single particle fragmentation results. Therefore, to accurately simulate the crushing process and the macroscopic crushing characteristics of a single particle, it is recommended that the loading speed in the BVBM for single particles does not exceed 0.05 m/s, which corresponds to a nominal strain rate of 4.05×10^{-2} s^{-1}.

2.3.2.3 Influence of the SBM Meso-parameters

The softening tensile strength factor γ and the softening factor ζ are two important mesoscopic parameters in SBM. In a single-particle numerical compression test, these parameters are set within the range of 0 to 1 and 0 to 50, respectively. The gradient of change for γ is 0.1, while for ζ it is 10 in which BVBM remains consistent.

(1) Softening tensile strength factor γ

The softening tensile strength factor γ is used to determine the minimum tensile force required for a bond to be broken during the softening stage of SBM (Fig. 2.20b). A higher value of γ indicates that the softening behavior of the bond is less significant [93]. Figure 2.25 illustrates the variations of peak force and displacement for BVBM under different softening tensile strength factors. Both peak force and displacement decrease exponentially as γ increases, with the correlation coefficient R^2 of 0.999 and 0.997 respectively (Fig. 2.25a, b). Furthermore, the standard errors of peak force and displacement under different values of γ are relatively small, indicating that the robustness of BVBM simulation results is not affected by the softening tensile strength factor γ in SBM.

Figure 2.26 presents the force–displacement curves with various softening tensile strength factors and the same random seed. Each pre-peak force–displacement curve, under different softening tensile strength factors, shows a similar slope. This indicates that the elastic modulus of the individual particle remains unaffected by the softening tensile strength factor. This is due to the fact that, during the elastic compression stage, the majority of bonds in BVBM do not undergo tensile forces that exceed their tensile strength. As a result, the softening behavior of the bonds is not activated. As the softening tensile strength factor increases, both the peak force and the pre-peak displacement increase. However, the strain softening characteristics of the force–displacement curve in the post-peak plastic failure stage remain relatively unchanged. This can be explained by the fact that, as the compression process progresses, more and more bonds in BVBM experience tensile forces that exceed their tensile strength,

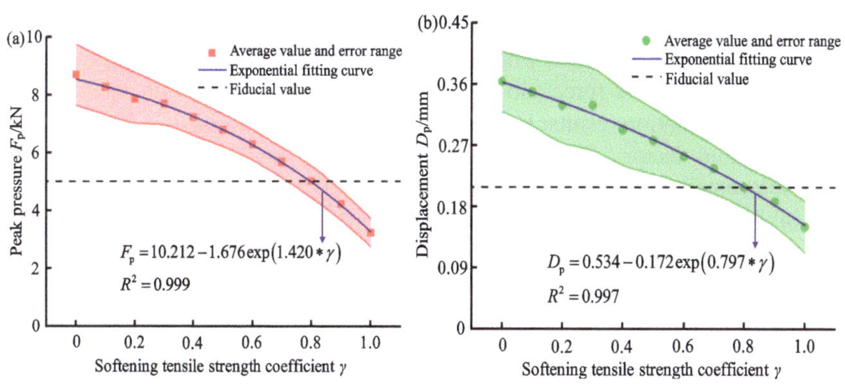

(a) Peak pressure (b) Displacement

Fig. 2.25 Evolutions of peak pressure and displacement of single particle under the influence of different softening tensile strength coefficients

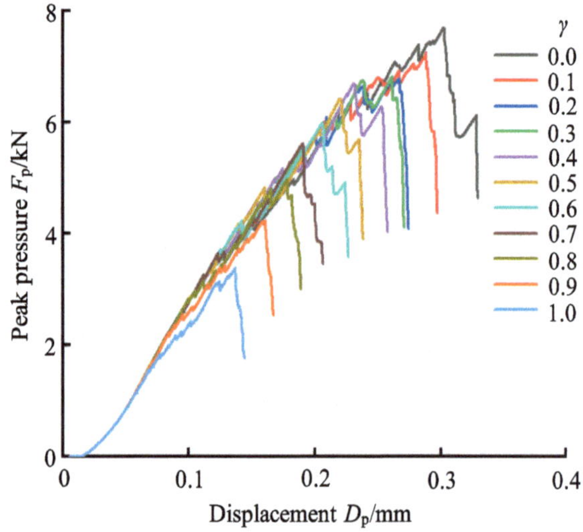

Fig. 2.26 Load–displacement curves of different softening tensile strength coefficients under the same random seeds

leading to a more pronounced manifestation of softening behavior. Therefore, a smaller value of the softening tensile strength factor γ results in a prolonged softening stage, necessitating greater force and displacement for BVBM failure.

(2) Softening factor ζ

The softening rate of SBM is controlled by the softening factor ζ. A larger softening factor ζ results in a smaller absolute value of the softening rate [93]. Figure 2.27 illustrates the evolutions of peak force and displacement for BVBM under different softening factors. Both the peak force and displacement increase linearly with the increase of the softening factor ζ. The correlation coefficient R^2 is 0.998 and 0.995 for peak force and displacement, respectively. Furthermore, the light blue zones in Fig. 2.27a, b demonstrate that when the softening factor ζ ranges from 0 to 10, the standard errors of peak force and displacement are small, indicating robustness in the BVBM simulation results. However, in the light-yellow zones in Fig. 2.27a, b, as the softening factor ζ increases from 10 to 50, the standard errors of peak force and displacement exhibit a significant increase. This observation indicates a higher level of variability in the results of the BVBM simulations.

Figure 2.28 illustrates the force–displacement curves with varying softening factors while maintaining the same random seed. The slope of each pre-peak force–displacement curve remains consistent across different softening factors, indicating that the elastic modulus of the single particle is unaffected by these factors. As the softening factor increases, both the peak force and pre-peak displacement exhibit significant increases. Moreover, the force–displacement curve in the post-peak plastic failure stage no longer shows an instantaneous drop. This observation suggests that BVBM transitions gradually from brittle to ductile behavior and exhibits strain softening characteristics. This phenomenon can be attributed to the fact that a larger

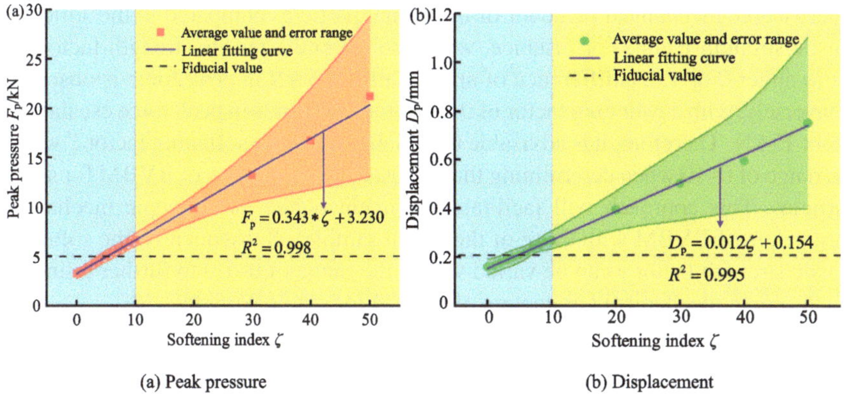

(a) Peak pressure

(b) Displacement

Fig. 2.27 Evolutions of peak pressure and displacement of single particle under the influence of different softening factors

softening factor ζ results in a smaller absolute value of the softening rate, leading to more pronounced softening behavior of the SBM bond. Consequently, the bond can withstand tension for a longer duration during compression and is less likely to fail. Therefore, increasing the softening factor significantly enhances both the peak force and displacement of BVBM, underscoring its more pronounced ductility characteristics.

The analysis of Figs. 2.25, 2.26, 2.27, 2.28 reveals the following key findings: (1) Within the examined value range, the BVBM demonstrates macroscopic mechanical behaviors, such as peak force and displacement, that closely resemble those observed in hard caved ore and rock [94]. This resemblance is particularly evident when the BVBM is subjected to various softening tensile strength factors and softening factors. (2) The softening factor exerts a more significant influence on the

Fig. 2.28 Load–displacement curves under different softening factors with the same random seed

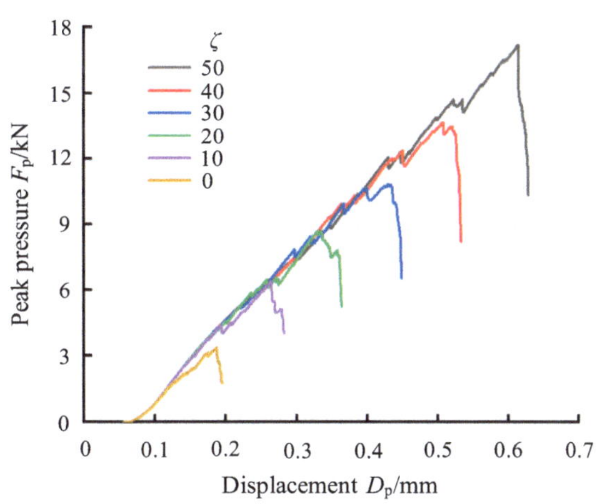

macroscopic mechanical behavior of individual particles compared to the softening tensile strength factor. For instance, when the softening tensile strength factor γ is set to either 0 or 1.0, a difference of approximately 5 kN in peak force is observed. Conversely, with a softening factor of 0 or 50, the difference in peak force escalates to about 15 kN. Therefore, it is advisable to initially adjust the softening factor ζ within the range of 0–10 when determining the mesoscopic parameters of BVBM for single particles. This approach will facilitate alignment of the macroscopic mechanical parameters of BVBM with those of the actual samples. Subsequently, the softening tensile strength factor γ can be varied within the range of 0–1.0 to further refine the macroscopic mechanical parameters of BVBM for single particle.

2.3.2.4 Influence of the BVBM Sizes

The size of caved ore and rock typically ranges from millimeters to meters. As particle size increases under uniaxial compression, the force required for fragmentation also escalates [73]. In the numerical simulation, the mesoscopic parameters of the BVBM and the ball diameter were kept constant, while the macro sizes of the blocks were adjusted according to the scaling factor α, which varies from 0.6 to 10. This results in a range of $\alpha \times S_p$ for the macroscopic short-axis dimension of the BVBM, spanning from 14.474 mm to 206.240 mm. It is crucial to adjust the distance between the upper and lower rigid walls in the simulation tests when varying the sizes of the BVBM. The actual loading speed can be determined using the axial strain rate and the distance between the two loading plates, as illustrated in Eq. (2.12) [90]:

$$|v| = \varepsilon h_0 / 2 \tag{2.12}$$

Here, v is the loading rate of two rigid walls, m/s; h_0 is the initial distance between the two rigid walls in the axial direction, m; ε is the axial strain rate, %.

Therefore, when adjusting the macroscopic short-axis size $\alpha \times S_p$ of the BVBM, the axial strain rate remains constant, and the loading rate v is determined based on the scaling factor α and Eq. (2.12). BVBM is generated using the real particle profile of the same geometric shape. The ball diameter, loading speed, softening tensile strength factor, and softening factor remain constant, with only the macro size of BVBM being altered. Figure 2.29 illustrates the evolutions of peak force and displacement under different BVBM sizes. Both peak force and displacement increase as the BVBM size increases, following a power function relationship. The correlation coefficient R^2 are 0.999 and 0.998 for peak force and displacement, respectively. Additionally, the standard errors of peak force and displacement under different BVBM sizes are small, indicating that the robustness and reliability of BVBM are not affected by its macroscopic size. It is important to note that the number of blocks within each BVBM increases as the macroscopic size of the BVBM increases, while the sizes of the balls remain constant. For instance, in this study, the numbers of blocks ranged from 10 to 10^4. In summary, combined with the results of Sect. 2.3.2.1, it is recommended to comprehensively consider the accuracy and computational

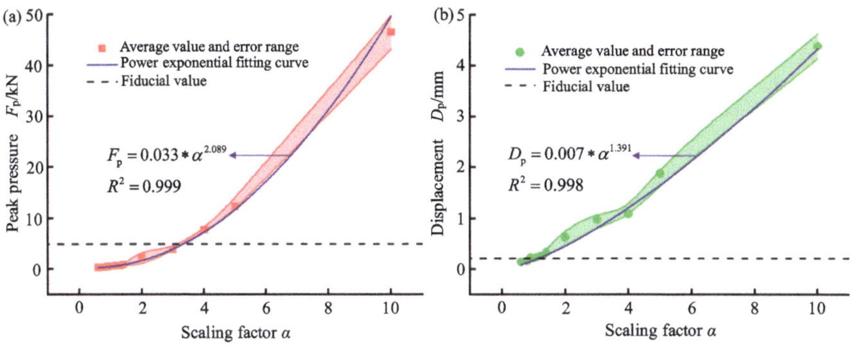

(a) Peak pressure (b) Displacement

Fig. 2.29 Evolutions of peak pressure and displacement of single particle under the influence of different BVBM size

efficiency of the simulation in the single-particle BVBM of caved ore and rock to determine the appropriate macroscopic size of the BVBM and the size (number) of the spherical particles (blocks) contained therein.

2.3.2.5 Sensitivity Analysis of Five Factors

The results summarized in Table 2.7 present the recommended value ranges for five factors: ball diameter, loading speed, softening tensile strength factor, softening factor, and BVBM size, based on the analyses conducted in Sects. 2.3.2.1–2.3.2.4. To quantitatively assess the influence of these five factors on the macro-mechanical properties of single particles BVBM, such as peak pressure and displacement, and to enhance the efficiency of matching mesoscopic parameters, a sensitivity analysis is necessary for each of the aforementioned factors. Given that the trends in peak pressure and displacement derived from the analyses in Sects. 2.3.2.1–2.3.2.4 are generally consistent, we use peak pressure as a representative example. A single-factor sensitivity analysis method [95] is employed to investigate the degree of influence of these factors. This involves analyzing the correlation between the relative change rates of each factor and the relative change rate of peak pressure, represented by sensitivity γ_{i}. The sensitivity γ_{i} of the ith influencing factor is expressed as shown in Eq. (2.13):

$$\gamma_i = \frac{\Delta F_{\mathrm{p}i}}{\Delta \beta_i} \tag{2.13}$$

Here, i is the ith sensitive factor, γ_i is the sensitivity coefficient of the peak force to the ith influencing factor, ΔF_i is the change rate of the peak force when the sensitive factor β_i changes, and $\Delta \beta_i$ is the change rate of the ith influencing factor.

Table 2.7 Suggested value range of BVBM parameters of single particle of caved ore and rock

Ratio of short axis size to the average diameter of balls $S_p/<d>$	Loading speed v/ m s^{-1}	Softening tensile strength factor γ	Softening factor ζ	Scaling factor α
5–10	0.001–0.050	0.0–1.0	0–10	–

The sensitivity coefficient γ_i indicates the magnitude of influence that the ith factor has on the peak force.

To ensure consistency across different dimensions, the data within the value range of each factor in Table 2.7 is normalized using Eq. (2.14) [96]. This normalization process ensures that the range of values falls within the range of [0.1, 0.9]:

$$H = 0.1 + 0.8\frac{Q - \min Q}{\max Q - \min Q} \tag{2.14}$$

Here, Q and H are the values of the data corresponding to each factor after and before normalization, respectively. The maximum and minimum values of the data corresponding to each factor are denoted as max$Q$ and min$Q$, respectively.

Here, the ratio of the short axis size of BVBM to the average diameter of the balls is expressed as the average diameter of the balls $<d>$, which is convenient for data processing. In the single-factor sensitivity analysis, the range of the macro-size scaling factor, α, is established between 0.6 and 5.0. It is important to note that the selection of this range does not significantly affect the results.

Figure 2.30 shows the evolutions of sensitivity coefficient of five influencing factors. The fluctuation range of the single factor sensitivity on the peak force of BVBM is within 10% for all five factors. The specific order of single factor sensitivity is as follows: BVBM size $\alpha \times S_p$ > softening factor ζ > softening tensile strength factor γ > loading speed $v \approx$ ball diameter $<d>$. Based on the analysis presented, the proposed methodology for modeling and determining the mesoscopic parameters for the BVBM of a single particle is articulated as follows: First, the macroscopic dimensions of the BVBM must be aligned with the actual particle size of the caved ore and rock. Subsequently, the loading speed and ball diameter, as illustrated in Table 2.7, are established, taking into account both the accuracy and computational efficiency of the simulation. In the process of determining the mesoscopic parameters, it is essential to initially adjust the softening factor, followed by a fine-tuning of the softening tensile strength factor. This sequence is imperative for enhancing the efficiency of matching the macro-mesoscopic parameters, particularly regarding the alignment of peak force and displacement.

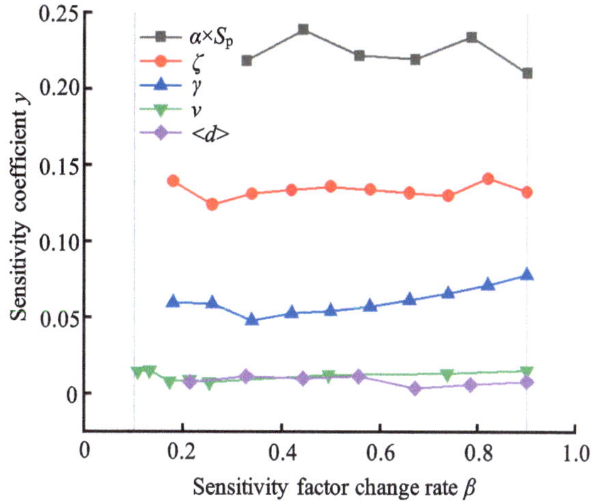

Fig. 2.30 Evolutions of sensitivity coefficient of five influencing factors

2.3.3 *Meso-crushing Mechanism of Single Particle*

Generally, the deformation and failure of standard rock samples in physical tests are attributed to two primary mechanisms: shear and tension, with single-particle compressive failure being a result of tensile failure [92]. The failure of a single particle parallels the failure process of standard rock samples; both fundamentally involve the propagation and coalescence of microcracks generated internally as the load increases, ultimately leading to the formation of macroscopic cracks. In examining the effects of various factors on the single-particle crushing characteristics of caved ore and rock, the BVBM simulation not only necessitates the acquisition of peak pressure and displacement but also requires the observation of the mesoscopic fracture mechanisms of the particle numerical model. The distribution of tensile and compressive stresses following BVBM failure, as illustrated in Fig. 2.31, indicates that failure predominantly occurs at the center of the particles, primarily through bond breakage leading to tensile failure. The bonds that rupture due to tension aggregate into surfaces, resulting in the complete failure of particles as tensile cracks develop. In contrast to PBM, there are specific conditions regarding the state of bonds in SBM; from the initial state to failure, an additional softening process is observed. Based on the changes in the state of bonds during the compression process of BVBM, as recorded in Sect. 2.3.2 under various influencing factors, it is evident that the softening factor plays a crucial role in the evolution of cracks.

As shown in Fig. 2.32, a series of simulations were conducted using BVBM with softening factors of 5 and 50. The spatial coordinates of the bonds are represented by the positions of disks, where red disks indicate broken bonds and blue disks represent bonding links that are in the softening stage. After compression, all particle models reached central failure; however, the order of bond breaks varied across different BVBMs during the testing process. By observing the locations and quantities of

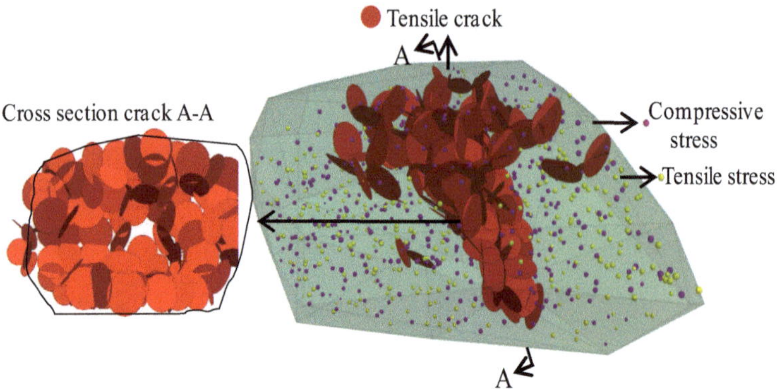

Fig. 2.31 Maps of tensile/compressive stresses and tensile-crack distribution in the BVBM model

bond breaks over time, it can be concluded that the failure mechanisms of BVBM can be categorized into two main types. Mechanism I: The fractures of bonds first appear at both ends of the rigid wall, gradually clustering into two clusters, which then extend towards the center over time while also diffusing outward from the center of the particles. Mechanism II: The fractures of bonds initially cluster at the top of the rigid wall, and as time progresses, the bonds in the softening stage extend outward, causing the broken links to expand downward. There exists a transitional scenario between Mechanisms I and II, where both ends extend and one end extends simultaneously. Mechanisms I and II can be considered model failures induced by indentation mechanisms, where the failure mode is formed by cracks generated in the shattered zone beneath the rigid wall. Mechanism I is observed in the BVBM with a lower softening factor; when the softening factor is low, it is clearly evident that after the hard rock fails at both ends, the failure rapidly progresses. Mechanism II primarily occurs when the softening factor is higher, resulting in greater ductility of the particles. Following the localized compression by the upper rigid wall, plastic deformation failure first occurs at the top and then extends downward. The appearance time of the bonds indicated by the red and blue disks demonstrates the significant ductility of such soft rock particles.

Further observation of the changes in the bonds of the BVBM model, as shown in Fig. 2.33, indicates that with a softening factor ζ of 5 or 20, the development of

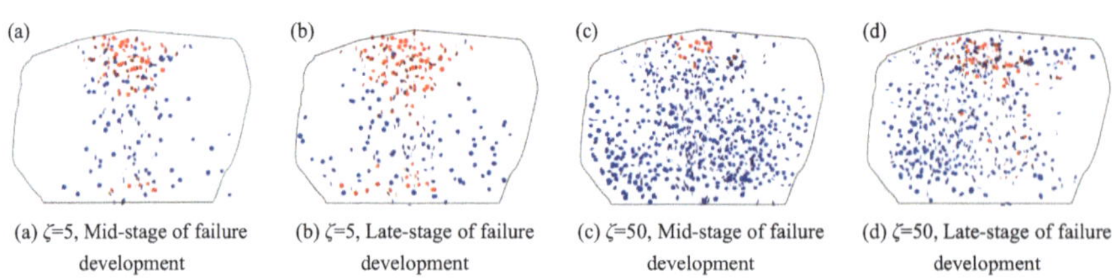

Fig. 2.32 Mesoscopic failure process of single particle under different softening factors

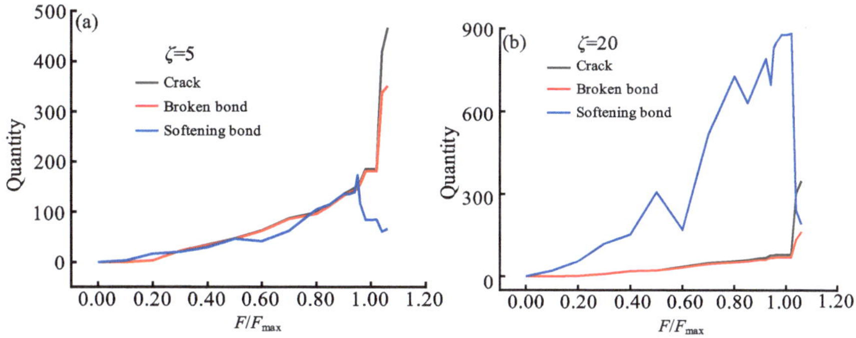

(a) Softening factor $\zeta = 5$ (b) Softening factor $\zeta = 20$

Fig. 2.33 Crack propagation and bond evolution within particles

broken bonds and cracks caused by stretching is similar, suggesting that the BVBM model is prone to tensile failure. By comparing the number of broken bonds and those in the softening stage, it can be concluded that as the softening factor ζ increases, the ductility of the particles gradually increases. The curves in Fig. 2.34, which illustrate the changes in the number of particle cracks, broken bonds, and bonds in the softening stage under different softening factors ζ, indicate that the fracture evolution of the BVBM models with different softening factors ζ is similar. The horizontal axis is uniformly defined as the ratio of the applied axial load F to the peak pressure F_p at that time. The number of cracks, broken bonding bonds, and bonds in the softening stage all increases with the increase of load, and experiences a sudden change when the particles fail, confirming that this model is applicable to hard brittle particles and soft rock particles such as coal rock. Notably, a smaller softening factor for the particles correlates with fewer bonds in the softening stage and a greater number of cracks and broken bonds, indicating that the particles fail earlier and exhibit lower strength, which aligns with the conclusions drawn in Sect. 2.3.2.3.

The BVBM construction method presented above offers significant advantages over traditional spherical particles and PBM. First, regarding particle shape, the irregular blocks generated by Voronoi tessellation resemble the irregular crystal grains of minerals within the rock mass more closely, thereby aligning the particle shape in the model with reality. Second, the load–displacement curve: the inter-particle contact behavior in the SBM utilized in the model accurately reflects the actual mechanical behavior of rock masses. This includes the softening phase during the failure process, which can simulate the post-peak ductility behavior observed in the stress–strain curve of real rock masses, thus aligning more closely with actual failure processes. Third, concerning particle crushing mode, the superior inter-particle contact behavior of SBM in BVBM enhances the representativeness of the simulated particle crushing mode [76, 85]. As indicated in Sect. 2.3.3 regarding the influence of the softening factor on particle failure mechanisms, this BVBM modeling method is not only applicable to hard and brittle particles like rock masses but will also expand its research scope to soft rock particles such as coal [97, 98]. On this basis, further modeling

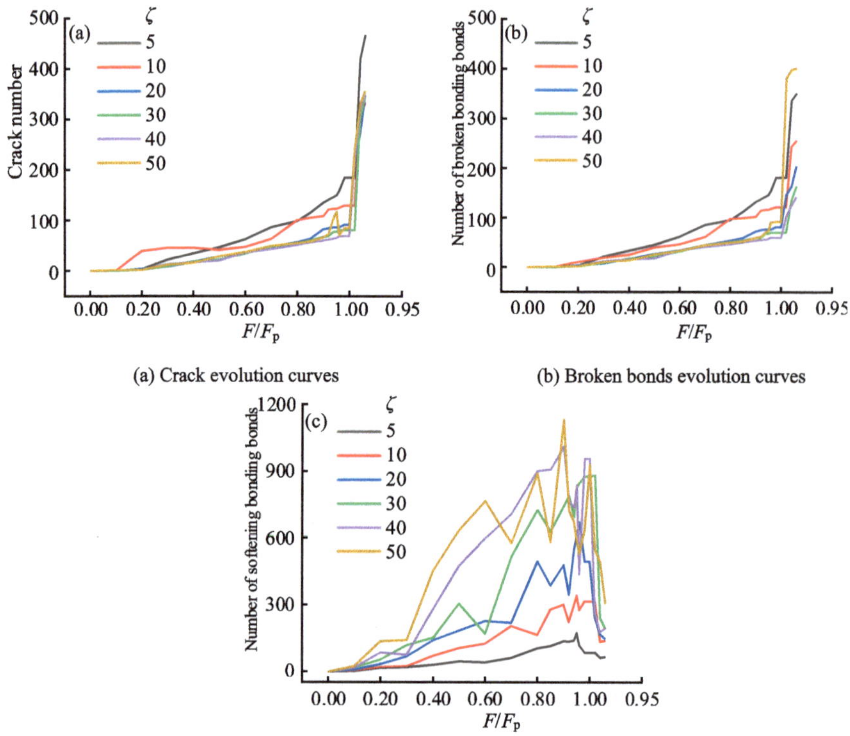

(a) Crack evolution curves

(b) Broken bonds evolution curves

(c) Bonds evolution curve in softening stage

Fig. 2.34 Effect of different softening factor ζ on particle crack, broken bond and number of bonds in softening stage

optimizations can be conducted in areas such as the concave particle construction method [99] and multi-particle modeling methods combined with image recognition technology [100, 101], while also considering the efficiency of parameter tuning for more meso-parameters in multi-particle tests.

2.4 Conclusions

This chapter [102, 103] begins with the execution of compression crushing experiments on individual particles utilizing three-dimensional white light scanning technology. It examines both the macro and micro characteristics of the compression crushing process, focusing on aspects such as the distribution of particle crushing strength, modes of particle crushing, distribution of fragment sizes, and the surface morphology of the crushing fractures. Building on these findings, a numerical simulation method for the compression crushing of single particles, based on the BVBM approach, is proposed, alongside a macro-meso analytical method for assessing particle crushing characteristics. This research systematically investigates the influence patterns and mechanisms of various parameters on the crushing behavior of single particles, providing recommended ranges for each parameter. Furthermore, the crushing mechanisms of soft and hard rock particles are analyzed from a mesoscopic perspective. The main research conclusions are as follows:

(1) As brittle particles, the strength distribution of caved ore and rock conforms to the Weibull function model. In this study, the Weibull parameter m for the rock particles obtained is 2.173, and the characteristic strength F_0 is 7195.925 N.
(2) The crushing modes of caved ore and rock can be categorized into four types: mid-span crushing, edge attrition, through-crack crushing, and random cracking, with mid-span crushing identified as the primary mode, accounting for 0.433. The flatness and elongation of the particles exert a greater influence on mid-span crushing than sphericity. When the lengths of the three principal dimensions of the particles are similar, the likelihood of mid-span crushing diminishes. Furthermore, the mass distribution of the 'first size fragments' and 'second size fragments' generated from mid-span crushing particles follows a normal distribution, with means of approximately 0.65 and 0.30, respectively.
(3) The internal crack propagation within caved ore and rock results in the formation of external macroscopic cracks. The roughness of the fracture surface exhibits fractal characteristics, with the fractal dimension D decreasing as the cross-sectional area increases, gradually approaching equilibrium. When performing fractal analysis on the fracture surface of the particles, a cross-sectional size greater than 36 mm^2 yields a more stable fractal dimension D. The shape, equivalent particle size, and crushing energy of the particles significantly influence the average fractal dimension $\overline{D}$ of the fracture surface, with sphericity having the most significant effect. The influences of flatness, energy, angularity, and equivalent particle size are similar and secondary, while elongation shows no statistical difference.
(4) BVBM is constructed using three-dimensional white light scanning and reconstruction technology, employing Voronoi block models and SBM from PFC to develop a breakable irregular block model. This model exhibits a three-dimensional shape that closely resembles real irregular rock particles and is capable of fracturing. The SBM utilized incorporates a softening phase and a particle failure mode, which aligns more closely with the actual failure behavior of rocks. This method of constructing a numerically modeled single particle of breakable rock with a realistic shape can also be applied to fields related to granular materials, such as rockfill dams.
(5) A parametric study was conducted focusing on the size of spherical particles, loading speed, softening tensile strength factor, softening factor, and the macro size of BVBM. The results indicate: (1) The ratio of the macro short axis size of a single particle in BVBM to the average diameter of spherical particles (S_p/$<d>$) should be approximately in the range of 5–10. (2) The loading speed for BVBM should not exceed 0.05 m/s, corresponding to a nominal strain rate of $4.05 \times 10^{-2}\ s^{-1}$. (3) During the process of matching the mesoscopic parameters

of SBM in the BVBM model, the softening factor should first be adjusted within the range of 0–10, followed by the adjustment of the softening tensile strength factor within the range of 0–1.0. (4) In the BVBM model, both the accuracy of the simulation and computational efficiency must be comprehensively considered to determine the appropriate macro size of BVBM and the sizes of the included spherical particles.

(6) The results of the single-factor sensitivity analysis indicate that the sensitivity fluctuation range of five influencing factors on the peak pressure of BVBM remains within 10%. The specific order is as follows: macro size $\alpha \times S_p >$ softening factor $\zeta >$ softening tensile strength factor $\gamma >$ loading rate $v \approx$ spherical particle size $\langle d \rangle$. Accordingly, the modeling and meso-parameter matching method for individual particles of BVBM has been adjusted.

(7) The tensile failure of the particle model is induced by indentation, and its failure mechanism can be categorized into Mechanism I (failure propagating from both ends to the center) and Mechanism II (failure propagating from the top end to the bottom end), with a transitional phase existing between these two mechanisms. The particle model characterized by a low softening factor aligns with Mechanism I, whereas the model with a high softening factor corresponds to Mechanism II. The development of broken bonds and cracks resulting from tensile stress occurs in a similar manner, indicating that the model particles experience tensile failure. A comparison of the number of broken bonds with those present during the softening phase suggests the presence of ductility. Furthermore, the rupture evolution of the BVBM under varying softening factors ζ exhibits similarities, thereby validating the applicability of the BVBM model to both hard brittle particles and softer rock particles, such as coal.

Note. Figures 2.17, 2.18, 2.19, 2.20, 2.21, 2.22, 2.23, 2.24, 2.25, 2.26, 2.27, 2.28, 2.29 and 2.30 and Tables 2.6, 2.7 in Sect. 2.3 are reprinted from the authors' previously published work [103]:

Jin AB, Li MY, Sun H, et al. Numerical modeling of the crushing characteristics of single ore particle based on breakable Voronoi block model [J]. Powder Technology, 2024, 445: 120,114.

Permission has been obtained from the publisher for the reuse of these materials.

References

1. Kou BQ (2018) X-ray Imaging study of dynamic properties of three-dimensional granular system. Shanghai Jiao Tong University, Shanghai
2. Hu JQ (2017) Mechanical analysis and research on the transport process of granular materials. Shenyang Jianzhu University, Shenyang
3. Yang YH (2015) Theoretical study on force transmission and scaling law of drag force in granular materials. Lanzhou University, Lanzhou
4. Lu KQ, Liu JX (2004) Static and dynamic properties of granular matter. Physics 33(09):629–635

5. Ministry of Natural Resources of the People's Republic of China. Global Mining Development Report 2020–2021[EB/OL]. https://www.mnr.gov.cn/dt/ywbb/202110/t20211025_2700039.html, 2022–02–15.
6. Ministry of Natural Resources of the People's Republic of China. China Mineral Resources Report 2022[EB/OL]. https://www.mnr.gov.cn/sj/sjfw/kc_19263/zgkczybg/202209/t20220921_2759600.html. 2023–02–21
7. Sun H (2019) Study on migration and evolution mechanism of caved ore and rock based on the particle flow theory. University of Science and Technology Beijing, Beijing
8. Li T (2018) Study on migration law of granular ore-rock under draw by caving method. University of Science and Technology Beijing, Beijing
9. Dorador L, Eberhardt E, Elmo D (2014) Influence of secondary fragmentation and column height on block size distribution and fines migration reaching drawpoints. In: Proceedings of the 3rd international symposium on block and sublevel caving, Santiago, pp 128–137
10. Sun H, Gao YT, Davide E et al (2019) A study of gravity flow based on the upside-down drop shape theory and considering rock shape and fragmentation. Springer Vienna 52(3):881–893
11. Brown ET (2007) Block caving geomechanics. Julius Kruttschnitt Mineral Research Centre, The University of Queensland, Brisbane
12. Chitombo GP (2010) Cave mining: 16 years after Laubscher's 1994 paper 'Cave mining a state of the art. In: Proceedings of the 2nd international symposium on block and sublevel caving, Perth, pp 45–61
13. Dorador L (2016) Experimental investigation of the effect of broken ore properties on secondary fragmentation during block caving. The University of British Columbia, Vancouver
14. Laubscher DH (2000) Block caving manual. The University of Queensland, Brisbane
15. Valdes JR, Leleu SL (2008) Influence of mineral composition on the simple shear response of sands: experimental study. J Geotech Geoenviron Eng 134(12):1820–1824
16. Castro RL, Fuenzalida MA, Lund F (2014) Experimental study of gravity flow under confined conditions. Int J Rock Mech Min Sci 67:164–169
17. Esterhuizen GS (1994) A program to predict block cave fragmentation. User's Guide, BCFV2.1
18. Esterhuizen GS (2005) A program to predict block cave fragmentation. User's Guide, BCFV3.05
19. Butcher RJ, Thin IGT (2007) The inputs and choices for predicting fragmentation in block cave projects. In: Proceedings of the1st international symposium on block and sub-level caving. Johannesburg, pp 35–49
20. Pierce ME (2010) A model for gravity flow of fragmented rock in block caving mines. The University of Queensland, Brisbane
21. Lorig LJ, Cundall PA (2000) A rapid gravity flow simulator. JKMRC and Itasca Consulting Group Inc., Brisbane
22. Kojovic T (2010) Application of the hybrid model to RWD. The University of Queensland, Brisbane
23. Rogers S, Elmo D, Webb G, Catalan A (2010) A discrete fracture network based approach to defining in situ, primary and secondary fragmentation distributions for the Cadia East panel cave. In: Proceedings of the 2nd international symposium on block and sublevel caving, Perth
24. Zhang YN (2019) Study on macro-mesoscopic mechanical properties of single particle crushing. Wuhan University, Wuhan
25. Deng XX (2018) The effects of the local constraints patterns on the fragmentation of a single grain. Wuhan University, Wuhan
26. Pan YY (2020) Study on the influence of particle morphology on shear fracture mechanical properties of coarse grained soils. Chengdu University of Technology, Chengdu
27. Guyon E, Troadec JP (1994) Du sac de billes au tas de sable. Editions Odile Jacob, Paris

28. Gorham DA, Salman AD, Pitt MJ (2003) Static and dynamic failure of PMMA spheres. Powder Technol 138(2–3):229–238
29. Antonyuk S, Palis S, Heinrich S (2011) Fragmentation behaviour of agglomerates and crystals by static loading and impact. Powder Technol 206(1–2):88–98
30. Zhang YD, Buscarnera G, Einav I (2016) Grain size dependence of yielding in granular soils interpreted using fracture mechanics, fragmentation mechanics and Weibull statistics. Géotechnique 66(2):149–160
31. Ma LJ, Li Z, Luo ZM et al (2019) Experimental study of strain rate effects on mechanical properties of coral particles. Rock Soil Mech 40(12):4637–4643
32. Wang YD (2017) Fractal theory of single particle crushing strength and compression deformation of coarse granular materials. Shanghai Jiao Tong University, Shanghai
33. Mc Dowell GR (2002) On the yielding and plastic compression of sand. Soils Found 42(1):139–145
34. Cheng YP, Bolton MD, Nakata Y (2005) Grain crushing and critical states observed in DEM simulations. Powders Grains 2:1393–1397
35. Lim WL, Mc Dowell GR, Collop AC (2004) The application of Weibull statistics to the strength of railway ballast. Granular Matter 6(4):229–237
36. Huang J, Xu S, Yi H et al (2014) Size effect on the compression fragmentation strengths of glass particles. Powder Technol 268(1):86–94
37. Oskooei PR, Mohammadinia A, Arulrajah A et al (2021) DEM modeling and experimental analysis of the fragmentation behavior of recycled crushed brick particles. Transp Geotech 100586
38. Meng MQ, Wang L, Jiang X et al (2020) Single-particle crushing test and numerical simulation of coarse grained soil based on size effect. Rock Soil Mech 41(09):2953–2962
39. Zhou HJ, Ma G, Yuan W et al (2017) Size effect on the crushing strengths of rock particles. Rock Soil Mech 38(08):2425–2433
40. Xu YF, Wang YD, Xi Y et al, Size effect on crushing of rock particles. J Eng Geol 22(06):1023–1027
41. Zhou MJ, Wen YF, Deng G et al (2019) Three-dimensional discrete element simulation of random breaking strength and size effect in single particle splitting test of rockfill. Rock Soil Mech 40(S1):503–510
42. Xu YF, Xie HP (1998) Study on the macro-mesoscopic mechanics of porous and fractured rock mass. Chinese J Geotech Eng 20(4):113–114
43. Mandelbrot BB (1983) The fractal geometry of nature. W. H. Freeman and Company, New York
44. Tyler SW, Wheatcraft SW (1992) Fractal scaling of soil particle-size distributions: analysis and limitations. Soil Sci Soc Am J 56(2):362–369
45. Xie HP (1992) Fractal geometry and its application to rock and soil materials. Chinese J Geotech Eng 14(1):14–24
46. Xie HP (1993) Recent advances in fractals and fractals in mechanics. Mech Eng 15(2):9–18
47. Xu YF, Liu SH, Dong P (2001) Microstructural model for granular soils. Rock Soil Mech 22(4):366–372
48. Jaeger JC (1967) Failure of rocks under tensile conditions. Int J Rock Mech Mining Sci Geomech Abstr 4(2):219–227
49. Hiramatsu Y, Oka Y (1966) Determination of the tensile strength of rock by a compression test of an irregular test piece. Int J Rock Mech Mining Sci 3(2):89–90
50. Nakata AFL, HYde M, Hyodo H et al (1999) A probabilistic approach to sand particle crushing in the triaxial test. Géotechnique 49(5):567–583
51. Sun YD (2019) Study on the mechanism of particle fragmentation of coral debris. Southeast University, Nanjing
52. Zhou Q, Pan YT, Xu HX (2020) Experimental research on strength distribution of coal rock particles under uniaxial compression. J China Univ Mining Technol 49(02):381–386+402
53. Jia XM, Chai HJ, Zheng YR (2010) Mesomechanics research of large direct shear test on soil and rock aggregate mixture with particle flow code simulation. Rock Soil Mech 31(9):2695–2703

54. Yan Y, Ji SY (2009) Effects of particle shape on shear strength of discrete media. Rock Soil Mech 30(S1):225–230
55. Shi DD, Zhou J, Liu WB et al (2010) Exploring macro- and micro-scale responses of sand in direct shear tests by numerical simulations using non-circular particles. Chin J Geotech Eng 32(10):1557–1565
56. Afshar T, Disfani MM, Arulrajah A et al (2017) Impact of particle shape on fragmentation of recycled construction and demolition aggregates. Powder Technol 308:1–12
57. Zhou W, Xu K, Ma G et al (2019) On the fragmentation function for constructing the fragment replacement modes. Particuology 44:207–217
58. Zhao XL, Zhu JG, Jia Y et al (2022) Experimental and numerical study of size effects on the crushing strength of rockfill particles. Int J Numer Anal Meth Geomech 46(11):2060–2086
59. Zhou B, Ku Q, Li CH et al (2022) Single-particle crushing behaviour of carbonate sands studied by X-Ray microtomography and a combined finite-discrete element method. Acta Geotech 17:3195–3209. https://doi.org/10.1007/s11440-022-01469-w
60. Bagherzadeh KA, Mirghasemi AA, Mohammadi S (2011) Numerical simulation of particle fragmentation of angular particles using combined DEM and FEM. Powder Technol 205(1–3):15–29
61. Ma G, Zhou W, Chang XL et al (2016) A hybrid approach for modeling of breakable granular materials using combined finite-discrete element method. Granular Matter 18:7
62. Ham TG, Nakata Y, Orense R et al (2010) Influence of water on the compression behavior of decomposed granite soil. J Geotech Geoenviron Eng 136(5):697–705
63. Druckrey AM, Alshibli KA (2016) 3D finite element modeling of sand particle fracture based on in situ X-Ray synchrotron imaging. Int J Numer Anal Meth Geomech 40(1):105–116
64. Ben-Nun O, Einav I (1910) The role of self-organization during confined comminution of granular materials. Philos Trans R Soc 2010(368):231–247
65. Zhou W, Yang LF, Ma G et al (2015) Macro-micro responses of crushable granular materials in simulated true triaxial tests. Granular Matter 17:497–509
66. Alonso EE, Tapias M, Gili J (2012) Scale effects in rockfill behavior. Géotechnique 2(3):155–160
67. Wang JF, Yan HB (2013) On the role of particle fragmentation in the shear failure behavior of granular soils by DEM. Int J Numer Anal Meth Geomech 37(8):832–854
68. Wang B, Martin U, Rapp S (2017) Discrete element modeling of the single-particle crushing test for ballast stones. Comput Geotech 88:61–73
69. Wang YH, Ma G, Mei JZ et al (2021) Machine learning reveals the influences of grain morphology on grain crushing strength. Acta Geotech 16:3617–3630
70. Ge H, Quezada JC, Houerou VL et al (2021) Three-dimensional simulation of asphalt mixture incorporating aggregate size and morphology distribution based on contact dynamics method. Constr Build Mater 302(2):124124
71. Blott SJ, Pye K (2010) Particle shape: a review and new methods of characterization and classification. Sedimentology 55(1):31–63
72. Tu YL, Liu XR, Ren QY et al (2020) Effects of rock contents and particle fragmentation on strength characteristics of soil-rock aggregate. Rock Soil Mech 41(12):3919–3928
73. Mandelbrot BB (1967) How long is the coast of Britain? Statistical self-similarity and fractional dimension. Science 156(3775):636–638
74. Mandelbrot BB (1975) Les objects fractals: forme, hasard et dimension. Flmmarion, Paris
75. Li H, Huang RQ (2014) Method of quantitative determination of joint roughness coefficient. Chin J Rock Mech Eng 33(S2):3489–3497
76. Clarke KC (1986) Computation of the fractal dimension of topographic surfaces using the triangular prism surface area method. Comput Geosci 12(5):713–722
77. Xie HP, Wang JA, Xie WH (1997) Fractal effects of surface roughness on the mechanical behavior of rock joints. Chaos, Solitons Fractals 8(2):221–252
78. Zhou W, Xie H (2008) Direct estimation of the fractal dimensions of a fracture surface of rock. Surf Rev Lett 10(05):751–762

79. Wang YL, Li LY, Deng SY et al (2006) Computation of the surface fractal dimension (D_S) from micro-images of sediment particles. Environ Chem 25(4):400–404
80. Zhang YH, Zhou HW, Xie HP (2005) Improved cubic covering method for fractal dimensions of a fracture surface of rock. Chin J Rock Mech Eng 24(17):3192–3196
81. Ge YF (2014) Research on roughness and peak shear strength for rock discontinuities based on BAP. China University of Geosciences, Wuhan
82. Man K, Liu XL, Song ZF et al (2021) Research on fracture surface morphology of rock with static and dynamic fracture toughness. J Central South Univ (Nat Sci Edition) 52(08):2876–2886
83. Liang CY, Wu SR, Li X (2015) Research on micro-meso characteristics of granite fracture under uniaxial compression at low and intermediate strain rates. Chin J Rock Mech Eng 34(S1):2977–2986
84. Al-Rousan T, Masad E, Tutumluer E (2007) Evaluation of image analysis techniques for quantifying aggregate shape characteristics. Constr Build Mater 21(5):978–990
85. Xiang DJ, Li HW, Liu XY (2005) Practical multivariate statistical analysis. China University of Geosciences Press
86. Itasca Consulting Group (2019) Particle flow code (PFC) version 6.0. Itasca Consulting Group Inc., Minneapolis
87. Sun H, Chen SJ, Jin AB et al (2021) Study on mechanical properties of sandstone based on Voronoi block model. J Northeast Univ (Nat Sci) 42(11):1600–1608
88. Kandasami RK, Murthy TG (2014) Effect of intermediate principal stress on the mechanical behavior of angular Sand. Soil Behav Geomech GSP236:406–415
89. Cho N, Martin CD, Sego DC (2007) A clumped particle model for rock. Int J Rock Mech Min Sci 44(7):997–1010
90. Zhang LP, Wong LNY (2012) Cracking processes in rock-like material containing a single flaw under uniaxial compression: a numerical study based on parallel bonded-particle model approach. Rock Mech Rock Eng 45(5):711–737
91. Sierakowski RL (1985) Dynamic effect in concrete materials. Application of fracture mechanics to cementitious composites, pp 535–557
92. Saeidi F, Yahyaei M, Powell M et al (2017) Investigating the effect of applied strain rate in a single fragmentation event. Miner Eng 100:211–222
93. Ma YF, Huang HY (2018) DEM analysis of failure mechanisms in the intact Brazilian test. Int J Rock Mech Min Sci 102:109–119
94. Okubo S, Fukui K (1996) Complete stress-strain curves for various rock types in uniaxial tension. Int J Rock Mech Min Sci Geomech Abs 33(6):549–556
95. Cooper DF, Chapman CB (1987) Risk analysis for large projects: models, methods and cases. Wiley, New York
96. Ahmad A, Kotsovou G, Cotsovos DM et al (2018) Assessing the accuracy of RC design code predictions through the use of artificial neural networks. Int J Adv Struct Eng 10:349–365
97. Liu XW, Chen HX, Liu B et al (2022) Experimental and numerical study on failure characteristics and mechanism of coal under different quasi-static loading rates. Theoret Appl Fract Mech 121:103478
98. Ma SZ, Liu KW, Guo TF et al (2022) Experimental and numerical investigation on the mechanical characteristics and failure mechanism of cracked coal & rock-like combined sample under uniaxial compression. Theoret Appl Fract Mech 122:103583
99. Wu DT, Guo WB, Luo F et al (2023) Simulation experimental investigations on particle fragmentation mechanism and fractal characteristics of mixed size gangue backfill materials. J Mater Res Technol 24:125–142

100. Li JM, Huang YL, Pu H et al (2021) Influence of block shape on macroscopic deformation response and meso-fabric evolution of crushed gangue under the triaxial compression. Powder Technol 384:112–124
101. Ma G, Xie YJ, Long GC et al (2022) Mesoscale investigation on concrete creep behaviors based on discrete element method. Constr Build Mater 342:127957
102. Jin AB, Li MY, Sun H et al (2023) Compression crushing characteristics of single ore particle with different shapes. J Central South Univ (Sci Technol) 54(9):3585–3596
103. Jin AB, Li MY, Sun H et al (2024) Numerical modeling of the crushing characteristics of single ore particle based on breakable Voronoi block model. Powder Technol 445:120114

Chapter 3
Shear Characteristics of the Particle System of Caved Ore and Rock

3.1 Brief Overview of Shear Characteristics Research in the Particle System

Granular materials are ubiquitous in nature, everyday life, and industrial processes. They encompass natural substances such as sand, soil, snow, and ice; agricultural products like corn, soybeans, and rice; pharmaceutical formulations including pills and granules; and ores and rock particles in the mining industry. Due to variations in their storage conditions and applications, systems of caved ore and rock formed by the interactions of numerous particles exhibit considerable diversity. Notably, the particle systems of caved ore and rock are prevalent in mining operations. Investigating the mechanisms of particle interactions and elucidating the shear mechanical properties and movement characteristics of these systems can provide valuable insights for engineering applications and enhance safety in mining production. The following sections will comprehensively review the current state of research on the shear mechanical properties and shear-induced movement behavior of particle systems of caved ore and rock.

3.1.1 Shear Mechanical Characteristics of the Particle System

The direct shear test effectively reveals the shear response of granular media under boundary displacement conditions. Shear patterns generated in large-scale boundary displacement problems, such as surface fault ruptures during earthquakes, can also be replicated in direct shear tests [1]. Consequently, the direct shear test has become one of the most widely used techniques for investigating the mechanical and movement properties of granular materials, with numerous scholars conducting relevant laboratory test and numerical simulations in this field. Among these scholars, Wang

H. Sun et al., *Study on Flow and Hang-up Characteristics of Caved Ore and Rock*,
https://doi.org/10.1007/978-981-92-0335-2_3

[2], Chen [3], Kong et al. [4], and Yang [5] investigated the stress–strain relationship, particle displacement, and variations in particle contact forces during the shear process of granular systems through consolidated direct shear laboratory test and numerical simulations. They explored the main factors influencing the shear resistance of granular systems. Yang et al. [6] examined the anisotropic characteristics of contact forces during shearing using numerical simulation methods, noting that contact force anisotropy develops significantly in the early stages of shearing and subsequently stabilizes, with its principal direction undergoing counterclockwise rotation. Liu et al. [7], Zhang et al. [8], Yang et al. [9], and Zhang et al. [10] conducted large-scale direct shear test on soil-rock mixture particles, demonstrating that an increase in coarse particle enhances the shear strength and dilatancy of the packing system, with energy dissipation in coarse particle primarily dominated by friction. In studies on the shear stress–strain relationship of bulk materials, phenomena such as a 'sharp rise' [11] and 'drop' [12] are observed in the later stages of the curve. Li and Aydin [13] performed Fourier transformation on fluctuations in the shear stress-displacement curve of granular systems via direct shear test, indicating that shear force fluctuations surpass the dilatancy component of shear resistance, and identifying particle size, normal stress, and shear rate as contributing factors to shear stress fluctuations. Sun et al. [14], Yang et al. [15], and Xue et al. [16] utilized laboratory direct shear tests and numerical simulations to examine the evolutions of force chains, stress–strain behavior, contact force anisotropy, and overall shear strength during the shearing process.

Numerous studies have investigated the factors influencing the shear mechanical properties of granular systems, including the effects of moisture content [17], vertical stress [18], particle fragmentation [19], and gradation [20] on shear strength characteristics. Additionally, Chai et al. [21] conducted direct shear tests demonstrating that the shear strength of calcareous sand systems initially decreases and then increases with rising shear rates. Zhao et al. [22] performed large-scale direct shear tests on phyllite, revealing that the orientation of rock blocks significantly reduces the shear strength of phyllite particles. Cheng et al. [23] carried out direct shear tests with various gradation patterns, indicating that continuously graded particle systems exhibit higher shear strength under identical particle size conditions. Lin et al. [24] simulated lunar soil using parameters such as roundness and elongation in numerical simulations, demonstrating that ellipsoid-like particles possess greater shear strength. Liu et al. [25] conducted Cone Penetration Tests (CPT) on quartz sand, showing that particles with higher roundness generate lower cone tip resistance. Yan and Ji [26] systematically investigated particle interlocking effects through numerical simulations involving particles of varying shapes and sizes.

3.1.2 Shear Movement Characteristics of the Particle System

In the study of shear movement characteristics in granular systems, phenomena can generally be categorized into particle motion on and above the shear surface, local strain, and shear bands. Dong and Chai [27] and Jia et al. [28], through physical experiments and numerical simulations respectively, observed that the shear surface exhibits significant roughness, which they characterized using fractal dimensions. Macaulay and Rognon [29] validated a frictional constitutive model for particles, demonstrating that the movement capacity of individual particles is weaker than the inertial movement capacity of the granular system as a whole. Cui and O'sullivan [30] proposed that particles undergo rotation during shearing and that such rotation plays a crucial role in the shear behavior of granular systems. Zhang and Thornton [31] suggested that rapidly rotating particles are primarily distributed along a diagonal band from the shear input zone to the shear output zone, and that this band is conjugate to the force chain band. Wang et al. [32] indicated that the number of rotating particles and their angular velocities follow a Gaussian distribution.

In the study of local strain in particle shear systems, Sun et al. [33] concluded through physical tests that, under high-friction conditions, the influence of disorder on system stability outweighs that of local structure. Both Angus et al. [34] and Zheng et al. [35] suggested that increased friction suppresses particle motion within localized strain regions. Regarding shear band research, Hong et al. [36], Jiang et al. [37], and Pan et al. [38] proposed that, under constant friction conditions, particle size significantly influences shear band formation within the system, with shear band thickness measuring 10–21 times the average particle size of the gradation. Gang et al. [39] investigated the evolution of shear bands and local strain using numerical simulations of direct shear tests, demonstrating a significant correlation between non-affine displacement and local strain, and indicating that shear bands result from irreversible particle redistribution. Li et al. [40] and Wang et al. [41] studied the 'reverse grading' phenomenon in granular slope landslides, noting the absence of particle size segregation during high-speed landslides, with only localized tendencies for differential movement trajectories between coarse and fine particles. Jin et al. [42] conducted quantitative research on the percolation of fine particles during the draw process. Lei et al. [43] performed monitoring and quantitative analysis of local strain and displacement fields in granular systems during shear, utilizing transparent shear devices and digital image correlation techniques. Jing et al. [44] employed numerical simulations to investigate the movement characteristics of particles of two different sizes by altering the direction of gravity, demonstrating that the particle size ratio and friction coefficient significantly influence the movement of coarse and fine particles. Van et al. [45] conducted oscillatory shear tests on bidisperse mixed granular systems, revealing a polarized phenomenon in the movement efficiency of coarse and fine particles across different media. Xu et al. [46] and John et al. [47] studied the movement characteristics of two and three different-sized particles, respectively, indicating that fine particles ultimately concentrate at the center of rotating granular systems, while coarse particles move toward the surface. Zheng et al. [48] investigated the movement characteristics of individual coarse particles during shear through torsional shear tests, proposing the concept of the 'buoyancy-lift factor' to explain the initiation mechanism of coarse particle movement in fine granular media. Trewhela et al. [49, 50] quantitatively studied the movement trajectories, rotation, and strain rate tensors of individual coarse particles using cyclic shear tests, revealing the movement mechanisms between coarse and fine particles.

3.2 Research on Shear Characteristics of Uniform Particle System of Caved Ore and Rock

Currently, research on the shear mechanical characteristics of the particle systems of caved ore and rock remains relatively limited, primarily relying on numerical simulations without sufficient validation through physical testing. Furthermore, investigations into the spatiotemporal evolution of particle movement require enhancement. Therefore, this study employs a self-developed visual shear apparatus in conjunction with Digital Image Correlation (DIC) technology to monitor the shear strength of uniform particle systems of caved ore and rock during shearing, as well as the movement states of these materials, local strain, and the evolution patterns of shear bands. The research aims to systematically investigate the shear mechanical properties and movement characteristics of caved ore and rock.

3.2.1 Physical Testing Procedure for Shear Behavior of the Uniform Particle System of Caved Ore and Rock

3.2.1.1 Test Apparatus and Materials

In traditional direct shear tests, particularly large-scale direct shear tests, researchers have consistently utilized physical shear models constructed entirely of steel materials. This construction complicates the effective monitoring of particle movement states during shear testing. Consequently, this experiment employs a YSZJ-50 rock direct shear apparatus equipped with a self-developed visual shear model to conduct direct shear tests on uniform particle systems of caved ore and rock. The self-developed visual shear model has internal dimensions of 300 mm × 250 mm × 200 mm, as illustrated in Fig. 3.1. Specifically, the visual shear box is made of stainless-steel casting and features a tempered glass front panel, thereby overcoming the visibility limitations of traditional shear boxes. The shear box is divided into upper and lower sections, as shown in Fig. 3.1a, where the lower section remains fixed while the upper section undergoes vertical consolidation loading and horizontal shear, with the shear direction illustrated in Fig. 3.1b. The introduction of DIC technology enables real-time monitoring of particle displacement, rotation, local strain, and shear band development during the shearing process, effectively expanding the application of DIC technology in studying the mechanical properties of particle systems of caved ore and rock.

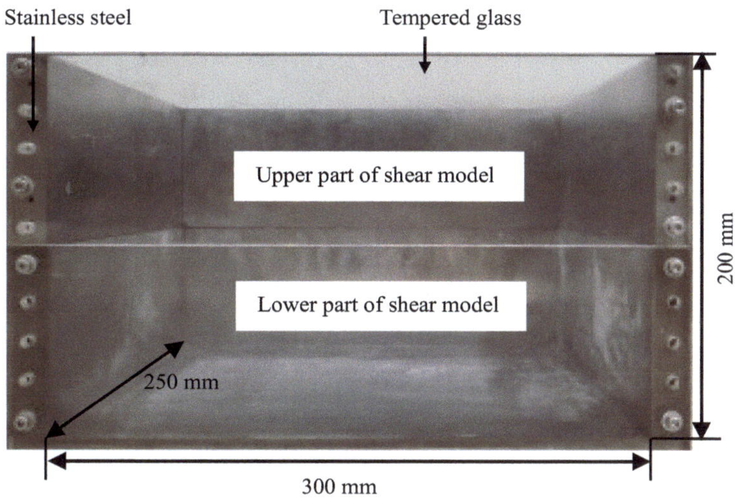

(a) Particle visualization shear model box

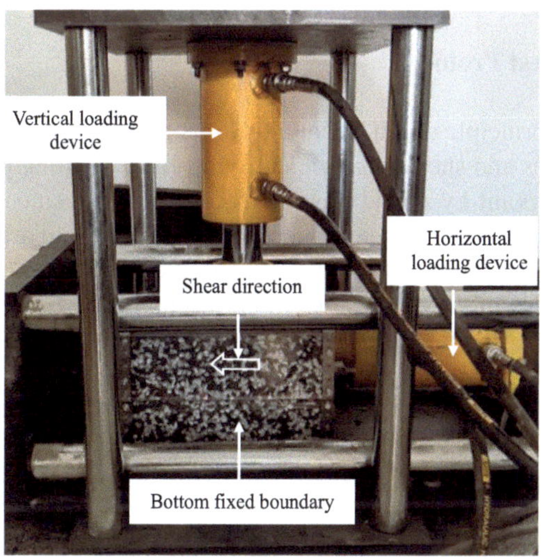

(b) Complete particle shear physical model

Fig. 3.1 Visualized shear model box for caved ore and rock and the complete physical shear model

The study utilized naturally occurring gray-black limestone particles, noted for their hard texture, uniaxial compressive strength of 72 MPa, and density of 3150 kg/m^3. To meet the testing requirements, the limestone particles were sieved using a square-mesh screen (Fig. 3.2), selecting particles with sizes ranging from 5–10 mm to 10–15 mm for the experiment.

Fig. 3.2 Sieved limestone particles

3.2.1.2 Shear Test Protocol

The fundamental principle of DIC technology involves segmenting the initial image into distinct regions and subsequently identifying corresponding positions in subsequent images on a point-by-point basis. The recorded displacement differences form a displacement field, and parameters such as the velocity field are calculated based on time intervals [51]. Since DIC relies on speckle pattern recognition, the direct spraying of speckles often yields suboptimal results for particle media. Therefore, the particle system was processed by spraying black and white speckles to create particles of two distinct colors. Prior to testing, the two-colored speckled particles were uniformly mixed and placed on the tempered glass side of the visual shear box.

The shear test research is divided into two main parts: (1) Study of shear mechanical properties: Particles sized between 5 and 10 mm are subjected to shear under vertical pressures of 100 kPa, 200 kPa, and 300 kPa to investigate their stress-displacement relationships. To further examine the influence of coarse particles on shear strength, particles in the size range of 10–15 mm are added with mass content ratios of fine to coarse particles of 1/5, 1/4, 1/3, 1/2, and 1. After uniform mixing, these particles are placed inside the shear model and tested under a vertical pressure of 200 kPa to investigate the effect of particle gradation on the shear mechanical properties of the particle system (see Table 3.1). (2) Study of shear motion characteristics: DIC technology is employed for real-time monitoring and analysis of particle displacement, local strain, and shear bands during the shearing process.

The standard procedure for conducting a direct shear test on particle systems comprises the following steps: (1) Installation of the shear model; (2) Filling the model with particles; (3) Application of vertical pressure, stopping the loading process upon reaching the predetermined pressure; (4) Calibration of the spatial

Table 3.1 Particle-size proportion and vertical-pressure settings in the shear-test protocol

Shearing scheme	Coarse-to-fine particle size ratio	Vertical pressure/kPa
	(5–10 mm): (10–15 mm)	
1	1:0	100
		200
		300
2	5:1	200
	4:1	
	3:1	
	2:1	
	1:1	

coordinates and dimensions of the DIC equipment; (5) Initiation of horizontal loading while simultaneously monitoring the shear process using DIC; (6) Analysis and processing of the test data to obtain particle movement data and elucidate the evolution patterns of local strain and shear bands.

3.2.2 *Shear Mechanical Characteristics of Uniform Particle Systems of Caved Ore and Rock*

In investigating the shear characteristics of granular materials, a comprehensive understanding of their fundamental mechanical properties is essential for analyzing shear resistance. Accordingly, for the particle system of caved ore and rock, various vertical pressures and particle gradations were configured to meet the test requirements (see Fig. 3.1) in order to examine the shear mechanical properties of the particle system. In this direct shear test, the shear process was conducted by fixing the lower shear box while applying a horizontal leftward shear stress to the upper shear box. The results of the direct shear test conducted under a vertical pressure of 100 kPa are illustrated in Fig. 3.3, which presents the front view of the particle system upon completion of the test.

During the testing phase, pressure sensors and displacement meters were employed to monitor and record the variations in stress and shear displacement throughout the shearing process. The evolution pattern of shear stress intensity during shearing can be summarized as follows: shear stress initially increases rapidly before gradually stabilizing. As vertical pressure increases, the shear strength exhibits a significant increase, with its variation pattern illustrated in Table 3.2. In alignment with the mechanical characteristics observed in other particle systems, the response of shear strength variation in the particle system of caved ore and rock to vertical pressure demonstrates an approximately linear growth trend.

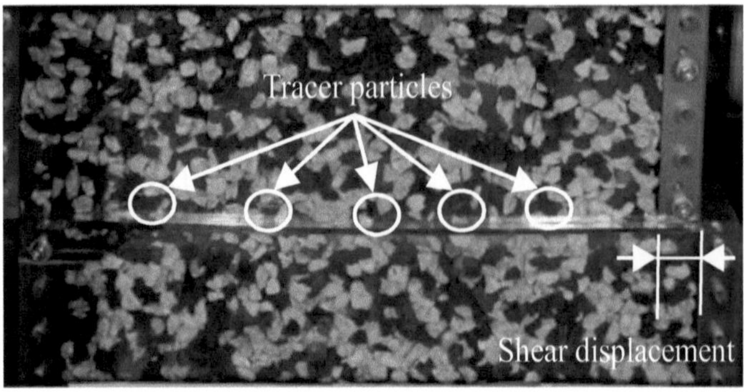

Fig. 3.3 Direct shear test results under 100 kPa vertical pressure

Table 3.2 Shear strength of particle systems under different vertical pressures

Vertical pressure/kPa	100	200	300
Shear strength/kPa	44.38	61.40	92.02

To further investigate the shear mechanical properties of the particle system, varying proportions of 10–15 mm coarse particles were incorporated into the fine particles. After ensuring uniform mixing, a vertical stress of 200 kPa was applied for direct shear testing. The data analysis reveals the variation pattern of the shear strength of the particle system in relation to coarse particle content, as illustrated in Fig. 3.4. Specifically, as the coarse particle content increases, the shear strength is observed to improve. This phenomenon can be attributed to two primary factors: first, the increased coarse particle content enhances interparticle interlocking; second, the inclusion of coarse particles improves the overall compactness of the particle system.

Through linear interpolation, it can be concluded that the relationship between coarse particle content and shear strength follows a negative exponential growth function, as illustrated in Eq. (3.1):

$$\tau = 72438 - 27583 * \mathrm{e}^{-\frac{(x-0.1678)}{0.0772}} \tag{3.1}$$

Here, τ is the shear strength, Pa, x is the coarse particle content, %, and e is the base of the natural logarithm.

In summary, the analysis based on Fig. 3.4 and Eq. (3.1) indicates that as the content of coarse particles increases, the shear strength gradually improves; however, the rate of improvement progressively diminishes. This finding illustrates that while enhancing interparticle interlocking and compactness can increase shear resistance, its effectiveness becomes limited at higher coarse particle contents.

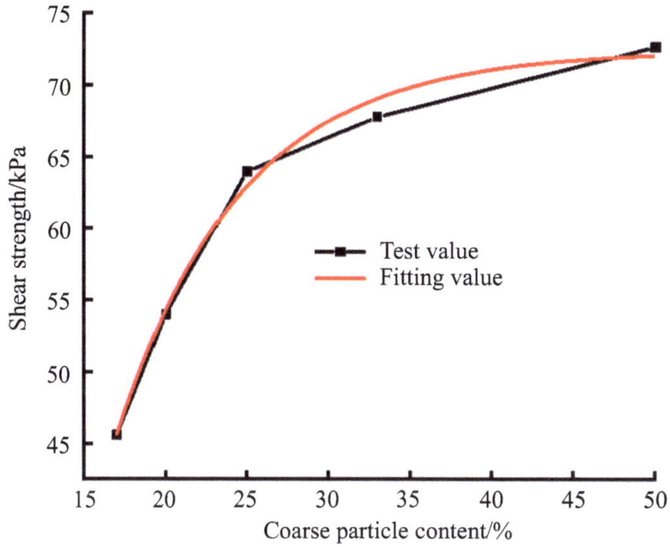

Fig. 3.4 Curve of shear strength vs coarse-particle content for the particle system under 200 kPa vertical stress

3.2.3 Shear Movement Characteristics of Particle System Based on DIC Technology

Based on DIC technology, this study investigates the characteristics of surface shear movement within the particle system of caved ore and rock. It specifically examines particle movement patterns at the shear surface, the development of local strain, and the evolution of shear bands.

3.2.3.1 Movement Characteristics of Particles Along the Shear Plane

During the shearing process, both individual particles and the entire particle system exhibit similar movement trends under varying pressure conditions. Therefore, the particle movement at a pressure of 100 kPa is selected as a representative example for analyzing the characteristics of particle movement during shearing.

Based on the analysis of tracer particles, it is observed that as shear displacement increases, the red tracer particles (Fig. 3.3) not only move in conjunction with the overall particle system but are also influenced by their surrounding particles. This interaction results in a displacement and rotation that are distinct from those of the particle system (Fig. 3.5). The evolution of displacement can be categorized into two scenarios: (1) Interparticle interactions lead to a situation where the actual displacement of the particles is less than that of the particle system, a phenomenon termed the lateral obstruction effect (Fig. 3.6). (2) Under the influence of dilatancy, particles experience displacement towards both the upper and lower sides, referred to as the vertical compression effect (Fig. 3.7). The magnitude of displacement and rotation varies slightly depending on the positions of the particles, as illustrated in Fig. 3.8.

Before and after shearing, the particles on the shear surface underwent three distinct types of movement, as illustrated in Fig. 3.8. The intensity and trends of these movements varied according to the positions of the particles. Overall, both lateral obstruction and vertical compression of the particles initially decreased and then increased with increasing distance from the shear inlet, while particle rotation exhibited the opposite trend, first increasing and then decreasing. This observation indicates a reversal in the sign of acceleration for both particle displacement and rotation at this specific point.

The analysis of the data presented in Fig. 3.8 indicates that the tracer particles on the shear surface exhibit a common abrupt change point in their motion trend, located 100 mm from the shear inlet, which corresponds to one-third of the total

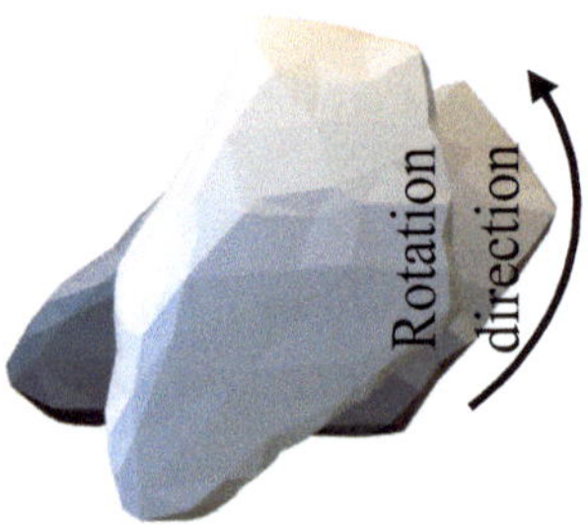

Fig. 3.5 Schematic of tracer particle rotation

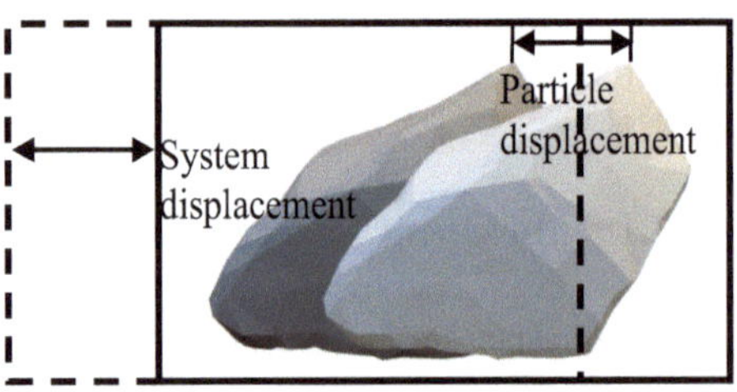

Fig. 3.6 Schematic of lateral particle obstruction (displacement difference)

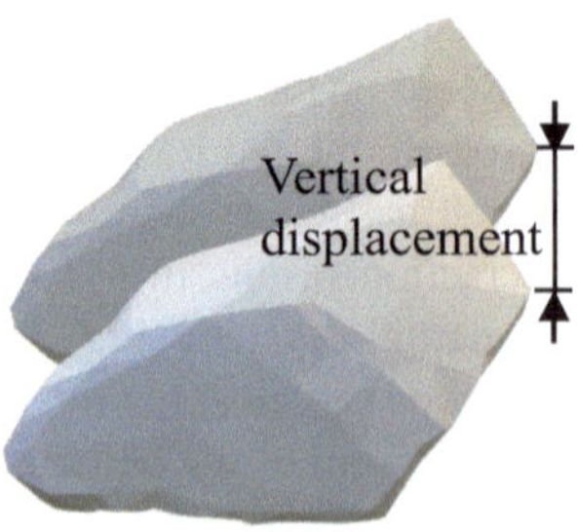

Fig. 3.7 Schematic of particle vertical compression

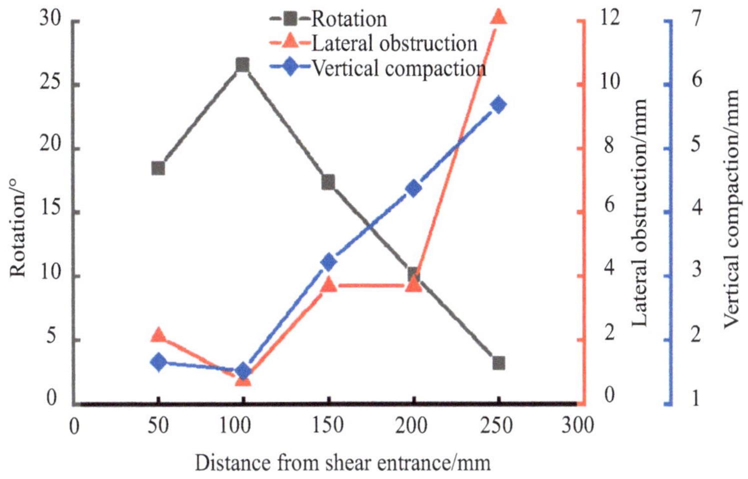

Fig. 3.8 Movement pattern of tracer particles on the shear plane

shear surface length. These particles, influenced by shear stress and constrained by the particle system, display movement characteristics consistent with the pattern illustrated in Fig. 3.8. At the junction of driving and restraining forces, the state of particle movement transitions, a phenomenon referred to as the driving-restraining compromise point. On the shear surface, lateral obstruction, vertical compression, and rotational changes in particle displacement reveal the dilatancy effect during shearing. This suggests that the impact of the dilatancy effect on the entire particle system during shearing is not uniform; rather, it fluctuates along the shear surface: initially decreasing and subsequently increasing with greater distance from the shear inlet, ultimately reaching its minimum effect at one-third of the shear surface length from the shear inlet.

During the shearing process, the external energy input induced by shear remains constant. This energy facilitates particle compaction, displacement, and rotation. Wang et al. [32] previously indicated that particle redistribution occurs during shearing: the structure evolves from the shear inlet zone to the shear outlet zone, initially simplifying and then becoming more complex, with structural stability first decreasing and then increasing. The right side (shear inlet) generates driving shear stress, exerting a compacting effect on particles, while the left side (shear outlet) acts as a fixed boundary that hinders particle movement, also inducing compaction. Consequently, the central region forms a relatively less compact zone compared to both sides. Higher compactness in local particle systems promotes the formation of stable local structures, thereby enhancing resistance to lateral particle movement. Moreover, as the hindering effect on lateral movement from the right side (shear inlet) is constrained by the horizontal load, the lateral obstruction initially decreases and then increases with growing distance from the shear inlet.

The stronger the lateral obstruction experienced by particles, the more pronounced their tendency for vertical compression becomes. Consequently, the vertical compression of particles demonstrates a trend of initial decrease followed by an increase, mirroring the pattern observed in lateral obstruction. The rotational capacity of particles diminishes with the increasing compactness of the particle system and is also related to stress transmission within the system. Therefore, the rotational effect exhibits a trend of initial increase followed by a decrease as the distance from the shear inlet grows.

3.2.3.2 Localized Shear-Strain Characteristics

Unlike the shear failure observed in rock, the shear failure process in particle systems occurs through the coordinated action of multiple local strains. Local strain refers to localized regions where shear strain concentration develops during the shearing process. The emergence of local strain is typically induced by opposing moment actions on local particles. In existing research, scholars have employed local strain to characterize the degree of non-affine deformation [52]. However, most studies on local strain rely on numerical simulation methods, which often lack physical validation and predominantly describe local strain post-shearing. Some researchers have also proposed criteria for local strain discrimination through theoretical calculations (Eq. (3.2)) [53, 54]. This study primarily employs DIC technology to monitor and analyze the shear strain field. During the shearing process of particle systems, local strain directly influences the shear process, thus necessitating an investigation into the temporal evolution patterns of local strain.

$$\det\left[n_j D_{ijkl}^{ep} - \frac{1}{2}\left(\sigma_{ik} + \sigma_{jk} n_j n_i - \sigma_{ij} n_j n_k - \sigma_{jl}\delta_{ik} n_j n_l\right)\right] = 0 \tag{3.2}$$

Here, n_i denotes the unit normal vector of the shear band, D_{ijkl}^{ep} represents the elastoplastic stiffness tensor, and σ_{ij} is the current stress tensor component.

As illustrated in Fig. 3.9, during the shearing process, local strain regions progressively converge from a random distribution toward the shear surface. In the initial stage of shearing, the impact of shear force on the particle system is minimal, with local strain primarily induced by vertical pressure. Since vertical pressure uniformly affects the entire area at the onset of shearing, the regions exhibiting shear strain demonstrate a globally irregular and random distribution, as depicted in Fig. 3.9a. As the shear force gradually intensifies, its influence on the evolution of local strain becomes more pronounced, leading to the aggregation of local strain from the shear inlet area (Fig. 3.9b) toward the shear surface (Fig. 3.9c). Eventually, the local strain becomes interconnected and stabilizes, with no further significant changes observed, as shown in Fig. 3.9d–f.

To further investigate the evolution patterns of local strain, we examined the temporal development of local strain, focusing on local strain values, the number of

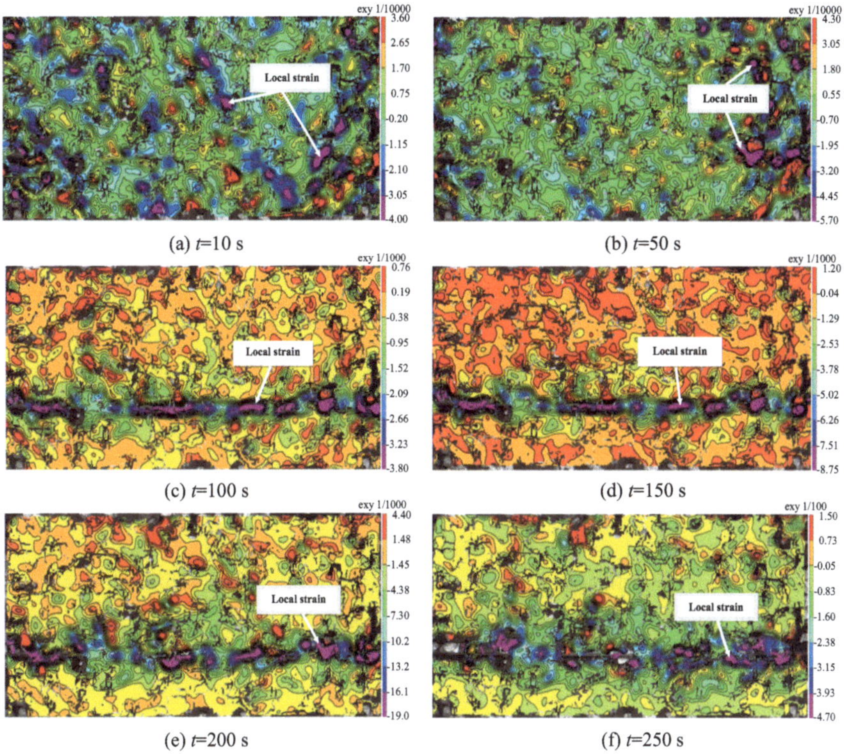

Fig. 3.9 Evolution of local strain during shearing

concentration areas, and local strain probability density. Within a defined interval, the likelihood of local strain occurrence is characterized as local strain probability density, which is inherently non-negative [55]. This probability density can be quantitatively assessed using DIC technology by analyzing shear strain. Based on the maximum thickness of the shear band, specific regions were delineated, and a statistical analysis of local strain within these regions was performed to elucidate the temporal evolution patterns of local strain, as illustrated in Fig. 3.10.

Near the shear surface, the number of local strain concentration areas exhibits random variations but generally remains within the range of 8–15. This indicates that, under constant particle size gradation, the quantity of local strain concentration areas fluctuates randomly within this range. Furthermore, the local strain value gradually increases over time with a slightly rising growth rate, demonstrating that the magnitude of local strain continuously grows with increasing shear displacement. Since local strain is induced by shearing, its temporal evolution transitions from global randomness to concentration near the shear surface. Although local strain in other regions does not disappear, it becomes systematically negligible due to the more intense local strain near the shear surface. Statistical analysis of local strain within the delineated regions reveals that the probability density of local strain follows a

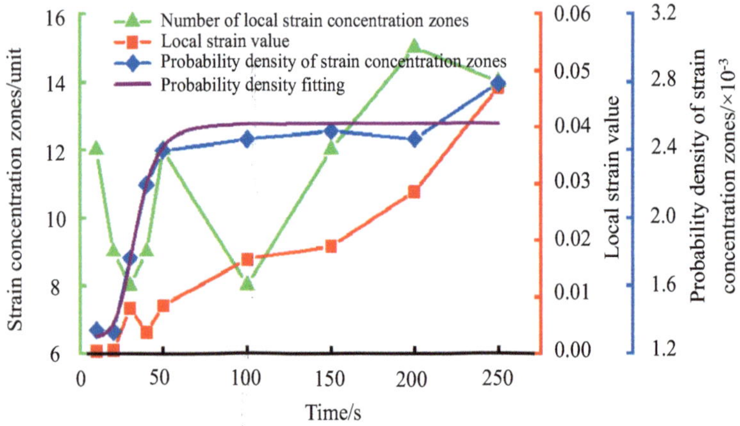

Fig. 3.10 Temporal evolution of local strain

pattern of initial rapid growth followed by stabilization, conforming to the logistic regression model. Further quantification yields the temporal evolution of local strain as shown in Eq. (3.3):

$$p = A_2 + \frac{A_1 - A_2}{1 + \left(\frac{t}{t_0}\right)^c} \tag{3.3}$$

Here, p is the local strain probability density; A_1 and A_2 are coefficients with values of $1.30 \times 10^{-3} \pm 1.11 \times 10^{-4}$ and $2.55 \times 10^{-3} \pm 6.30 \times 10^{-5}$, respectively; t_0 is the initial time; and c is the exponent, which has a value of 5.21 ± 1.84 in this model.

The analysis of Eq. (3.3) reveals that, as the shearing process progresses, the probability density of local strain increases rapidly during the initial stage. This indicates that, in the early phase, the particle system exhibits significant resistance to shearing, accompanied by a period of sharp growth. In the later stage, the probability density of local strain stabilizes with minor fluctuations, demonstrating that the likelihood of local strain occurrence has an upper limit and remains relatively stable within the shear band region.

3.2.3.3 Morphological Characteristics of Shear Band Based on Displacement Contours

During the shearing process of particle systems, shear bands are defined as regions where particles undergo relative displacement due to opposing force directions. These bands initiate at the onset of shearing and progressively develop, ultimately exhibiting stable morphological characteristics. Numerical simulations are frequently employed in shear band research, as they enable direct observation of particle movement both

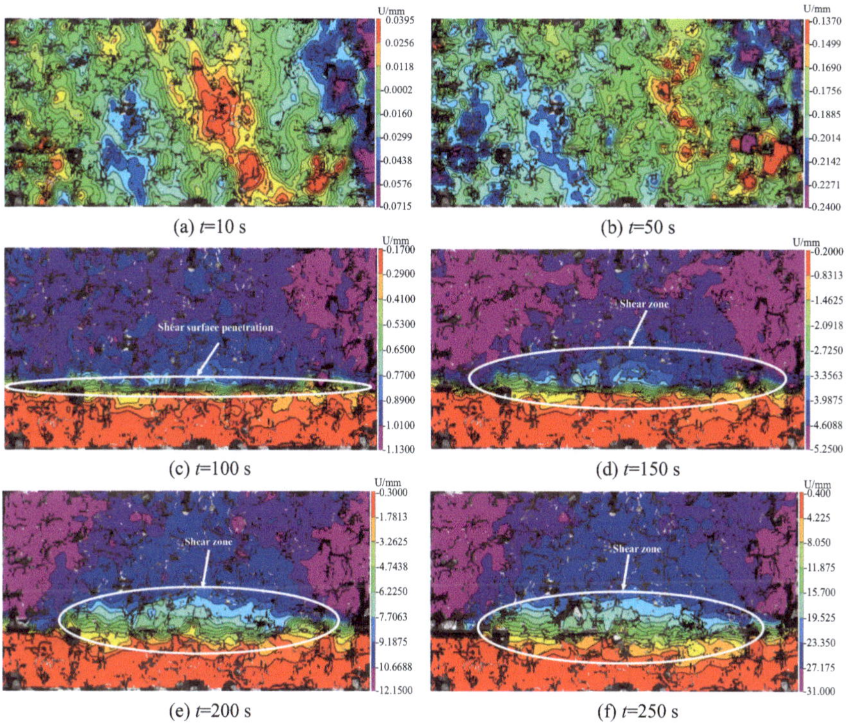

Fig. 3.11 Temporal evolution of shear band during shearing process

within and outside the shear bands, as well as local deformations of the particle system. In this study, DIC technology was utilized to monitor the evolution patterns of shear bands (Fig. 3.11).

By correlating the local strain observed during the shearing process (Fig. 3.9), we analyzed the evolution mechanism of shear bands (Fig. 3.11). In the initial shearing stage, when the shear force is relatively small, its effect on the particle system primarily manifests as localized particle compaction. During this phase, significant displacement occurs in localized regions throughout the entire area (Fig. 3.11a). The corresponding shear strain at the same time instance, as shown in Fig. 3.9a, demonstrates the development of local strains of varying magnitudes across the entire region. As the shearing process progresses, the particle system becomes increasingly compacted. The continuous increase in shear stress induces displacement of particles near the shear inlet (Fig. 3.11b). The corresponding shear strain, illustrated in Fig. 3.9b, indicates that local strain becomes predominantly concentrated at the shear initiation location. With the progressive increase in shearing duration, particles become interconnected along the shear surface, and displacement contours align parallel to this surface. A distinct discrepancy emerges between the displacement contours of the upper and lower sections of the shear box (Fig. 3.11c). At this stage, the initial formation of the shear band takes shape, with the corresponding local strain

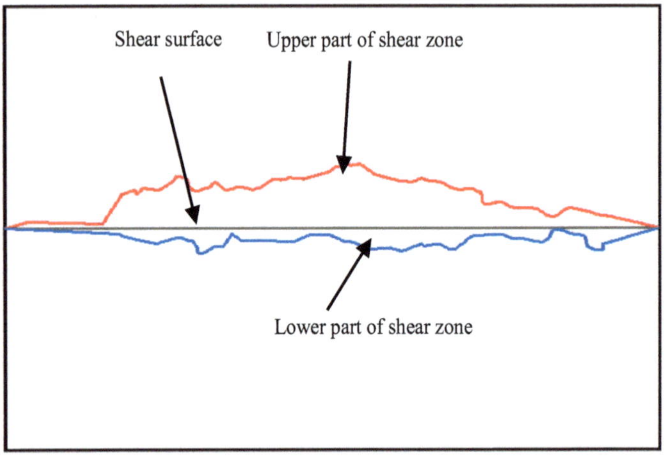

Fig. 3.12 Schematic of the developed shear-band geometry

(Fig. 3.9c) concentrating in the vicinity of the shear surface. Subsequently, the shear band gradually expands in thickness from the shear surface, extending toward both sides (Fig. 3.11d–f), while the local strain simultaneously stabilizes (Fig. 3.9d–f). A comparison between Fig. 3.11e, f reveals that following its formation, the shear band progressively stabilizes, exhibiting no significant changes in its maximum thickness or substantial alterations in its morphology, which indicates a state of relative stability. In summary, the penetration of local strain through the shear surface results in the formation of a shear band resembling a 'semi-spindle' shape, characterized by an arc-shaped surface in its upper portion. A schematic representation of the formed shear band configuration is illustrated in Fig. 3.12.

Distinct morphological differences are evident between the upper and lower portions of the shear band. The upper section displays a relatively regular configuration, whereas the lower portion appears more irregular. Furthermore, in the upper shear band, the growth pattern from the shear surface to the maximum shear band thickness demonstrates a gradual increase near the shear inlet, reaching its maximum thickness before decreasing back toward the shear surface. This phenomenon occurs because, as the shearing process progresses, the particle system gradually compacts and forms a stable structure. While the right-side experiences shear force, the left end of the shear box acts similarly to a fixed boundary, impeding particle displacement. Consequently, the compaction effect becomes particularly pronounced at the terminal section of the shear box. The formation of a stable structure through compaction leads to the observed morphological asymmetry between the left and right sides of the shear band.

To further investigate the temporal evolution of shear bands, a statistical analysis was conducted on the maximum thickness of the shear bands at selected time intervals. The results obtained from DIC processing indicate that the evolutionary patterns differ between the upper and lower portions of the shear band throughout the

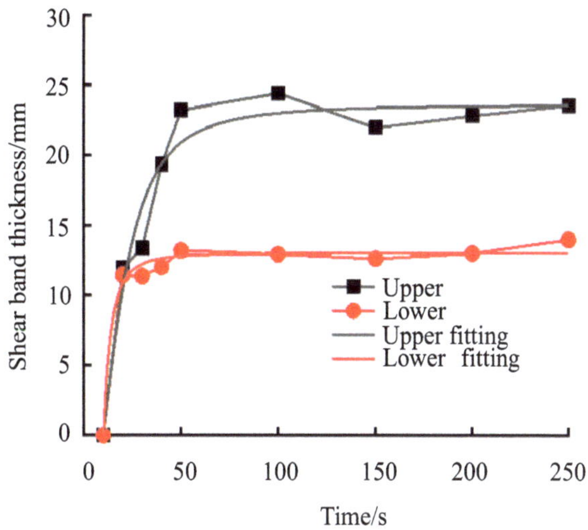

Fig. 3.13 Time-evolution curve of maximum shear-band thickness

shearing process. For a more detailed examination, the shear band was divided into upper and lower sections using the shear surface as the dividing plane, with specific results illustrated in Fig. 3.13.

The analysis of Fig. 3.13 indicates that the thickness of the shear band develops rapidly during the initial stage and subsequently stabilizes. This pattern is observed in both the upper and lower sections of the shear band. Furthermore, statistical analysis reveals that, similar to the probability density of local strain, both sections conform to the Logistic model, as represented in Eq. (3.4):

$$d = k_2 + \frac{(k_1 - k_2)}{1 + \left(\frac{t}{t_0}\right)^n} \quad (3.4)$$

Here, k_1 and k_2 are coefficients, taking values of -0.58 ± 1.02, -415.01 ± 31.85 and 2.37 ± 0.13, 1.31 ± 0.04, respectively; t_0 is the initial time; and n is the exponent, assigned values of 2.26 ± 1.09, 2.56 ± 0.28, respectively.

Similarly, based on the fluctuation ranges of the upper and lower shear band sections depicted in Fig. 3.12, along with the quantified maximum thicknesses of these sections illustrated in Fig. 3.13, analysis indicates that the maximum thickness of the lower shear band is approximately two-fifths that of the upper section. The particle size range utilized in this test is 5–10 mm. When comparing the thickness difference between the upper and lower shear band sections with the average particle size, it is observed that the thickness difference is approximately 2–3 times the average particle size. The incorporation of coarse particles contributes to a certain degree of increase in the shear band thickness; however, the maximum thickness difference between the stabilized upper and lower shear band sections still adheres to the ratio of 2–3 times the average particle size. Previous studies have investigated the influence of particle gradation on shear band thickness, yet the disparity between the upper and lower sections is influenced not only by particle gradation but also by the dilatancy effect. During the shearing process, the particle system experiences external influences from the shear box, as well as internal interactions among the particles. The elimination of rigid body displacement for the entire particle system results in the instantaneous particle shear displacement vector diagram, as illustrated in Fig. 3.14. Notably, the displacement vectors of particles in the lower region exhibit a clear tendency toward the upper-left direction. This observation suggests that, after removing rigid body displacement from the entire system, the particles experience upward compression due to the dilatancy effect, particularly in light of the fixed bottom boundary. This mechanism explains why the maximum thickness of the lower shear band, as depicted in Fig. 3.12, is less than that of the upper shear band. Furthermore, as previously discussed, the spatial distribution of the dilatancy effect is non-uniform, manifesting as variations in particle movement patterns. Additionally, the particle vector diagram, with rigid displacement removed and presented in Fig. 3.14, indicates that while vectors at different locations generally trend toward the upper left, their angles and magnitudes vary. This finding demonstrates that particles at various positions are subjected to differing degrees of influence from the dilatancy effect during shearing, thereby further elucidating the irregular morphology observed in the lower portion of the shear band in Fig. 3.12.

Furthermore, during the shearing process, particle displacements exhibit coordinated behavior, characterized by the following: while particles experience both translation and rotation, the heterogeneity in force distribution across various particle positions results in variations in interparticle interaction effects. Near the right side (shear inlet), the dilatancy effect on particles is mitigated by the shear stress exerted by the shear box, which causes particle movement to be primarily influenced by lateral shear stress. As the distance from the shear inlet increases, the inhibiting effect of shear stress on the dilatancy effect lessens, allowing the dilatancy effect to become

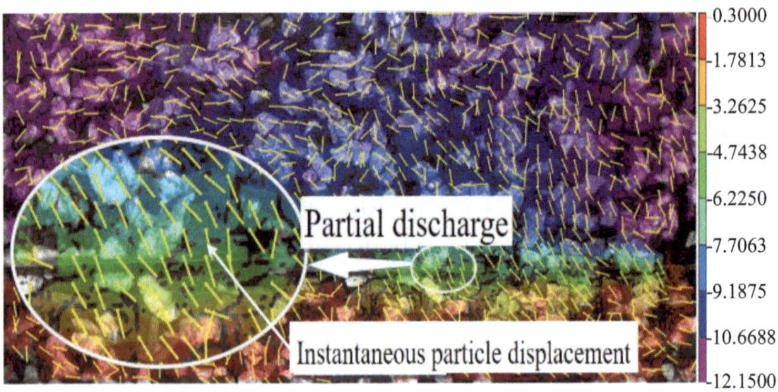

Fig. 3.14 Shear-displacement vectors with rigid motion removed

more pronounced. Consequently, this leads to the observed patterns in lateral obstruction displacement, vertical compression effects, and rotational capacity, as illustrated in Fig. 3.8.

3.3 Study on Shear Characteristics of Particle System Under Influence of Single Coarse Particle

3.3.1 Physical Testing Procedure for Shear Behavior of Particle System Considering Single Coarse Particle

To investigate the influence of single coarse particles on the shear mechanical properties and movement characteristics of the particle system of caved ore and rock, a series of direct shear test were conducted on particle systems under various conditions including coarse particle size, shape, placement location, and different vertical pressures.

3.3.1.1 Direct Shear Test System and Test Materials

This direct shear test employed a self-developed non-contact shear characteristic monitoring and analysis system specifically designed for particle systems (see Fig. 3.15). The system consists of two primary subsystems: a shear image acquisition subsystem and a shear image processing subsystem.

The shear image acquisition subsystem primarily utilizes a visual shear model and a high-definition camera to monitor the particle movement process. Frame sequences are exported via MATLAB software to create a comprehensive image sequence. Specifically: (1) The visual shear model measures 300 mm × 250 mm × 200 mm (length × width × height). Except for the front side, which is constructed from 12 mm thick transparent tempered glass, the remainder of the model is made of stainless steel. All inner walls are polished to a smooth finish to minimize boundary effects on particle movement. Tests confirm that under vertical pressures below 300 kPa, the tempered glass exhibits negligible deformation, making it suitable for monitoring particle movement states on the end surface during shearing. (2) The high-definition camera employed is a Canon 90D, capable of recording 4 K high-definition video at 100 fps, with an effective resolution of 32.5 megapixels. (3) Frame sequences are generated using MATLAB software to crop non-essential areas from the 4 K video recorded by the camera, retaining only the image sequences necessary for the specific identification of coarse particles and surrounding fine particles.

The shear image processing subsystem utilizes Image Pro Plus (IPP) image identification software to dynamically track particle trajectories and analyze movement characteristics during the shearing process. The IPP software enables dynamic object identification based on contours or color ranges and incorporates multiple filters

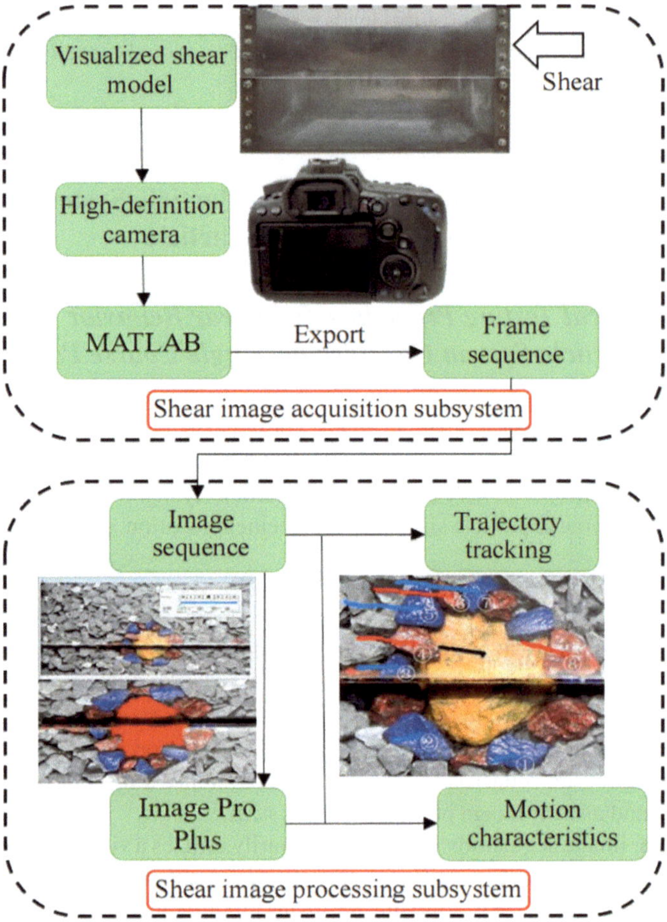

Fig. 3.15 Non-contact shear characteristic monitoring and analysis system for the particle system of caved ore and rock

to enhance object contrast. Furthermore, it permits manual adjustments of image lighting to mitigate the effects of overexposure on image identification. Capable of performing trajectory tracking for particulate materials and quantitatively processing various motion-related parameters, IPP software has been extensively applied in fields such as medicine [56], biology [57], and materials science [58, 59]. Given that caved ore and rock exhibit distinct contours and high contrast, which meet the fundamental characteristics required for object recognition, the IPP software is particularly well-suited for investigating the shear movement properties of caved ore and rock.

All particulate materials used in this study were composed of naturally occurring gray-black limestone (Fig. 3.16a), exhibiting a density of 3150 kg/m^3. Prior to testing, the particles were sieved, selecting fine particles within the size ranges of 1–5 mm, 5–10 mm, and 10–15 mm. Coarse particles with varying coarse-to-fine particle size

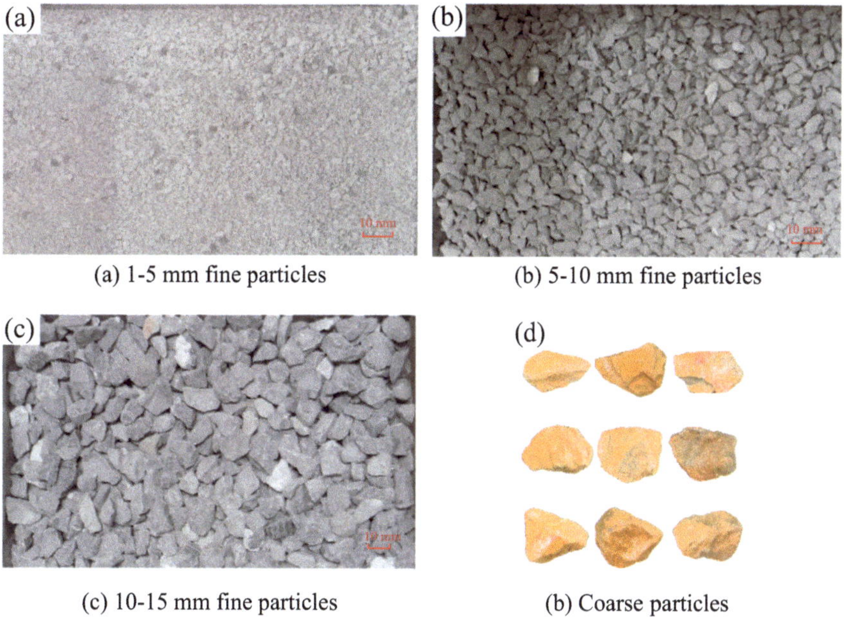

(a) 1-5 mm fine particles (b) 5-10 mm fine particles (c) 10-15 mm fine particles (b) Coarse particles

Fig. 3.16 Fine and coarse particles used in the tests

ratios d_c/d_f, based on the 5–10 mm fine particle range, were designated as influencing factors for the test. The primary test subjects were the 5–10 mm particles, while the 1–5 mm and 10–15 mm particles were employed solely to investigate the influence of individual coarse particles on fine particles across different size ranges. Before testing, all coarse and fine particle materials were washed and sun-dried to mitigate the impact of surface dust on the accuracy of image acquisition. Furthermore, to enhance the clarity of distinguishing coarse particles from surrounding fine particles in images, the corresponding coarse and fine particles were dyed, as depicted in Fig. 3.16b.

3.3.1.2 Test Plans

This research comprises three primary components: (1) An investigation into the particle size ratio d_c/d_f between coarse and fine particles, which significantly influences the shear strength of the particle system under conditions dominated by individual coarse particles. (2) An examination of the effects of coarse particle shape and vertical pressure p_v on the shear strength and movement characteristics of the particle system. (3) An analysis of how the placement location of coarse particles affects the shear strength and movement characteristics of the particle system.

In this test, the shape factor r_v of coarse particles was set to values of 0.201, 0.304, 0.409, 0.507, 0.540, 0.745, and 0.939. The first four shape factors were incrementally increased to investigate the approximate range of coarse particle shape factors that influence overall shear strength. The latter three shape factors were similarly increased to study their effects on the particle system. During the experiments, coarse particles were centrally positioned on the visible surface. The specific experimental scheme is detailed in Table 3.3, while a schematic diagram illustrating the placement of coarse particles is presented in Fig. 3.17. To ensure consistency, the maximum projection surface of the coarse particles was maintained parallel to the visible surface during placement. Each arrangement was compared with the initial coarse particle placement photograph to minimize errors arising from manual setup. Furthermore, all particles employed in each filling were sourced from the same batch and possessed identical mass to mitigate variations in average compactness and void ratio across multiple fillings. Coarse particles labeled A, B, and C-2 were centrally located on the visible surface. Among the 39 test groups, the fine particle size consistently measured between 5 and 10 mm, with the exceptions of A-8 and A-9. Variations in the coarse-to-fine particle size ratio d_c/d_f were achieved by adjusting the coarse particle sizes during the test. The sizes of coarse particle d_c were 2.10, 3.07, 4.23, 5.11, 6.07, and 7.04 times the average size of the fine particle d_f.

The specific test procedure is outlined as follows: (1) Install the visual model and camera device. (2) Uniformly mix the fine particles and fill the model to half capacity. Subsequently, place the coarse particles, which are distinctly colored from the fine particles, at the specified positions indicated in Table 3.3 and Fig. 3.17. Cover these coarse particles with fine particles and continue filling the model box until it is completely full. (3) Apply the appropriate vertical pressure above the filled visual device. (4) Conduct the direct shear test while monitoring and recording the movement of both coarse and fine particles, as well as the shear strength of the particle system. (5) In accordance with the test plan, sequentially vary parameters such as the coarse-to-fine particle size ratio, coarse particle shape, placement location, and vertical pressure, repeating the aforementioned steps.

3.3.2 *Influence of Single Coarse Particle on Shear Strength of Particle System*

Based on the results of the aforementioned 39 direct shear tests, the influences of various factors—including the coarse-to-fine particle size ratio d_c/d_f, coarse particle shape factor, coarse particle placement location, and vertical pressure—on the shear strength of the particle system were analyzed.

Table 3.3 Test plan for investigating the effect of coarse particles on the shear behavior of particle system

No	Particle size ratio (d_c/d_f)	Particle size range/mm	Shape factor (r_v)	Vertical pressure/kPa	Coarse particle location
A-1	1.00	5–10	0.745	100	Center of visible surface
A-2	2.10				
A-3	3.07				
A-4	4.23				
A-5	5.11				
A-6	6.07				
A-7	7.04				
A-8	12.69	1–5			
A-9	2.53	10–15			
B-1	4.23	5–10	0.201	100	Center of visible surface
B-2			0.304		
B-3			0.409		
B-4			0.507		
B-5			0.201	200	
B-6			0.304		
B-7			0.409		
B-8			0.507		
B-9			0.201	300	
B-10			0.304		
B-11			0.409		
B-12			0.507	100	
B-13			0.540		
B-14			0.745		
B-15			0.939		
B-16			0.540	200	
B-17			0.745		
B-18			0.939		
B-19			0.540	300	
B-20			0.745		
B-21			0.939		
C-1	4.23	5–10	0.745	100	75 mm left of visible surface center
C-2					Visible surface center
C-3					75 mm right of visible surface center

(continued)

Table 3.3 (continued)

No	Particle size ratio (d_c/d_f)	Particle size range/mm	Shape factor (r_v)	Vertical pressure/kPa	Coarse particle location
C-4					50 mm above visible surface center
C-5					50 mm below visible surface center
C-6					75 mm left and 50 mm above visible surface center
C-7					75 mm right and 50 mm above visible surface center
C-8					75 mm left and 50 mm below visible surface center
C-9					75 mm right and 50 mm below visible surface center

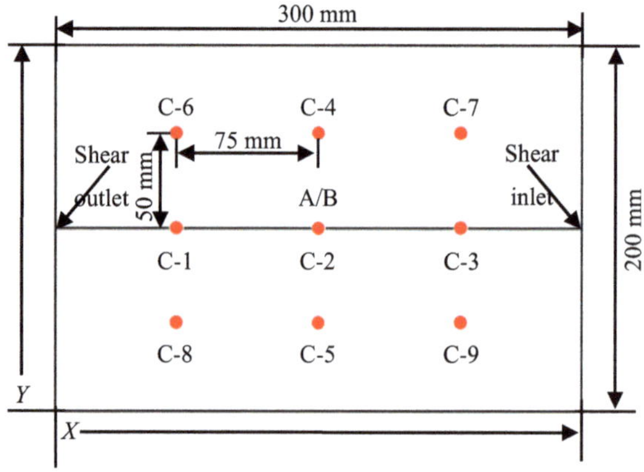

Fig. 3.17 Schematic of coarse-particle placement positions

3.3.2.1 Influence of Single Coarse Particle on Shear Strength of Fine Particles with Identical Size Range

For tests A-1 to A-9, with the coarse particle shape factor held constant $r_v = 0.745$, the influence of varying coarse-to-fine particle size ratios d_c/d_f on the overall shear strength (peak strength) of the particle system under single coarse particle conditions was analyzed. As illustrated in Fig. 3.18, when the coarse-to-fine particle size ratio d_c/d_f is within the range of 1–3.07, the shear strength of the particle system increases gradually and linearly; however, the increments remain below 1.64%, indicating a negligible effect. Only when the coarse-to-fine particle size ratio d_c/d_f exceeds 4.23 does the shear strength of the particle system experience a significant increase, achieving an 11.60% improvement compared to the shear strength observed without

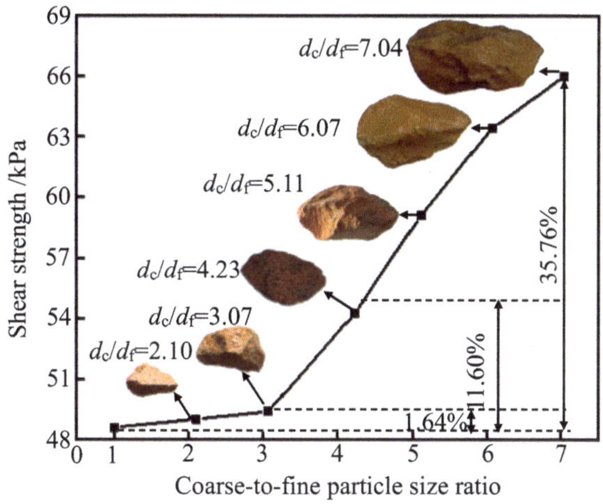

Fig. 3.18 Shear strength of particle system under the influence of different coarse particle sizes

coarse particles $d_c/d_f = 1.0$. When d_c/d_f exceeds 7.04, the improvement can reach as high as 35.76%.

Analysis of Fig. 3.18 indicates that under individual coarse particle conditions, there is a specific threshold for the coarse-to-fine particle size ratio d_c/d_f ranging from 3.07 to 4.23. Upon reaching this threshold, d_c/d_f significantly influences the shear strength of the particle system.

3.3.2.2 Influence of Single Coarse Particle on Shear Strength of Fine Particles with Different Size Ranges

As the size range of fine particles (Fig. 3.16) varies, the influence of the same coarse particle on shear strength also changes. Coarse particles, which are 4.23 times larger than the average particle size in the 5–10 mm range from tests A-1 to A-7, were selected and subjected to shear tests alongside fine particles sized between 1–5 mm and 10–15 mm (tests A-8 and A-9). In this context, the coarse-to-fine particle size ratios d_c/d_f were 12.69 and 2.53, respectively. The placement of these particles was consistent with that of the coarse particles in tests A-1 to A-7, and the comparison of their shear strengths is illustrated in Fig. 3.19.

Figure 3.19 illustrates a comparison of shear strength among fine particles across various size ranges while maintaining consistent coarse particle conditions. The results demonstrate that the overall shear strength of the particle system increases with larger particle sizes. Notably, the same coarse particle exerts differing degrees of influence on fine particles of varying size ranges, characterized by distinct d_c/d_f values. The 5–10 mm fine particles, paired with a single coarse particle, serve as the control group. When the coarse particle is introduced to the 1–5 mm fine particles, the system's shear strength increases by 8.60%, which is slightly lower than the 11.62% enhancement observed in the 5–10 mm system, characterized by a d_c/d_f of 12.69. To

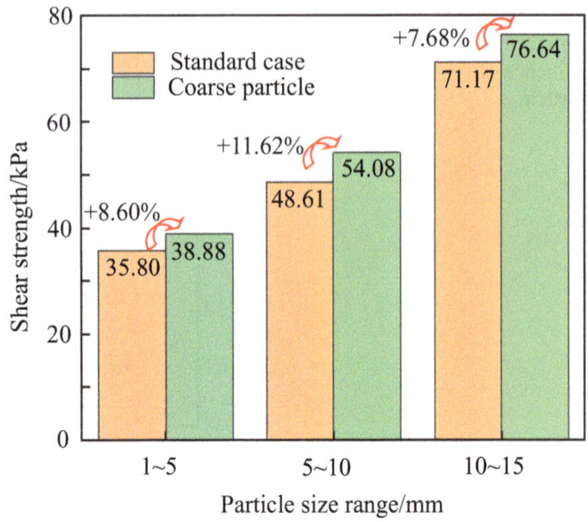

Fig. 3.19 Comparison of shear strength among fine particles in different size ranges

achieve a comparable shear strength improvement in the 1–5 mm system, an increase in the coarse particle size is necessary. Similarly, the 10–15 mm fine particle system exhibits a 7.68% improvement with a d_c/d_f of 2.53, also necessitating larger coarse particles to match the enhancement level of the 5–10 mm system. In summary, the same coarse particle influences fine particle systems of varying size ranges differently. Altering the fine particle size simultaneously changes the coarse-to-fine particle size ratio d_c/d_f, but the required ratio varies across different particle size ranges. For fine particles with smaller average sizes, a larger coarse-to-fine particle size ratio d_c/d_f is necessary to achieve shear strength improvements that are comparable to those of fine particles with larger average sizes. Conversely, in fine particle systems with larger average sizes, only coarse particles with smaller coarse-to-fine particle size ratios are required to attain equivalent shear strength enhancements as those in fine particle systems with smaller average sizes.

3.3.2.3 Influence of Coarse Particle Shape Factor and Vertical Pressure on Shear Strength

As established in the previous section, there exists a critical value for the coarse-to-fine particle size ratio d_c/d_f between 3.07 and 4.23, beyond which it begins to significantly influence the shear strength of the particle system. Concurrently, coarse particles with different shape factors also exert varying effects on the shear strength of the particle system. Based on this, while maintaining a constant coarse-to-fine particle size ratio $d_c/d_f = 4.23$, tests B-1 to B-9 were conducted to analyze the effects of coarse particles with different shape factors and vertical pressures on the shear strength of the particle system.

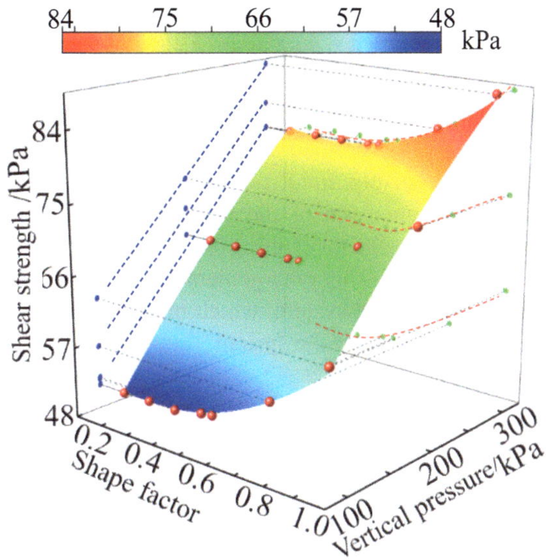

Fig. 3.20 Shear strength of particle system under the influence of different coarse particle shapes and vertical pressures

Based on the shear strength data obtained from tests B-1 to B-9, which examined different coarse particle shapes and vertical pressures, various formulas were selected to model the relationships among the three factors. Analysis revealed that the Levenberg–Marquardt iterative algorithm yielded the highest correlation coefficient R^2. Consequently, a three-dimensional surface plot was generated to illustrate the influence of coarse particle shape and vertical pressure on the shear strength of the particle system (Fig. 3.20).

As shown in Fig. 3.20, the shear strength of the particle system exhibits clear regularity with increasing coarse particle shape factor and vertical pressure. The projection of shear strength on the *xz* plane represents its variation with the shape factor, while the projection on the *yz* plane illustrates its variation with vertical pressure. Analysis of Fig. 3.20 reveals that, within the studied value range, the shear strength of the particle system containing a single coarse particle increases approximately linearly with rising vertical pressure. When $r_v < 0.5$, the shear strength projections almost overlap as the coarse particle shape factor increases, indicating negligible change in shear strength, which can be interpreted as normal experimental error. When $r_v \geq 0.5$, the shear strength increases gradually, as indicated by the red dashed line in Fig. 3.20, and the rate of increase with vertical pressure (blue dashed line) is higher than that with the coarse particle shape factor (red dashed line). The contour map of shear strength variation with shape factor and vertical pressure satisfies the polynomial fitting equation for the shear strength of the particle system (Eq. (3.5)), with a correlation coefficient $R^2 = 0.990$.

$$\tau = 24.95 - 16.64r_v + 0.23p_v + 29.96r_v^2 = 0.0002p_v^2 - 0.021r_vp_v \tag{3.5}$$

Here, τ denotes the shear strength of the particle system, r_v represents the particle shape factor, and p_v is the vertical pressure.

When the shape factor is small, coarse particles exhibit pronounced platy characteristics. Under identical sieving conditions, the projected area perpendicular to the shear stress, defined as the product of the intermediate and minimum dimensions, is relatively small for coarse particles. As the shape factor increases, the minimum dimension of the particle increases while the intermediate dimension remains unchanged, resulting in an enlarged projected area perpendicular to the shear direction. This enlargement leads to an increased effective interaction area within the particle system, which macroscopically manifests as an enhancement in the shear strength of the system.

3.3.2.4 Influence of Single Coarse-Particle Position on Shear Strength

When the coarse-to-fine particle size ratio d_c/d_f is 4.23 and the coarse particle shape factor r_v exceeds 0.5, a single coarse particle significantly enhances the shear strength of the particle system. To investigate the influence of different coarse particle placement locations on the shear strength under these conditions, Tests C-1 to C-9 were conducted while maintaining constant parameters of $d_c/d_f = 4.23$ and $r_v = 0.745$, with an applied vertical pressure of 100 kPa (Fig. 3.17). The shear strength data obtained from these tests under a vertical pressure of 100 kPa were analyzed using the Levenberg–Marquardt iterative algorithm for fitting. This analysis resulted in a three-dimensional surface plot that illustrates the effect of coarse particle placement on the shear strength of the particle system (Fig. 3.21).

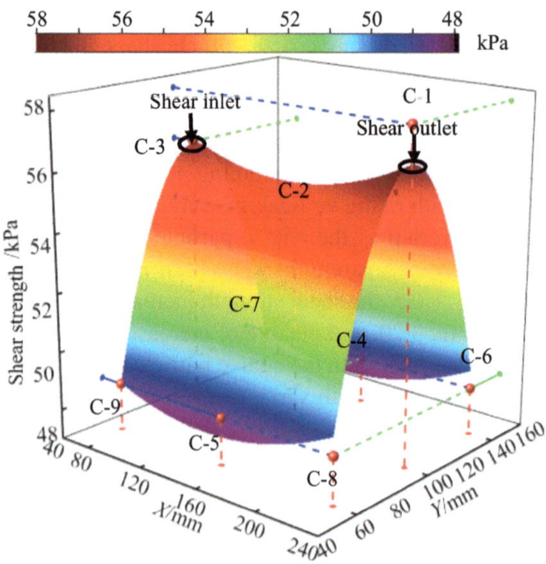

Fig. 3.21 Shear strength of the particle system under 100 kPa vertical stress for different single coarse-particle positions

The contour map illustrating the shear strength of the particle system, influenced by the varying placements of coarse particles, generally resembles a saddle surface. This observation is consistent with the polynomial fitting equation for the shear strength of the particle system (Eq. (3.6)), which exhibits a correlation coefficient of $R^2 = 0.962$. The shear strength of the particle system reaches its maximum when the coarse particle is located at the shear outlet on the shear surface, followed by the shear inlet, while the minimum strength is observed when the coarse particle is positioned at the center of the visible surface.

$$\tau = 28.48 - 0.48X + 6.09Y + 0.02X^2 - 0.3Y^2 + 1.2XY \tag{3.6}$$

Here, τ denotes the shear strength, and X and Y represent the horizontal and vertical positions of the particle, respectively.

In contrast to Tests C-1 through C-3, where coarse particles were positioned on the shear surface, the placement of coarse particles in tests C-4 to C-9 did not result in a significant enhancement of the shear strength of the particle system. This lack of effect can be attributed to the fact that these coarse particles were located outside the shear band (within a width range of 10–15 d_f [52]), where they do not influence the shear strength of the particle system.

3.3.3 *Influence of Single Coarse Particle on Movement Characteristics of Particle System*

Based on the investigation into the influence of single coarse particles on the shear strength of the particle system, further research was conducted on the movement characteristics of coarse particles and their surrounding fine particles. This study considers the effects of different shapes, positions of coarse particles, and varying vertical pressures.

3.3.3.1 Coarse-Particle Trajectory Tracking Procedure and IPP Reliability Verification

To enhance image recognition and track the movement trajectories of coarse particles, Test A-4 serves as an illustrative example. Initially, color differentiation was applied (Fig. 3.22a). Subsequently, the color selection tool in IPP software was employed to effectively isolate the coarse particles (Fig. 3.22b), with the delineation of particles depicted in Fig. 3.22c. The software automatically identified the positions of identical particles throughout the entire sequence, sequentially connecting these positions to construct a comprehensive movement trajectory for each particle (Fig. 3.22d). The horizontal and vertical displacements of the coarse particles were measured using a vernier caliper and IPP image recognition, respectively (Fig. 3.22e).

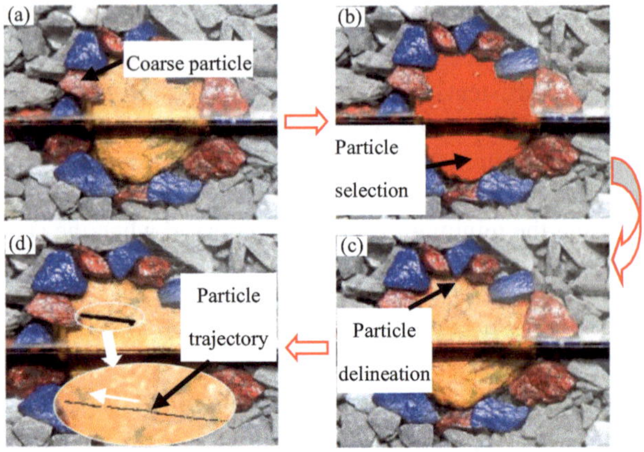

(a) Color differentiation; (b) Particle selection; (c) Particle delineation; (d) Trajectory tracking

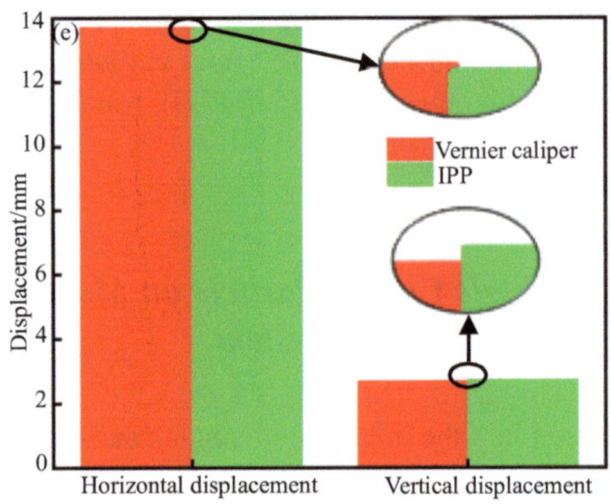

(e) Manual measurement and IPP identification

Fig. 3.22 Trajectory tracking procedure and IPP reliability verification

As shown in Fig. 3.22d, during the shearing process, the coarse particle gradually moves toward the upper left, which is the direction of the shear outlet. This movement deviates from the shear surface and exhibits a significant dilatancy effect. Furthermore, as illustrated in the magnified view within Fig. 3.22d, the trajectory of the coarse particle is not a perfectly smooth curve; rather, it displays minor fluctuations. These fluctuations can be interpreted as noise [47], which is associated with the non-uniform transmission of shear forces within the particle system, as well as the size of the fine particles surrounding the coarse particle and the compactness of the fine particle system structure. In this test group, the horizontal displacement of the coarse particle, measured using a vernier caliper, was 13.70 mm, while the vertical displacement was 2.70 mm. In contrast, the coarse particle movement trajectory monitored through IPP image recognition and spatial coordinate calibration yielded a horizontal displacement of 13.69235 mm and a vertical displacement of 2.73802 mm. A comparison of the two methods reveals that the accuracy of vernier caliper readings is limited by visual perception and mechanical precision, resulting in restricted precision in particle displacement identification. The data measured by IPP image recognition are largely consistent with manual measurements, and IPP offers higher recognition precision, capable of identifying particle displacements based on single pixels. This capability validates the reliability of IPP for particle image recognition. Additionally, IPP identifies particles based on color ranges, enabling automatic recognition of numerous particles and rapid calculation of the movement states of multiple particles, along with dynamic trajectory tracking, demonstrating significant advantages over manual measurements.

3.3.3.2 Analysis of Coarse Particle Displacement Under Different Shapes and Vertical Pressures

To further investigate the trajectory characteristics of coarse particles, we compiled statistics on their trajectories under the influence of various shapes and vertical stresses. The coarse particle shapes studied were those that significantly impact the overall shear strength. We analyzed the evolution patterns of vertical and horizontal displacements of coarse particles concerning their shape (using $p_v = 200$ kPa as an example) and vertical pressure (using $r_v = 0.939$ as an example) as illustrated in Fig. 3.23.

As illustrated in Fig. 3.23a, when the vertical pressure is maintained at a constant value of 200 kPa, an increase in the shape factor r_v of the coarse particle from 0.540 to 0.939 results in a corresponding increase in horizontal displacement by 19.56% and 39.02%, respectively, while vertical displacement increases by 24.90% and 112.81%, respectively. This observation suggests that, under consistent vertical pressure conditions, both horizontal and vertical displacements of the coarse particle are positively correlated with an increase in its shape factor. When r_v is held constant at 0.939 and vertical pressure is increased, horizontal displacement decreases by 19.10% and 33.65%, respectively, while vertical displacement decreases by 39.99% and 45.55%, respectively.

Furthermore, under the same coarse particle shape conditions, both horizontal and vertical displacements gradually decrease with increasing vertical stress. This observation indicates that the particle system becomes more compact under vertical pressure, enhancing the interlocking force between coarse and fine particles while reducing the mobility of the coarse particles. Furthermore, by comparing the variations in horizontal and vertical displacements of the coarse particles as illustrated in Fig. 3.23a, b, it can be concluded that, within the studied value range, the shape factor exerts a more significant influence on the vertical displacement of the coarse particles than on their horizontal displacement.

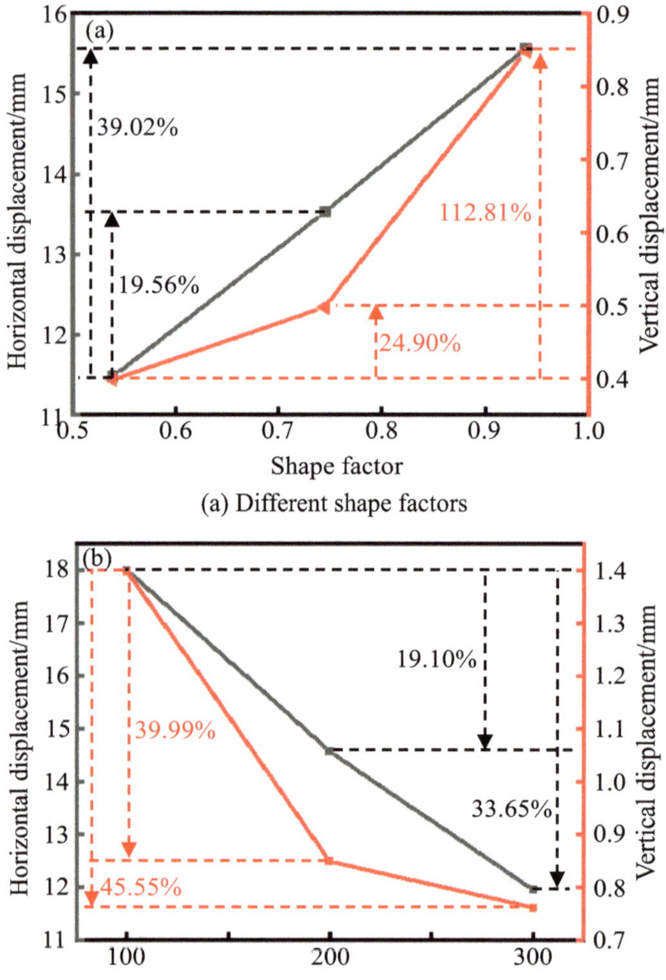

(a) Different shape factors

(b) Different vertical pressures

Fig. 3.23 Influence of different shape factors and vertical pressures on coarse particle displacement

3.3.3.3 Analysis of Coarse-Particle Trajectories Under Different Placement Positions

For tests C-1 to C-9, a quantitative analysis was conducted on the movement trajectories of coarse particles influenced by various placement positions. Since the coarse particles installed outside the shear band (C-4 to C-9) did not significantly affect the shear strength of the particle system, C-4 (above the shear band) and C-5 (below the shear band) were selected as examples. Alongside C-1 to C-3, which notably impacted the shear strength of the particle system from within the shear band, the starting points of the movement trajectories for these five groups of coarse particles were initialized (see Fig. 3.24). This initialization involved translating each trajectory so that its starting point coordinate was set to 0, calculating the relative positions of the particles, and analyzing the characteristics of the movement trajectories of coarse particles at different installation positions under a consistent stress state (100 kPa) and shear displacement (25 mm). The positive and negative signs in all particle movement trajectory plots indicate the direction of particle movement and are not related to the magnitude of displacement.

As shown in Fig. 3.24, the movement trajectory of coarse particle C-2 exhibits a local concave downward trend due to its interaction with a larger fine particle

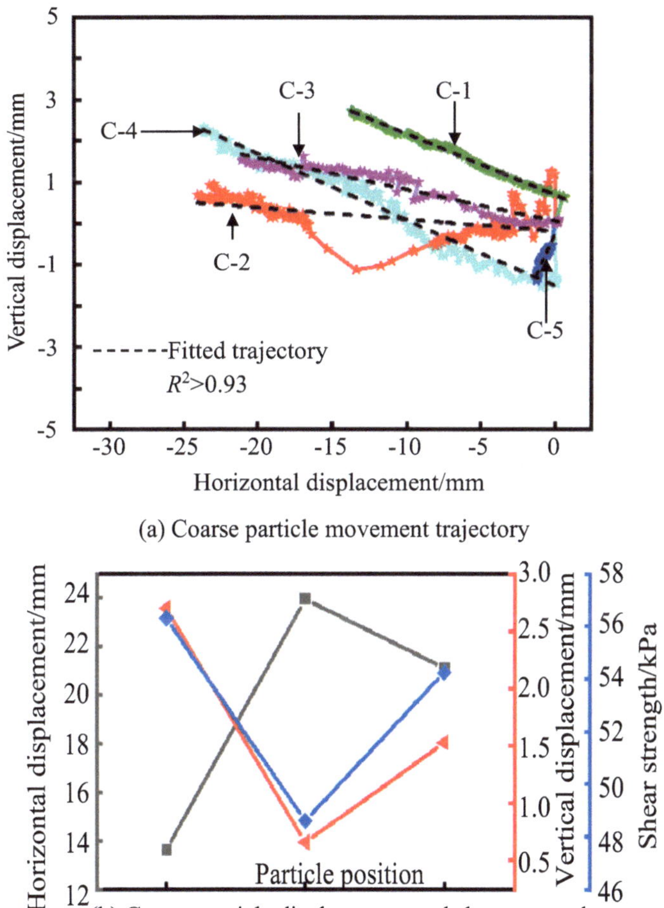

Fig. 3.24 Comparison of coarse-particle trajectories at identical shear displacement for different initial positions

during shearing. In contrast, the trajectories of the other coarse particles are approximately linear, with correlation coefficient R^2 values all exceeding 0.93. Furthermore, under identical shear displacements, the coarse particles at different positions demonstrate slight variations in both horizontal and vertical displacements: (1) Longitudinal comparisons indicate that horizontal displacements follow the order C-2≈C-4 > C-5, while on the shear plane, the order is C-1 > C-3 > C-2. Figure 3.24b illustrates a strong correlation between the displacement of coarse particles and the shear strength of the particle system, wherein shear strength is negatively correlated with horizontal displacement and positively correlated with vertical displacement during shearing. Within the particle system, coarse particles that are surrounded by stronger interlocking forces or tighter cluster structures inherently exhibit a reduced capacity for horizontal displacement, thereby impeding shear deformation. A smaller horizontal displacement indicates a stronger obstruction to the overall particle system, while a reduction in horizontal displacement enhances dilatancy effects, resulting in increased vertical displacement.

3.3.4 *Analysis of Surrounding Fine-Particle Movement Trajectories Under the Influence of a Single Coarse Particle*

Based on the analysis of coarse particle movement characteristics, this study further investigates the variation features of the movement trajectories of fine particles influenced by a single coarse particle.

3.3.4.1 Movement Laws of Surrounding Fine Particles Under the Influence of a Single Coarse Particle

The presence of a coarse particle significantly influences the movement trajectories of surrounding fine particles. Figure 3.25 illustrates the identified movement trajectory curves of fine particles in systems both with and without a coarse particle, as well as the particle movement trajectories and fitted curves after initializing the starting points of the fine particle displacements.

The movement trajectories of fine particles, in the absence of coarse particles, exhibit approximately linear patterns (Fig. 3.25a, b), with correlation coefficient R^2 values exceeding 0.95. Additionally, the movement trajectories of fine particles in the vicinity of coarse particles (Fig. 3.25c, d) conform to the Boltzmann distribution (Eq. (3.7)), with correlation coefficient R^2 values surpassing 0.98.

$$y = m + \frac{(n - m)}{1 + e^{\frac{x - x_0}{l}}} \tag{3.7}$$

Here, y is the vertical coordinate of the particle; m, n, l, and x_0 are constants; e is the natural constant; and x is the horizontal coordinate of the particle.

The analysis of Fig. 3.25b indicates that: (1) In the absence of coarse particles, the movement trajectories of fine particles are approximately linear. In contrast, the trajectories of fine particles at various positions around a coarse particle exhibit notable variations (see Fig. 3.25c). After initializing the fine particle movement trajectories, numerous trajectories appear close to or overlapping, such as those labeled (1) and (2), (3) and (4), and (7) and (8) in Fig. 3.25c. Notably, the maximum overlap of fine particle trajectories occurs when the horizontal displacement is less than 10 mm; distinct differences in the fine particle movement trajectories become apparent only when the horizontal displacement exceeds 10 mm. This observation suggests that under small shear displacements, the fine particle movement trajectories exhibit significant similarity, which can be attributed to the weaker dilatancy effect during the initial shearing stage. Furthermore, the fine particle movement trajectories

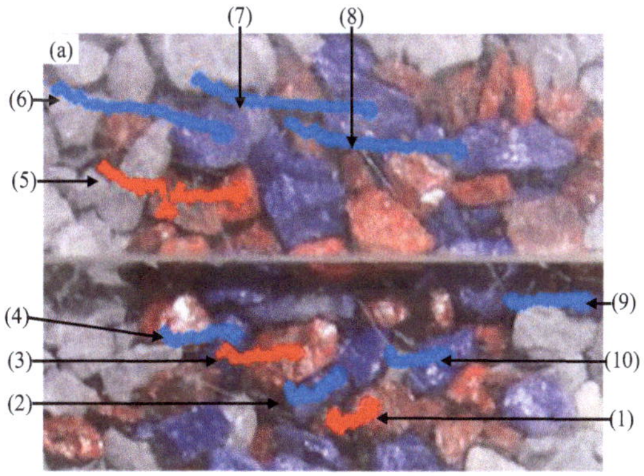

(a) Image-based identification results of fine-particle migration trajectories in the absence of coarse particles

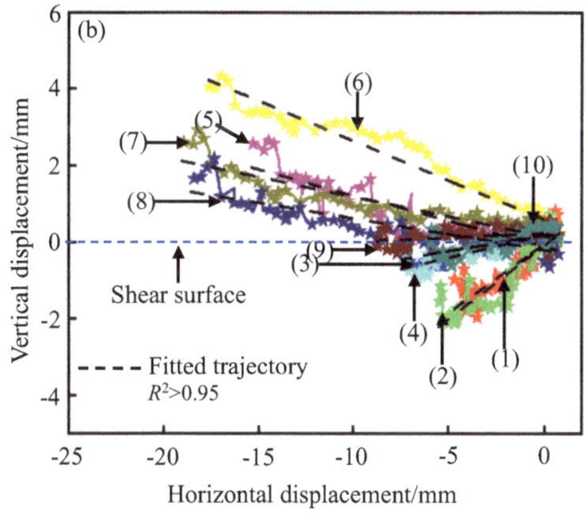

(b) Fine-particle movement trajectories and their fitted curves in the absence of coarse particles

Fig. 3.25 Movement trajectories of fine particles and their fitted curves

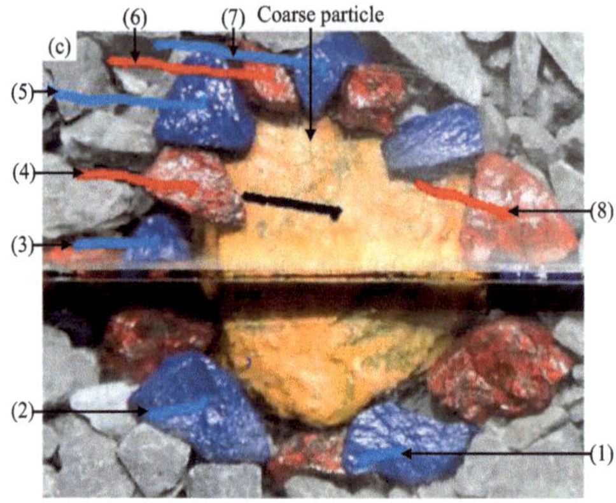

(c) Fine-particle trajectory identification results in the presence of coarse particles

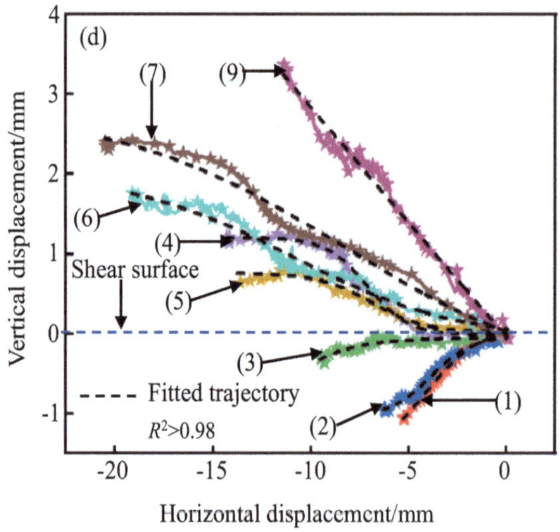

(d) Fine-particle movement trajectories and their fitted curves in the presence of coarse particles

Fig. 3.25 (continued)

can generally be categorized into two states: particles above the shear plane move upward, while those below move downward. (2) Under conditions involving coarse particles, the movement trajectories of fine particles become more dispersed due to the influence of the coarse particle, and these trajectories conform to the Boltzmann distribution (see Fig. 3.25d). The presence of the coarse particle results in a near absence of overlapping sections in the fine particle movement trajectories, indicating that the coarse particle reduces the similarity of these trajectories. Overall, the horizontal displacement of fine particles is positively correlated with their vertical height, while the vertical displacement is positively correlated with the distance from the shear plane.

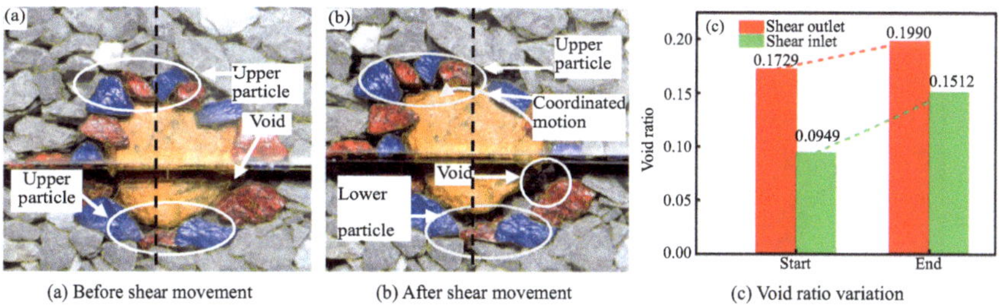

Fig. 3.26 Coordinated movement process of coarse and fine particles

The movement trajectories of fine particles result from the coordinated movement of both coarse and fine particles. As illustrated in Fig. 3.26, the area surrounding the coarse particle is divided into left and right sides, delineated by the black dashed line. Monitoring and statistical analysis of the voids around the coarse particle, conducted through particle image recognition, indicate that prior to the initiation of shearing, the void ratio in the shear-out direction is 0.1729, while in the shear-in direction, it is 0.0949. At the conclusion of shearing, the void ratio in the shear-out direction increases to 0.1990, reflecting a 15.07% increase compared to the initial measurement. Conversely, the void ratio in the shear-in direction rises to 0.1512, demonstrating a 59.24% increase from the pre-shearing value. This suggests that the average change in void ratio around the coarse particle is more pronounced in the shear-in direction. During the coordinated movement of coarse and fine particles, fine particles positioned beneath the coarse particle detach from its influence zone due to gravity, leading to relatively minor positional changes. Meanwhile, the coarse particle traverses the underlying fine particles, creating significant voids behind it (on the right side). As shearing progresses, these voids expand, and the fine particles behind gradually fill the voids under shear action. In contrast, fine particles situated above the coarse particle circumvent it during the coordinated movement.

3.3.4.2 Influence of Different Coarse Particle Shapes on Surrounding Fine Particle Movement Trajectories

To further investigate the influence of coarse particle shape and vertical pressure on the movement trajectories of surrounding fine particles, Tests B-1 to B-9 were analyzed, using 200 kPa as a reference to examine the effects of various coarse particle shapes on the movement trajectories of adjacent fine particles (Fig. 3.27).

As shown in Fig. 3.27, the movement trajectories of fine particles around coarse particles with different shapes all conform to the Boltzmann distribution (with correlation coefficient R^2 values all greater than 0.90). The vertical displacement exhibits a trend of initial slow growth followed by a rapid increase as the horizontal displacement (absolute value) increases. Specifically, as shown in Fig. 3.27a, for a coarse particle with $r_v = 0.540$, the maximum vertical displacement l_m of the surrounding

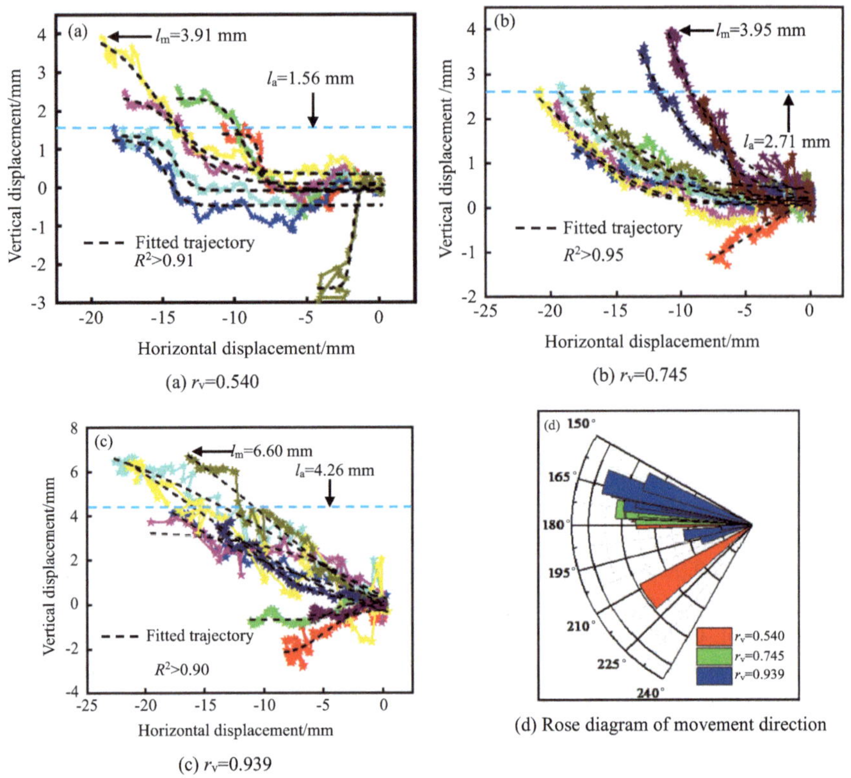

(a) r_v=0.540 (b) r_v=0.745 (c) r_v=0.939 (d) Rose diagram of movement direction

Fig. 3.27 Influence of coarse-particle shape on the movement trajectories of surrounding fine particles

fine particles is 3.91 mm, and the average vertical displacement l_a is 1.56 mm. In Fig. 3.27b, for a coarse particle with $r_v = 0.745$, the maximum vertical displacement l_m of the surrounding fine particles is 3.95 mm, and the average vertical displacement l_a is 2.71 mm, representing a 73.71% increase compared to $r_v = 0.540$. As shown in Fig. 3.27c, for a coarse particle with $r_v = 0.939$, the maximum vertical displacement l_m of the surrounding fine particles is 6.60 mm, and the average vertical displacement l_a is 4.26 mm, marking a 173.08% increase compared to $r_v = 0.540$, indicating a significant enhancement in the vertical displacement of fine particles. In summary, under a vertical pressure of 200 kPa, the shape factor of the coarse particle has a pronounced influence on the vertical displacement of the surrounding fine particles. Furthermore, analysis of the movement trajectories of fine particles around coarse particles under vertical pressures of 100 kPa and 300 kPa yields results consistent with those obtained under 200 kPa. Figure 3.27d presents rose diagrams illustrating the movement directions of fine particles around coarse particles with the three different shapes. Here, taking the shear-in direction as 0°, when $r_v = 0.540$, the range of fine particle movement directions is 150°–225°; as r_v gradually increases to 0.939, the fine particle movement directions progressively concentrate within 150°-200°. This indicates that as the shape factor of the coarse particle increases, the degree of order in the movement directions of the surrounding fine particles gradually enhances (i.e., the degree of anisotropy decreases). Integrating the findings on shear strength and coarse particle displacement, it can be concluded that the influence of coarse particle shape on the overall shear strength of the particle system and the movement trajectories of both coarse and fine particles is consistent.

3.3.4.3 Influence of Different Vertical Pressures on Surrounding Fine Particle Movement Trajectories

Furthermore, taking $r_v = 0.540$ as an example, the movement trajectories of fine particles around the coarse particle under the same shape factor but different vertical stress states are analyzed (Fig. 3.28).

As vertical pressure increases, the movement trajectories of fine particles continue to conform to the Boltzmann distribution, with all correlation coefficient R^2values exceeding 0.91. However, the vertical displacement of fine particles decreases, and the direction of this vertical displacement becomes more organized. As illustrated in Fig. 3.28a, at a vertical pressure of 100 kPa, the maximum vertical displacement l_m of fine particles surrounding the coarse particle is 5.95 mm, while the average vertical displacement l_a is 3.16 mm. In Fig. 3.28b, at a vertical pressure of 200 kPa, the maximum vertical displacement l_m is reduced to 3.91 mm, and the average vertical displacement l_a decreases to 1.56 mm, representing a 50.63% reduction compared to the 100 kPa condition. Figure 3.28c shows that at a vertical pressure of 300 kPa, the maximum vertical displacement l_m further declines to 3.36 mm, and the average vertical displacement l_a is 1.35 mm, indicating a 57.28% decrease relative to the 100 kPa vertical pressure. Figure 3.28d presents rose diagrams depicting the movement directions of surrounding fine particles under the three vertical pressures. Specifically, at $p_v = 100$ kPa, the range of fine particle movement directions spans from 105° to 225°. As p_v gradually increases to 300 kPa, the movement directions of fine particles progressively concentrate within the range of 150°–180°. This trend indicates that as vertical pressure rises, the packing structure of the particle system becomes denser, resulting in more organized movement directions of fine particles around the coarse particle.

3.3.4.4 Influence of Single Coarse Particle on Rotational Capacity of Surrounding Fine Particles

During the shearing process, the influence of a single coarse particle on the surrounding fine particles is manifested not only through particle displacement but also in the rotational capacity of the fine particles. To further investigate the rotational capacity of fine particles around a coarse particle, we collected statistics on the fine particles surrounding the coarse particle at the shear plane. The statistical results of fine particle rotation angles are presented in Fig. 3.29. The calculation of particle rotation uses the coarse particle as the origin and adheres to the standards of the Euler coordinate system, with the horizontal right direction defined as 0°, counterclockwise as positive, and clockwise as negative. The position of a fine particle relative to the coarse particle is defined as the location adjacent to the edge of the coarse particle, corresponding to the angle illustrated in Fig. 3.29a.

As illustrated in Fig. 3.29, the horizontal axis, using the rightward direction as a reference, represents the angle formed by the connecting line between coarse and fine particles, indicating the position of the fine particles. The vertical axis denotes

Fig. 3.28 Effect of vertical pressure on the movement trajectories of fine particles around coarse particles

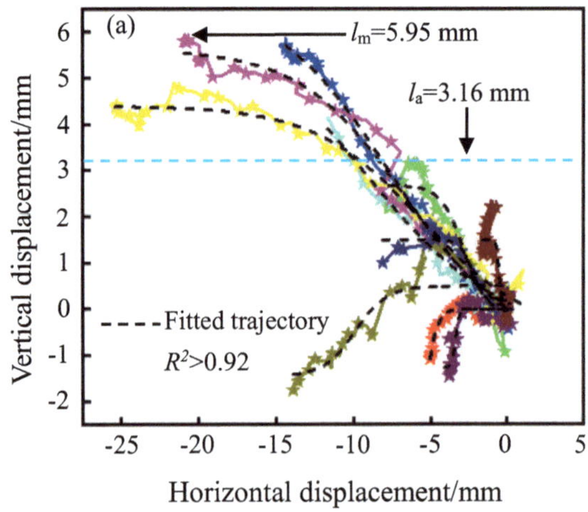

(a) p_v=100 kPa

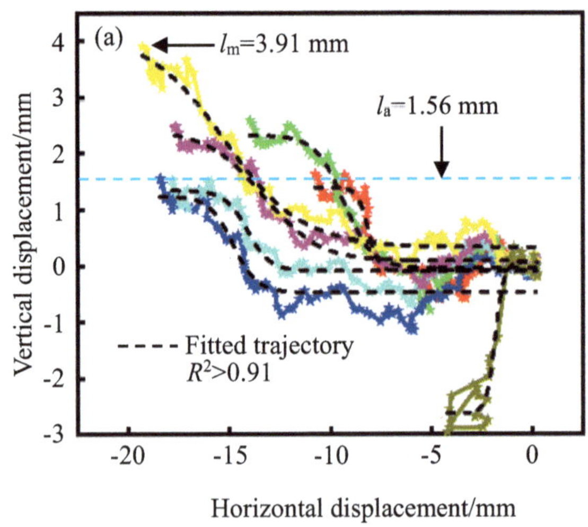

(b) p_v=200 kPa

Fig. 3.28 (continued)

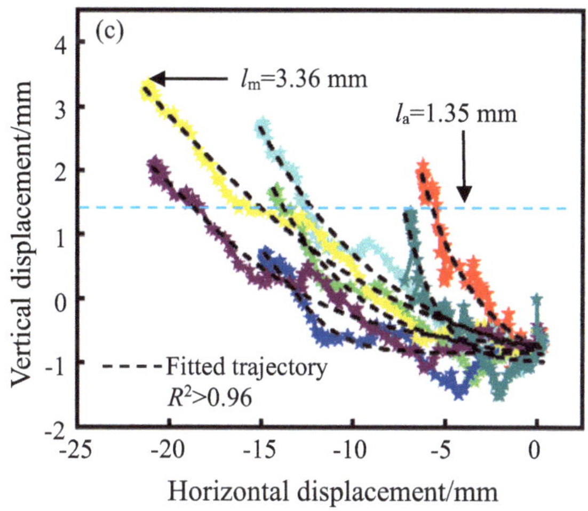

(c) p_v=300 kPa

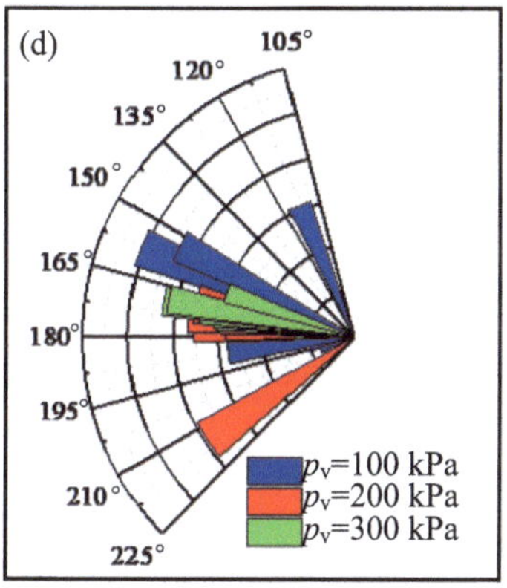

(d) Rose diagram of movement direction

the magnitude of the fine particle rotation angle. Overall, as the coarse particle is positioned at various locations on the shear plane, the rotation of the surrounding fine particles exhibits a consistent pattern: regardless of whether the particles are oriented vertically (top to bottom) or horizontally (left to right), the particle rotation initially decreases before subsequently increasing. Specifically, the rotational capacity of fine particles located directly above or below the coarse particle is slightly weaker than that of fine particles situated directly to the left or right of the coarse particle. For other positions, the rotation generally decreases compared to these specific orientations. A statistical analysis of this behavior reveals the rotation pattern of fine particles around the coarse particle, as expressed in Eq. (3.8):

$$\alpha = \alpha_0 + A \times \sin\left(\frac{(t-\varphi)\pi}{\omega}\right) \tag{3.8}$$

Here, α is the rotation angle of fine particles; α_0 is a coefficient with a value of 20.80 ± 2.46; A is the amplitude with a value of -16.87 ± 7.41; t is the position of fine particles relative to coarse particles; φ is the phase difference angle with a value of -16.87 ± 7.41; ω is a coefficient with a value of 42.94 ± 2.73.

During the shearing process, the presence of coarse particles not only increases the shear strength of the particle system and reduces its own displacement but also exerts a remarkably regular influence on the surrounding fine particles. As illustrated in Fig. 3.29 and Eq. (3.8), the rotation of fine particles around the coarse particle follows a sinusoidal variation pattern, where the particle rotation angle periodically increases or decreases depending on the position of the fine particle relative to the coarse

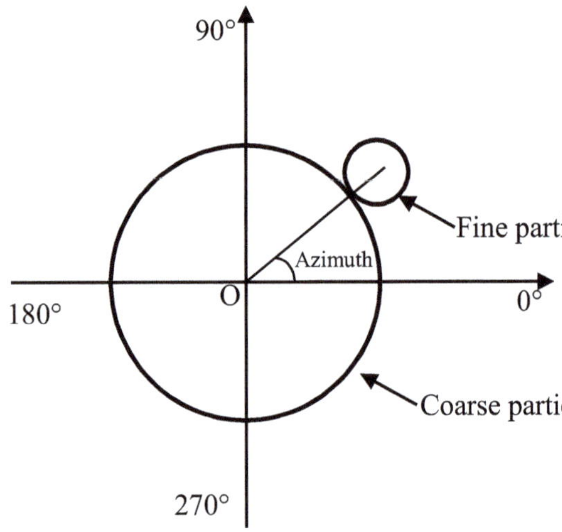

(a) Fine particle position relative to coarse particle

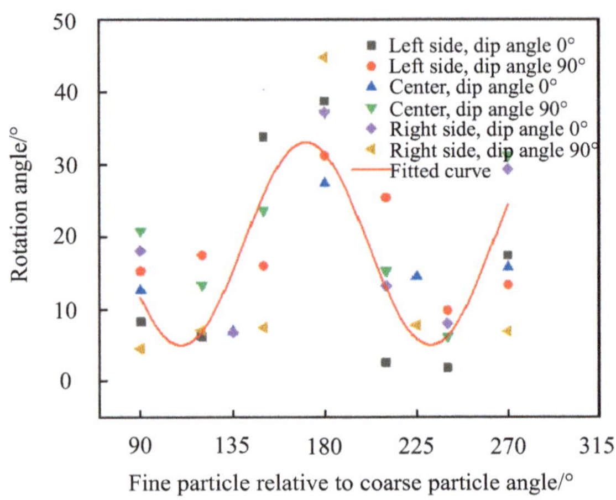

(b) Fine-particle distribution from top to bottom

Fig. 3.29 Statistical analysis of rotation angles of fine particles surrounding coarse particles

Fig. 3.29 (continued)

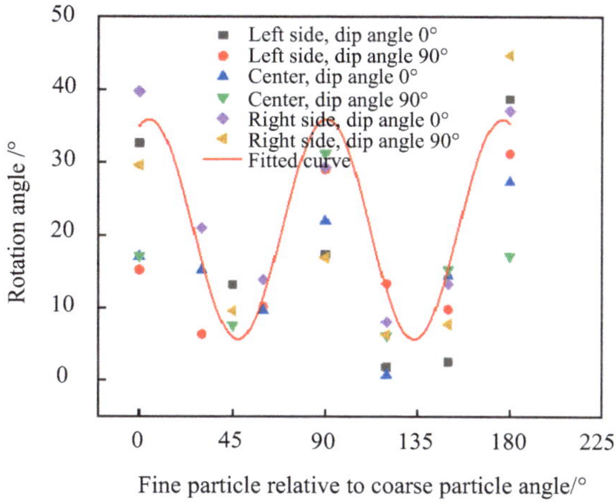

(c) Fine-particle distribution from left to right

particle. Taking 45° as one orientation, when the fine particle is positioned at this angle relative to the coarse particle, its rotation angle reaches an extreme value. Specifically, when the azimuthal angle of the fine particle relative to the coarse particle is 0°, 90°, 180°, or 270°, the fine particle experiences the strongest direct effect of shear force or vertical pressure, resulting in maximum rotation angles, while other azimuthal angles correspond to minimum rotation angles. The fine particle rotation angle fluctuates between these maximum and minimum points. Due to shearing, particles in the shear-out direction experience the greatest horizontal driving force from the coarse particle and undergo significant rotation, hindered by other particles in that direction. Fine particles directly facing the shear-in direction are most obstructed by the coarse particle and undergo intense rotation driven by both obstruction and shear forces. Fine particles above the coarse particle move across it under shear force, and due to the interlocking of particle edges, this crossing occurs through rotation. Particles below the coarse particle are influenced by the coarse particle moving across them, and the interaction between them also involves rotational movement. In other positions, the direct influence of the coarse particle is relatively weaker, resulting from the coordinated movement of coarse and fine particles during shearing.

3.4 Conclusions

Based on DIC technology and utilizing a self-designed visual shear model to address the low visualization limitations of traditional shear tests, direct shear tests were conducted on uniform particle systems of caved ore and rock [60, 61]. The results led to the following conclusions:

(1) The movement of particles on the shear plane can be categorized into three distinct types: transverse obstruction displacement, vertical compression, and rotation. Tracer particles on the shear plane exhibit varying movement states influenced by the dilatancy effect. Notably, the point located at one-third of the shear plane length from the shear-in entrance acts as a critical mutation point for movement trends, signifying the compromise between the dilatancy effect and transverse shear stress.
(2) The shear behavior of the particle system is governed by local strain, which transitions from a random distribution to a concentrated state along the shear plane as the shearing process progresses. The temporal evolution of its probability density near the shear plane adheres to the Logistic model.
(3) The local strain coalesces across the shear plane, resulting in the formation of a 'semi-spindle-shaped' shear band characterized by an arc-shaped upper portion. The evolution of both the maximum thickness of the upper part and the maximum thickness of the lower part of the shear band over time adheres to a Logistic pattern. The difference between the maximum thicknesses of the upper and lower sections of the shear band is approximately 2–3 times the average particle size.

A self-developed non-contact shear characteristic monitoring and analysis system for the particle system of caved ore and rock was employed to conduct experimental research on the influence of a single coarse particle on the shear strength and movement characteristics of the particle system. By comparing the traditional method based on vernier caliper measurements with the image identification method utilizing IPP software for measuring particle displacement, the reliability of IPP software in the domain of caved ore and rock research was established. Furthermore, in comparison to the traditional vernier caliper measurement method, IPP software demonstrates advantages such as enhanced accuracy and the capability for batch data processing. The primary research conclusions are as follows:

(1) When the shape factor of coarse particles r_v exceeds 0.5, the threshold value of the coarse-to-fine particle size ratio d_c/d_f that significantly enhances the overall shear strength of the 5–10 mm particle system ranges from 3.07 to 4.23. Furthermore, the shear strength of the particle system is maximized when the coarse particle is situated in the shear-out direction, followed by the shear-in direction, while it is minimized when the coarse particle is positioned at the center of the visible surface.
(2) Within the studied parameter range, the shear strength of the particle system exhibits a linear increase with the rise in vertical pressure. When the coarse particle shape factor r_v is less than 0.5, there is no significant change in shear strength with increasing r_v. However, when r_v is greater than or equal to 0.5, the shear strength gradually increases as the coarse particle shape factor increases. Notably, the vertical pressure has a more pronounced effect on the shear strength of the particle system compared to the coarse particle shape.
(3) Within the studied parameter range, the movement trajectories of fine particles around a coarse particle, influenced by varying coarse particle shapes and

vertical pressures, conform to the Boltzmann distribution. The vertical displacement of fine particles surrounding the coarse particle significantly increases with the coarse particle shape factor, while it gradually decreases with increasing vertical pressure. Furthermore, both the increase in the coarse particle shape factor and vertical pressure enhances the orderliness of the movement direction of the fine particles.

(4) During the shearing process, fine particles surrounding coarse particles undergo rotation due to the combined effects of the coarse particles and the shear force. The rotation angles of these fine particles exhibit a sinusoidal variation pattern. In the coordinated movement of coarse and fine particles, the positions directly above, below, to the left, and to the right of the coarse particle correspond to the maximum and minimum values of the fine particle rotation angles.

References

1. Garcia FE, Bray JD (2019) Modeling the shear response of granular materials with discrete element assemblages of sphere-clusters. Comput Geotech 106:99–107
2. Wang ZJ (2015) Study on mechanical properties of ballast based on discrete element method. Beijing Jiaotong University, Beijing
3. Chen F (2018) Research and application on strength characteristic of granular materials based on two-dimensional fractal particle size distribution model. Hunan University, Hunan
4. Kong L, Chen FX, Li J (2013) Meso-direct-shear test of sand based on digital image correlation method and its PFC numerical simulation. Rock Soil Mech 34(10):2971–2978
5. Yang S, Li XQ (2020) Simulation analysis of large direct shear test of soil and rock mixture based on PFC3D. Adv Eng Sci 52(3):78–85
6. Yang H, Xu WJ, Sun QC (2016) Study on the meso-structure development in direct shear tests of a granular material. Powder Technol 314:129–139
7. Liu ZQ, Xue YD, Huang HW et al (2012) Experimental research on shear behavior of colluvium. Rock Soil Mech 8:2349–2358
8. Zhang JM, Hu MJ, Hou GQ et al (2013) Dilatant behavior of intensely weathered coarse-grained soils in large-scale direct shear tests. Rock Soil Mech 34(S2):67–73
9. Yang ZP, Tian X, Lei XD et al (2021) Particle discrete element numerical study on factors of shear strength characteristics for soil-rock mixture. J Basic Sci Eng 29(1):135–146
10. Zhang ZP, Sheng Q, Fu XD et al (2021) Simulation of direct shear test for soil-rock mixtures based on particle discrete element method. J Basic Sci Eng 29(1):135–146
11. Liu XR, Tu YL, Wang P et al (2017) Particle breakage of soil-rock aggregate based on large-scale direct shear tests. Chin J Geotech Eng 39(8):1425–1434
12. Hu F, Li ZQ, Liu Q et al (2021) Mechanism of shear stress fluctuation and dropping of the soil-rock mixture. Hydrogeol Eng Geol 48(3):90–101
13. Li YR, Aydin A (2010) Behavior of rounded granular materials in direct shear: Mechanisms and quantification of fluctuations. Eng Geol 115(1–2):96–104
14. Sun ZZ, Ma G, Zhou W et al (2021) Influence of particle shape on size effect of crushing strength of rockfill particles. Rock Soil Mech 42(2):430–438
15. Yang H, Xu WJ, Zhang QB (2015) Macro- and meso-mechanism of strain localization in granular material. Chin J Rock Mech Eng 34(8):1692–1701
16. Xue YD, Liu ZQ, Huang HW (2012) Experimental analysis on shear strength characteristics of sand-gravel mixture. J Civil, Archit Environ Eng 34(6):75–79

17. Zhao Y, Zhou H, Feng XT et al (2013) Particle crushing and shear behaviour of an infilled joint soil under different conditions. Rock Soil Mech 34(1):13–22
18. Tong LH, Wen BQ, Xu CJ et al (2022) Theory of shear strength of granular materials based on rate-state model. Scientia Sinica Technologica 52(3):489–498
19. Wei HZ, Li XX, Zhang SD et al (2021) Influence of particle breakage on drained shear strength of calcareous sands. Int J Geomech 21(07):04021118
20. Hua WJ, Xiao YJ, Wang M et al (2021) Discrete element modeling (DEM) study on effect of gradation and morphology on shear strength behavior of rock debris as embankment fill materials. J Central South Univ (Sci Technol) 52(7):2332–2348
21. Chai W, Long ZL, Kuang DM et al (2019) Effect of shear rate on shear strength and deformation characteristics of calcareous sand in direct shear test. Rock Soil Mech 40(S1):359–366
22. Zhao XY, Wan YH, Zhang XB (2020) Experimental study of fragment orientation of phyllite talus at Whenchuan-Maerkang expressway. Rock Soil Mech 41(1):175–184
23. Cheng H, Wang X, Zhang JS et al (2018) Effects of particle size and gradation on shear behavior of interface between crushed stone and structure. J Central South Univ (Sci Technol) 49(4):925–932
24. Lin CX, Zhong SY, Ling DS (2016) Analysis of particle shape characteristics of lunar soil simulant and its effect on shear strength. J Northeastern Univ (Nat Sci) 37(11):1640–1644
25. Liu QB, Xiang W, Lehane BM et al (2011) Experimental study of effect of particle shapes on shear strength of sand and tip resistance of driven piles. Chin J Rock Mech Eng 30(2):400–410
26. Yan Y, Ji SY (2009) Effects of particle shape on shear strength of discrete media. Rock Soil Mech 30(S1):225–230+249
27. Dong Y, Chai HJ (2007) Experimental study on fractal character of shear surface of rock-soil aggregate mixture. Rock Soil Mech 28(5):1015–1020
28. Jia XM, Chai HJ, Zheng YR (2010) Mesomechanics research of large direct shear test on soil and rock aggregate mixture with particle flow code simulation. Rock Soil Mech 31(9):2695–2703
29. Macaulay M, Rognon P (2020) Two mechanisms of momentum transfer in granular flows. Phys Rev E 101(5–1):050901
30. Cui L, O'Sullivan C (2006) Exploring the macro-and micro-scale response of an idealised granular material in the direct shear apparatus. Géotechnique 56(7):455–468
31. Zhang L, Thornton C (2007) A numerical examination of the direct shear test. Géotechnique 57(4):343–354
32. Wang C, Deng A, Taheri A (2018) Three-dimensional discrete element modeling of direct shear test for granular rubber-sand. Comput Geotech 97:204–216
33. Sun X, Kob W, Blumenfeld R (2020) Friction-controlled entropy-stability competition in granular systems. Phys Rev Lett 125(26):268005
34. Angus A, Yahia LAA, Maione R et al (2020) Calibrating friction coefficients in discrete element method simulations with shear-cell experiments. Powder Technol 372:290–304
35. Zheng QJ, Luo Q, Yu AB (2020) Continuum modelling of primary and secondary granular flows in a torsional shear cell. Powder Technol 361:10–20
36. Hong Y, Li ZR, Tang SS et al (2020) Effect of average particle size on shear properties of sand and its mesomechanical analysis. J Jilin Univ (Earth Sci Ed) 50(6):1814–1822
37. Jiang MJ, Wang FZ, Zhu HH (2010) Shear band formation in ideal dense sand in direct shear test by discrete element analysis. Rock Soil Mech 31(1):253–257+298
38. Pan YY, Wei YF, Li YZ et al (2020) Study on microscopic deformation mechanism of sand soil shear zone in direct shear test. Hydro-Sci Eng 3:82–91
39. Gang M, Yza B, Yuan C et al (2021) Spatial correlation and temporal evolution of plastic heterogeneity in sheared granular materials. Powder Technol 378:263–273
40. Li K, Wang YF, Cheng QG et al (2021) Effects of fractal particle size distribution on segregation of granular flows. Chin J Rock Mech Eng 40(2):330–343
41. Wang YF, Cheng QG, Zhu Q (2012) Inverse grading analysis of deposit from rock avalanches. Chin J Rock Mech Eng 31(6):1089–1106
42. Jin AB, Chen SJ, Sun H et al (2020) Characteristics of particle percolation based on inhomogeneous particle size distribution. J Central South Univ (Sci Technol) 51(6):1673–1681

43. Lei D, Huang JF, Xu WX et al (2018) Deformation analysis of shear band in granular materials via a robust plane shear test and numerical simulation. Powder Technol 323:385–392
44. Jing L, Kwok CY, Leung YF (2017) Micromechanical origin of particle size segregation. Phys Rev Lett 118(11):118001
45. Van D, Gajjar P, Epely-Chauvin G et al (2015) Underlying asymmetry within particle size segregation. Phys Rev Lett 114(23):238001
46. Xu YP, Gao X, Li TW (2021) Numerical study of the bi-disperse particles segregation inside a spherical tumbler with discrete element method (DEM). Comput Math Appl 81:588–601
47. John MNTG, Parmesh G, Peter K (2015) Particle-size segregation in dense granular avalanches. C R Phys 16(01):73–85
48. Zheng L, Mo SP, Li YX et al (2019) Analysis of dynamic characteristics in binary granular segregation during thin shear. Acta Physica Sinica 68(16):260–267
49. Trewhela T, Ancey C, Gray JMNT (2021) An experimental scaling law for particle-size segregation in dense granular flows. J Fluid Mech 916(455):A55, 1–37
50. Trewhela T, Gray J, Ancey C (2021) Large particle segregation in two-dimensional sheared granular flows. Phys Rev Fluids 6(5):54302
51. Wang BX, Jin AB, Sun H, et al (2021) Study on fracture mechanism of specimens with 3D printed rough cross joints at different angles based on DIC. Rock Soil Mech 42(2):439–450+461
52. Guo N, Zhao J (2014) Local fluctuations and spatial correlations in granular flows under constant-volume quasistatic shear. Phys Rev E 89(04):042208
53. Qian JG, Huang MS (2005) An analytical solution for criterion of onset of strain localization of soils. Rock Soil Mech 26(3):432–437
54. Huang MS, Li XF, Qian JG (2012) Strain localization of anisotropic sand. Chin J Geotech Eng 34(10):1772–1780
55. Zhang Y, Cheng SZ (2017) Probability theory and mathematical statistics. Tsinghua University Press, Beijing
56. He YZ, Xie YB, Wu SY (2021) Clinical study of simple and non-invasive analysis of color changes of neonatal jaundice by Image-pro plus. Med Forum 25(02):168–170
57. Dhugal J, Atsushi L, Grossmann Y, et al (2019) Vertical profiles of marine particulates: a step towards global scale comparisons using an Autonomous Visual Plankton Recorder. Bull Plankton Soc Japan 61(01):72–81
58. Chen QL, Li W, Chen Z (2019) Analysis of microstructure characteristics of high sulfur steel based on computer image processing technology. Results Phys 12:392–397
59. Zhang X, Huang TH, Zhang YJ et al (2015) Image-Pro Plus analysis of pore structure of concrete. J Build Mater 18(01):177–182
60. Jin AB, Yao BS, Tang KL et al (2022) Analysis of characteristics of ore-rock granular shear motion based on DIC technology. J Central South Univ (Sci Technol) 53(8):3081–3091
61. Sun H, Tang KL, Jin AB et al (2022) Experimental study on influence of single coarse particle on shear properties of ore-rock particle system. J Zhejiang Univ (Eng Sci) 56(10):2037–2048

Chapter 4
Research on the Drop Shape Theory

The drop shape theory, recognized as one of the primary draw theories, has garnered increasing attention and practical application among mining engineering practitioners in recent years. This chapter will provide a detailed introduction to the validation of the upside-down drop shape theory in gravity flow, as well as the optimization of an empirical equation related to the drop shape theory.

4.1 Confirmation of the Upside-Down Drop Shape Theory in Gravity Flow

As underground mining progresses to greater depths and mining technologies advance, the caving mining method has gained prominence in metal extraction. A key characteristic of this method is its dependence on the extraction of caved ores located among surrounding waste rock, which are subsequently retrieved from designated drawpoints. However, this approach frequently leads to substantial ore loss and dilution rates, potentially diminishing the overall efficiency of mining operations [1]. Inefficient management of the draw process or inaccurate calculations of structural parameters can result in significant waste of valuable mineral resources and a decline in the economic performance of mining companies. To effectively implement the caving method and minimize ore loss and dilution, it is essential to grasp the fundamental principles governing the draw process. This understanding encompasses the behavior of caved ore and rock during extraction operations. Consequently, extensive research has been devoted to investigating the flow characteristics of both caved ore and rock. By examining the dynamics of the extraction zone (EZ), which refers to the area from which material has been removed in the mining model [2], and the movement zone (MZ), where material has shifted due to extraction activities [2], researchers can establish a foundation for understanding these principles. This research not only enhances and refines existing theories related to draw operations but

H. Sun et al., *Study on Flow and Hang-up Characteristics of Caved Ore and Rock*,
https://doi.org/10.1007/978-981-92-0335-2_4

also aids in predicting dilution and loss rates. Furthermore, it supports the optimization of various draw control measures and the design of structural parameters, such as the spacing and arrangement of drawpoints within caving mines. These objectives underscore the primary focus and driving force behind the current study.

Currently, three prominent methods are employed to study draw issues: experimental, theoretical, and numerical analysis. Among these approaches, experimental analysis includes both in-situ and model draw experiments. In-situ draw experiments [3] accurately reflect the actual draw conditions present in the field; however, they have significant drawbacks. These experiments tend to be time-consuming, incur higher costs, present greater management challenges, and their findings often lack broad applicability beyond the specific context in which they were conducted. Consequently, model draw experiments have gained more widespread acceptance and application in practical scenarios. Model draw experiments are crucial for advancing our understanding of draw problems, and extensive research has been conducted utilizing this methodology. Notable studies include those conducted by Kvapil [4, 5], McCormick [6], Watson [7], Laubscher [8], Power [9], and Castro [10, 11], who investigated the behaviors and mechanisms of gravity flow under various conditions, including isolated drawpoints, multiple drawpoints, and inclined boundaries. These studies utilized model draw tests with sand or gravel as mediums, significantly contributing to the rapid advancement of research in this field. In China, researchers Tao [12] and Wang [13] conducted draw experiments to investigate how various factors—such as particle size, drawpoint size, and water content—affect gravity flow. Their findings have established a critical foundation for the design of structural parameters in practical applications.

The method of theoretical analysis encompasses two primary approaches: the ellipsoid draw theory and the stochastic medium draw theory. The ellipsoid draw theory [14–16], initially developed to explain flow dynamics within bunkers and silos, emerged under conditions relatively close to gravity flow. This theoretical framework was subsequently adapted for application in the mining sector, where it has remained dominant for an extended period. The ellipsoid theory conceptualizes the EZ and MZ as ellipsoids, demonstrating significant practicality in the theoretical study of draw. Conversely, the stochastic medium draw theory [17–19] examines particle movement from a distinct perspective. It treats particles as a continuous medium and employs probability theory to derive the governing probability density equations for their movement. This approach facilitates a detailed analysis of particle movement patterns, showcasing considerable efficiency in calculations. Moreover, the outcomes generated through this theory closely align with results obtained from experimental models, thereby affirming its validity and reliability.

Numerical analysis techniques [10, 20–27] primarily utilize the finite difference approach (FDM), finite element techniques (FEM), DEM, or cellular automata theory (CA) to create numerical drawing models and address real-world draw challenges. These approaches are efficient, rapid, highly flexible, and repeatable, demonstrating significant potential for application in scientific research. However, it is crucial to validate these numerical models through experimental tests or full-scale evaluations.

In 1970, Fröström [28] conducted model tests at a scale of 1:50 at the Royal Institute of Technology in Stockholm, where he meticulously examined the flow characteristics using various size distributions of fragments. His research revealed a significant finding: the maximum width of the draw zone consistently occurred in the upper half, indicating that the draw zone is not symmetric about its midpoint. The shape resembles an upside-down water droplet [29]. This observation marks a substantial departure from previous understandings of draw zone shapes. Fröström concluded that the drop shape theory offers a more accurate representation of the draw zone's geometry compared to the traditional ellipsoid theory. This realization represents a significant paradigm shift in the theoretical framework governing gravity flow. Following this foundational work, Bergmark and Roos developed a mathematical equation for the drop shape, derived from fundamental natural laws [30]. Furthermore, the isolated draw tests conducted as part of this study reaffirmed the earlier conclusion that the shape of the EZ resembles an upside-down droplet rather than a conventional ellipsoid. To further explore the characteristics of EZ and MZ, a novel approach was proposed, effectively analyzing their shapes and the trends in their transformations. This research ultimately led to the formulation of new empirical equations aimed at accurately calculating the shapes of the EZ and MZ, contributing valuable insights to the field.

4.1.1 Description of Isolated Draw Experiment

4.1.1.1 Experimental Model and Materials

Castro [10] conducted a model test to investigate the flow characteristics of caved ore and rock within the context of the caving method. This research utilized two distinct geometrical scale models, specifically at ratios of 1:100 and 1:30, to observe and analyze the behavior of materials under realistic conditions. Castro R determined that the geometrical scale employed in the tests had an insignificant impact on the shape of the EZ. Given this finding, and after considering factors such as time efficiency and cost-effectiveness, it was decided to utilize the model (Fig. 4.1) with a geometrical scale of 1:100 for the isolated draw tests. This choice was made to streamline the testing process while ensuring reliable results. The outcomes of these tests are expected to provide critical insights and direction for the Pulang mining company, located in Yunnan Province, China, which is currently engaged in constructing necessary infrastructure. By establishing optimal structural parameters—such as appropriate spacing and layout of drawpoints—these results will aid in enhancing operational efficiency and effectiveness for mining operations.

The model utilized for the experiments was constructed with dimensions of 0.5 m in length, 0.5 m in width, and 1.2 m in height, resulting in a total volume conducive to controlled testing conditions. The drawpoint, strategically positioned at the center of the model's base, was specifically designed to prevent any intersection between the flow zones and the model's walls, thereby facilitating more accurate observations and

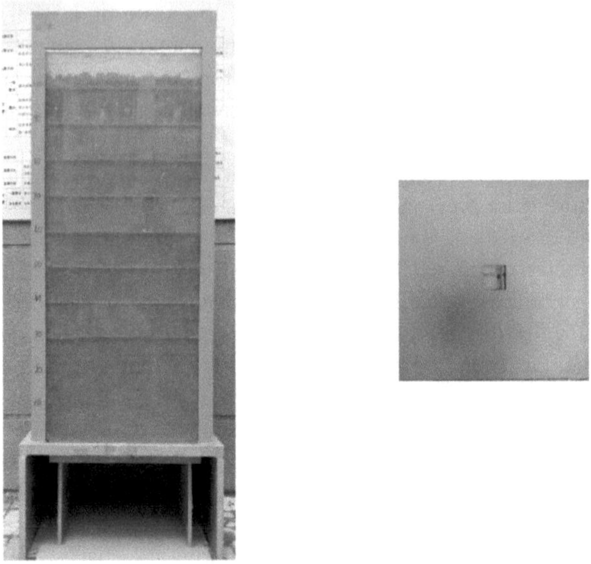

(a) Front view of model (b) Drawpoint in bottom of model

Fig. 4.1 Model in draw physical tests

results during the experiments. The tests employed magnetite ores sourced from the Meishan Iron Mine, as illustrated in Fig. 4.2. The characteristics of the ore were well-defined, with a solid density recorded at 3190 kg/m^3 and a bulk density of 1855 kg/m^3. In the context of isolated draw tests, several influential factors were considered, including particle size, the dimensions of the drawpoint, and the height of the column. To effectively categorize the materials, size classifications were established: materials measuring between 5 and 10 mm were designated as fine particles, those ranging from 10 to 15 mm were termed medium particles, and particles measuring from 15 to 20 mm were classified as coarse particles. A comprehensive approach was adopted in the testing procedure, leading to the design of three-dimensional tests that evaluated these three factors at three distinct levels, as detailed in Table 4.1. This structured framework enabled the researchers to systematically analyze the effects of each variable on the outcomes of the draw tests.

4.1.1.2 Experimental Processes

The tracers with labels were arranged radially at intervals of 50 mm along the height axis, with the center of radiation positioned on the axis of the drawpoint. As illustrated in Fig. 4.3, the two adjacent rows of labeled tracers formed a 30-degree angle, while the spacing between two consecutive tracers within the same row was 40 mm. Each labeled marker serves as a coordinate reference within the model. During each draw,

Fig. 4.2 Ores (left) and labeled tracers (right)

Table 4.1 Parameters in three dimensional experiments of 3 factors and 3 levels in isolated draw tests

No	Particle size/mm	Drawpoint size/m^2	Column height/mm
1	5–10	60 × 60	900
2	5–10	80 × 80	1050
3	5–10	100 × 100	1200
4	10–15	80 × 80	1200
5	10–15	100 × 100	900
6	10–15	60 × 60	1050
7	15–20	100 × 100	1050
8	15–20	60 × 60	1200
9	15–20	80 × 80	900

between 0.5 and 1.5 kg of ore was extracted from the drawpoint, and the quantity of labeled tracers drawn was recorded. The drawing process concluded once the EZ height reached 800 mm.

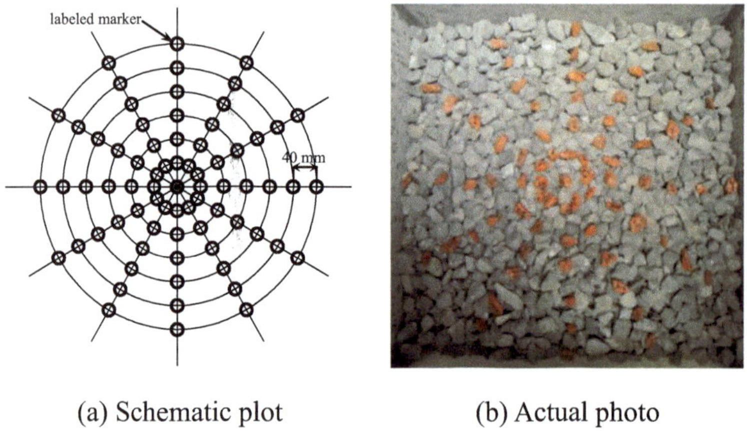

Fig. 4.3 Layout of labeled tracers

4.1.2 Analysis of Experimental Results

4.1.2.1 EZ's Shape

By measuring the mass extracted and documenting the number of labeled tracers utilized during each iteration, the shape of the EZ under infinite boundary conditions can be defined through an interpolation method. This process establishes a correlation between the height of the EZ and the mass drawn. Figure 4.4 illustrates a representative outcome of the EZ, corresponding to a vertical cross-section that traverses the center of the drawpoint at the model's base.

The analysis presented in Fig. 4.4 clearly demonstrates that the shape of the EZ exhibits longitudinal asymmetry and deviates from a conventional ellipsoidal form. Consequently, it is imperative to investigate the EZ's shape and its evolving characteristics through a thorough quantitative analysis of the experimental data collected. This understanding will provide deeper insights into the behavior and structure of the EZ in relation to the experimental conditions.

4.1.2.2 Relationship Between the Mass Drawn and Height of the EZ

Based on the non-linear Levenberg–Marquardt regression method, the experimental data in the 3D tests were fitted using Eq. (4.1):

$$m = k_1 * H^{k_2} \quad (4.1)$$

The constants obtained from the fitting process and the correlation coefficients are presented in Table 4.2. For instance, in the case of Test 6, the fitting outcomes

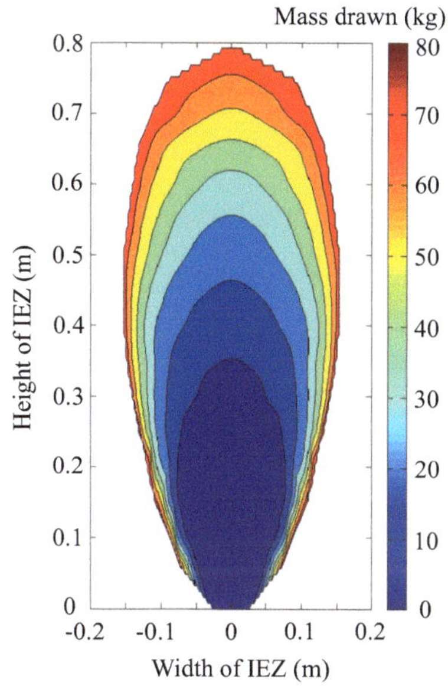

Fig. 4.4 Vertical section of IEZ for the Test 6

depicting the correlation between the mass drawn m and the height H of the EZ are illustrated in Fig. 4.5. The correlation coefficient value, R^2, is 0.998, indicating that there is a power function relationship between m and H as described by Eq. (4.1), with the constants being $k_1 = 131.249$ and $k_2 = 2.743$. Throughout this paper, the MKSA system is consistently utilized.

Taking the logarithm of both sides of Eq. (4.1) gives

$$\ln m = k_2 * lnH + \ln k_1 \quad (4.2)$$

which shows that ln$m$ and ln$H$ have a linear relationship. In Eq. (4.2), lnk_1 is the slope, and k_2 is the intercept.

Table 4.2 Equation (4.1) parameters in isolated draw tests

No	k_1	k_2	R^2
1	132.972	2.572	0.995
2	138.214	2.352	0.997
3	134.208	2.114	0.997
4	142.499	2.462	0.998
5	145.593	2.347	0.994
6	131.249	2.743	0.998
7	129.070	2.087	0.993
8	125.244	2.273	0.997
9	142.586	2.450	0.988

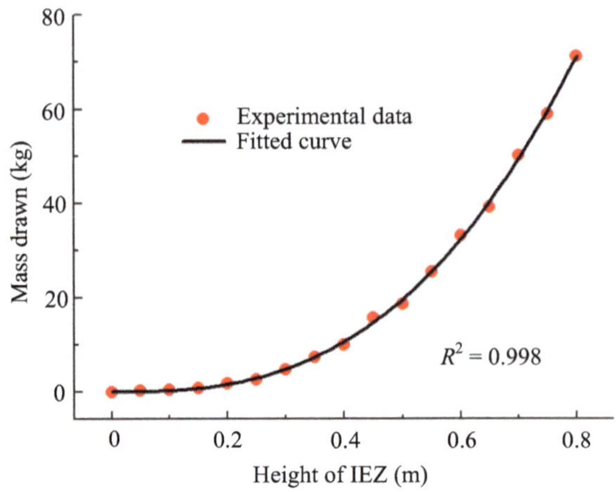

Fig. 4.5 Relationship between the mass drawn and the height of IEZ

4.1.2.3 Relationship Between Mass Drawn and Radius of EZ

Based on the Test 6 data, Fig. 4.6 illustrates the relationship between the mass drawn m and the radius of the EZ y at various heights of the EZ, ranging from 300 to 700 mm.

Given that the variables ln$m$ and ln$H$ exhibit a linear relationship, it becomes advantageous to establish a functional relationship involving the radius of each layer of the EZ. This approach will facilitate a more straightforward description of the problem at hand. Furthermore, it is important to note that the EZ is symmetrical around the axis of the drawpoint in scenarios of isolated draw. This inherent symmetry suggests that the relationship between ln$m$ and the vertical coordinate y must include even terms of the EZ's radius, such as y^2 or y^4, among others. For the purposes of this analysis, we will initially consider the term y^2.

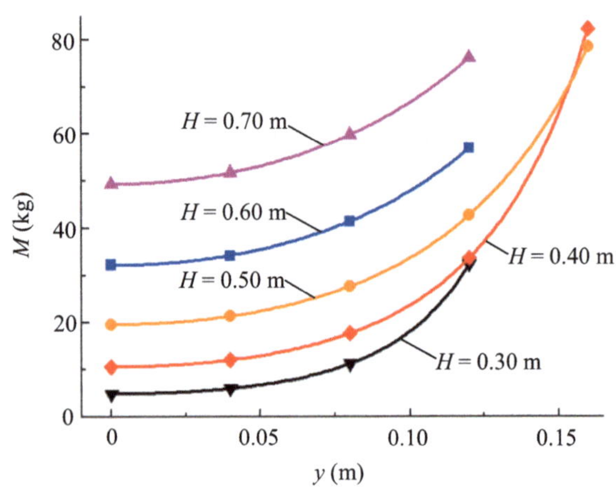

Fig. 4.6 Relationship between the mass drawn and the radius of the IEZ at different IEZ's heights

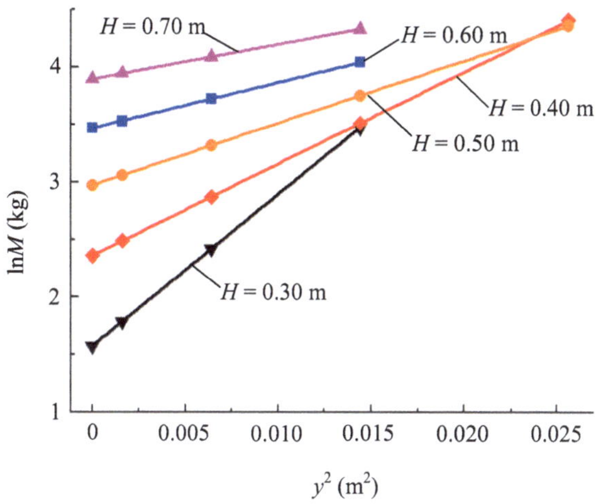

Fig. 4.7 Relationship between ln*m* and y^2 at different IEZ's heights

Table 4.3 Fitted results of coefficients *a* and *b*

H/m	*a*	*b*
0.30	1.574	132.034
0.40	2.363	79.960
0.50	2.976	54.191
0.60	3.471	39.718
0.70	3.899	30.121

The data collected from the tests were meticulously processed in accordance with the aforementioned analysis. The results depicting the relationship between ln*m* and y^2 across various heights of the EZ are illustrated in Fig. 4.7. The findings reveal a consistent linear relationship between ln*m* and y^2 at all heights of the EZ, with correlation coefficient values that are remarkably close to 1. These fitting results are documented in Table 4.3, underscoring the robustness of the observed linear relationship between the variables. Therefore

$$\text{In}m = b * y^2 + a \tag{4.3}$$

In Eq. (4.3), *a* and *b* are the intercept and slope, respectively, and they vary regularly with EZ's height.

4.1.2.4 Relationship of *a*, *b* and Vertical Height

(1) Relationship between *a* and vertical height

It can be known from Eq. (4.3) that when $y = 0$, $\ln m = a$; and, combining Eq. (4.2), gives

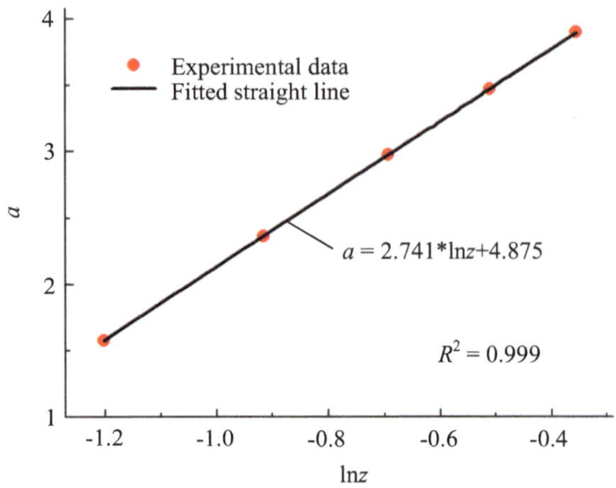

Fig. 4.8 Relationship between *a* and ln*z*

$$a = k_2 * lnH + \ln k_1 \tag{4.4}$$

It can be known from Eq. (4.4) that any EZ's height *H* corresponds to only one value of *a*, thus Eq. (4.4) can be extended to

$$a = k_2 * \ln z + \ln k_1 \tag{4.5}$$

Here, *z* is the vertical height.

In order to verify Eq. (4.4), based upon the data in Table 4.3, an *a*-ln*z* form was linearly fitted. As it is shown in Fig. 4.8, *a* and ln*z* satisfy the linear relationship, and the fitting constants $k_1 = 130.974$, $k_2 = 2.741$, which are almost the same as those in Fig. 4.5. Therefore, the correctness of Eq. (4.4) is verified.

(2) Relationship between *b* and vertical height

Based upon the data in Table 4.3, the ln*b*-ln*z* form was linearly fitted. As shown in Fig. 4.9, ln*b* and ln*z* satisfy the linear relationship:

$$b = 16.242 * z^{-1.740} = k_3 * z^{-k_4} \tag{4.6}$$

4.1.3 *Mathematical Description of a Drop Shape*

4.1.3.1 Equation Derivation

Combining Eqs. (4.3), (4.5) and (4.6) gives

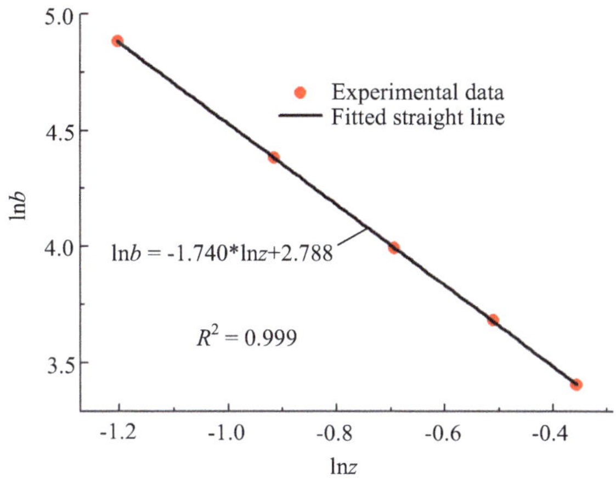

Fig. 4.9 Relationship between ln*b* and ln*z*

$$m = k_1 * z^{k_2} * \exp\left(\frac{k_3 * y^2}{z^{k_4}}\right) \tag{4.7}$$

and substituting Eq. (4.1) into Eq. (4.7) gives

$$y^2 = \frac{k_2}{k_3} * z^{k_4} * \ln\left(\frac{H}{z}\right) \tag{4.8}$$

Equation (4.8) represents the mathematical expression of the drop shape; however, it is not the final result and requires verification of its constant.

4.1.3.2 Coefficient Checking

Assuming that V is the volume of the drop shape at EZ's height H, then

$$V = \int_0^H \left(\pi * y^2\right) * dz = \frac{\pi * k_2}{k_3} * \int_0^H \left[z^{k_4} * \ln\left(\frac{H}{z}\right)\right] * dz$$

Solution of the equation gives

$$V = \frac{\pi * k_2}{k_3 * (k_4 + 1)^2} * H^{k_4 + 1} \tag{4.9}$$

and based on the relationship between the mass drawn and the volume, this gives

$$V = \frac{m}{\gamma} = \frac{k_1 * H^{k_2}}{\gamma} \tag{4.10}$$

Therefore

$$\gamma = \frac{k_1 * k_3 * (k_4 + 1)^2}{\pi * k_2} * H^{k_2 - (k_4 + 1)} \tag{4.11}$$

In the original loose condition, the bulk density γ can be thought of as not varying with height, namely γ has nothing to do with EZ's height H, thus

$$k_2 - (k_4 + 1) = 0$$

That is

$$k_4 = k_2 - 1 \tag{4.12}$$

From Fig. 4.5, $k_2 = 2.743$ and from Fig. 4.9, $k_4 = 1.740 \approx k_2 - 1 = 1.743$, which verifies the correctness of Eq. (4.12).

Equations (4.10) and (4.11) can be simplified by substituting Eq. (4.12):

$$V = \frac{\pi}{k_2 * k_3} * H^{k_2} \tag{4.13}$$

$$\gamma = \frac{k_1 * k_2 * k_3}{\pi} \tag{4.14}$$

4.1.3.3 Ultimate Expression of a Drop Shape

Combining Eqs. (4.8) and (4.12) gives

$$y^2 = \frac{k_2}{k_3} * z^{k_2 - 1} * \ln\left(\frac{H}{z}\right) \tag{4.15}$$

Equation (4.15) is the expression of a drop shape in a plane form, and its spatial form is

$$y^2 + x^2 = \frac{k_2}{k_3} * z^{k_2 - 1} * \ln\left(\frac{H}{z}\right) \tag{4.16}$$

Combining Eqs. (4.7) and (4.12) gives the relationship between the mass drawn and the spatial location of any point of the EZ as

$$m = k_1 * z^{k_2} * \exp\left(\frac{k_3 * y^2}{z^{k_2 - 1}}\right) \tag{4.17}$$

In summary, the comprehensive mathematical derivation concerning the drop shape has been successfully completed, resulting in the identification of three constants: k_1, k_2 and k_3. These constants are instrumental in illustrating how various factors influence the shape of the EZ. The derived equations indicate that determining the EZ's shape under isolated draw conditions is a straightforward process. It involves measuring the mass drawn m at varying heights H of the EZ using marked indicators, followed by conducting a fitting analysis of these measurements. Through this approach, the constants k_1 and k_2 can be accurately ascertained. Subsequently, by integrating these constants with the bulk density γ of the caved ore and rock, one can calculate the value of k_3 using Eq. (4.14). This systematic procedure allows for a cohesive understanding of the EZ shape. Finally, by substituting the known values of k_2 and k_3 into Eq. (4.15), the shape of the EZ can be derived with precision. This method offers a marked improvement over traditional techniques for measuring and charting the EZ's shape; it is not only more accurate but also remarkably quicker, significantly reducing the need for extensive experimentation and data processing.

4.1.4 Comparison Between Drop-Shape and Ellipsoidal EZs

According to the theory of ellipsoidal draw, the shape of the EZ is regarded as an ellipsoid, based on direct observations from experimental results. However, the equation suggests that the EZ resembles an inverted droplet shape, which relates the mass drawn to the height and width of the EZ determined through testing. Utilizing model test data, this study conducts a comprehensive quantitative evaluation of the advantages and disadvantages of both droplet-shaped EZ and ellipsoidal EZ. This analysis includes: (1) a variance examination comparing the radius of the actual EZ, the radius of the ellipsoidal EZ, and the radius of the droplet-shaped EZ across various heights; (2) a comparison of the trends in changes among the actual EZ, ellipsoidal EZ, and droplet-shaped EZ, including aspects such as the height at which the maximum width of the EZ occurs and the symmetry fit of both the ellipsoidal and droplet-shaped EZ relative to the actual EZ.

4.1.4.1 Variance Analysis

Initially, an interpolation method is employed to accurately assess the shape of the actual EZ. Following this measurement, both drop-shape and ellipsoidal EZ calculation methods are utilized to determine the radius of the EZ at varying heights. It is crucial to emphasize that the ellipsoidal models considered in this study include both truncated and non-truncated forms, with the latter referred to as the complete ellipsoid. The distinct shapes of the ellipsoidal EZ and the drop-shape EZ are illustrated in Fig. 4.10, where the variables y_{nte}, y_{te} and y_s correspond to the radius of the non-truncated ellipsoidal EZ, the truncated ellipsoidal EZ, and the drop-shape EZ, respectively.

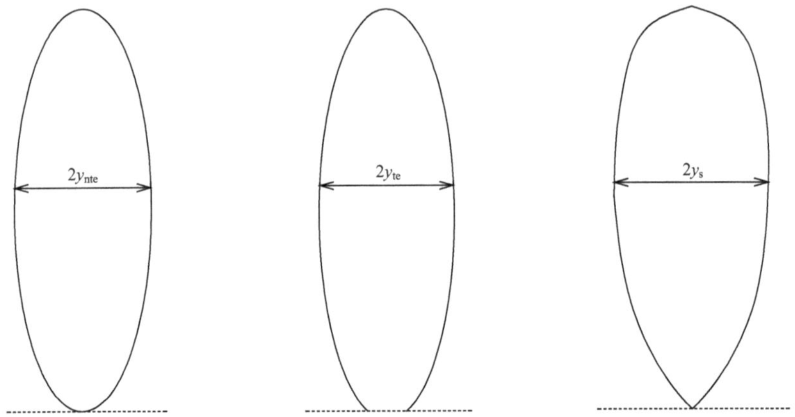

(a) Non-truncated ellipsoidal IEZ (b) Truncated ellipsoidal IEZ (c) Drop-shape IEZ

Fig. 4.10 Shapes of ellipsoidal IEZ and drop-shape IEZ

The formula for analysis of variance is

$$\sigma = \sqrt{\frac{1}{n-2} * \sum_{i=1}^{n} \left(y_i - y_i'\right)^2} \tag{4.18}$$

Here, σ represents the residual standard deviation, n is the number of layers in the comparison, y_{i} is the radius of the actual EZ at the ith layer, and y_i' is the radius of the drop-shape EZ or ellipsoidal EZ at the ith layer.

For the Test 6, when the height H is set at 800 mm, various measurements related to the EZ are presented in Table 4.4. Specifically, the actual radius of the drop-shaped EZ, denoted as y_{a}, alongside the radii for the non-truncated ellipsoidal EZ y_{nte} and the truncated ellipsoidal EZ y_{te}, are detailed. Utilizing Eq. (4.18), calculations reveal that the residual standard deviation for the drop-shaped EZ is $\sigma_{\mathrm{s}} = 8.366 \times 10^{-3}$ m, for the non-truncated ellipsoidal EZ, it is $\sigma_{\mathrm{nte}} = 1.582 \times 10^{-2}$ m, and for the truncated ellipsoidal EZ, it is $\sigma_{\mathrm{te}} = 1.518 \times 10^{-2}$ m. Furthermore, a similar analysis can be conducted when the height H is adjusted to 700 mm. In this case, the calculations yield a residual standard deviation of $\sigma_{\mathrm{s}} = 9.113 \times 10^{-3}$ m for the drop-shaped EZ, $\sigma_{\mathrm{nte}} = 1.545 \times 10^{-2}$ m for the non-truncated ellipsoidal EZ, and $\sigma_{\mathrm{te}} = 1.455 \times 10^{-2}$ m for the truncated ellipsoidal EZ. These figures demonstrate the variation in standard deviations of the different EZ shapes in response to changes in height.

The calculations presented indicate that, when comparing the residual standard deviations associated with the non-truncated ellipsoidal EZ and the truncated ellipsoidal EZ, the drop-shape EZ exhibits the lowest residual standard deviation. This finding underscores its superior fit in modeling. Furthermore, an examination of various heights for the EZ, including 600 mm, 500 mm, and 400 mm, consistently demonstrates a similar trend. This consistency suggests that the drop-shape EZ

Table 4.4 Comparison of the radii among actual EZ, drop-shape EZ and ellipsoidal EZ

z (mm)	y_a (mm)	y_s (mm)	y_{nte} (mm)	y_{te} (mm)
800	0.000	0.000	0.000	0.000
750	90.456	81.393	73.253	72.554
700	114.591	110.243	100.083	99.165
650	132.942	128.872	118.117	117.086
600	141.357	141.468	131.039	129.962
550	148.056	149.658	140.269	139.200
500	155.418	154.252	146.506	145.495
450	156.427	155.691	150.124	149.220
400	154.454	154.211	151.311	150.566
350	150.230	149.907	150.124	149.596
300	141.886	142.755	146.506	146.266
250	131.167	132.614	140.269	140.407
200	117.185	119.186	131.039	131.681
150	99.443	101.921	118.117	119.464
100	77.000	79.774	100.083	102.512
50	47.698	50.341	73.253	77.792
0	30.000	0.000	0.000	30.000

closely approximates the actual shape of the EZ, reaffirming its effectiveness as a more accurate representation compared to its ellipsoidal counterparts.

4.1.4.2 Comparison of Shape Change Trends

(1) Relationship between the height and maximal radius of the drop-shape EZ

From Eq. (4.15), set

$$\mathrm{f}(z) = \mathrm{y}^2 = \frac{k_2}{k_3} * z^{k_2 - 1} * \ln\left(\frac{H}{z}\right)$$

which yields the derivative

$$\mathrm{f}\prime(z) = \frac{k_2}{k_3} * \left[(k_2 - 1) * z^{k_2 - 1} * \ln\left(\frac{H}{z}\right) - z^{k_2 - 2}\right]$$

Setting

$$\mathrm{f}'(z) = 0$$

Gives

$$z = H * \exp\left(\frac{1}{1 - k_2}\right) \tag{4.19}$$

At this point, y^2 has a maximal value, thus y has a maximal value, and maximal EZ $y_{\max}$ is

$$y_{\max} = \sqrt{\frac{k_2}{\mathrm{e} * (k_2 - 1) * k_3}} * H^{\frac{k_2 - 1}{2}} \tag{4.20}$$

Here, e is the base of natural logarithm.

It can be shown from Eq. (4.20) that any drop-shape EZ's height H corresponds to only one maximal drop-shape EZ's radius $y_{\max}$. Thus Eq. (4.20) can be extended to:

$$y_{\max} = \sqrt{\frac{k_2}{\mathrm{e} * (k_2 - 1) * k_3}} * z^{\frac{k_2 - 1}{2}} \tag{4.21}$$

which shows that the relationship between the height and the maximal radius of the drop-shape EZ is in accordance with that of the power function shown in Eq. (4.21).

(2) Asymmetry of the drop-shape EZ

When maximal width of the drop-shape EZ is not at its half height $H/2$, clearly it is not symmetrical longitudinally; and when it is at half height, i.e. $x = H/2$, it can be deduced from Eq. (4.19) that $k_2 = 2.4427$. Therefore, set:

$$x_1 = \left(\frac{1}{2} + \alpha\right) * H$$

$$x_2 = \left(\frac{1}{2} - \alpha\right) * H$$

Here, x_1 and x_2 are two random points symmetrical about the midpoint of drop-shape EZ's height. When $k_2 = 2.4427$, set $x = x_1$, $x = x_2$, respectively, and set $\alpha = 0.20$, and substitute this into Eq. (4.15) for the spindle.

When $x = x_1$,

$$\mathrm{y}_1^2 = 0.2132 * \left(\frac{k_2}{k_3}\right) * H$$

When $x = x_2$,

$$\mathrm{y}_2^2 = 0.2119 * \left(\frac{k_2}{k_3}\right) * H$$

It can be demonstrated that $y_1 > y_2$, indicating that the drop-shaped EZ does not exhibit symmetry, even when the maximum radius is located at the midpoint of its height. This observation underscores a significant deviation from the predictions made by the ellipsoid drawing theory. Furthermore, the findings are consistent with the principles of upside-down drop shape theory, reinforcing the notion that the geometry of such drops is inherently asymmetrical.

Table 4.4 illustrates that at an EZ height of 800 mm, the maximum radii of both the non-truncated and truncated ellipsoidal EZs occur at the midpoint of the height. In contrast, the maximum radii for the actual and drop-shaped EZs are approximately 450 mm, which does not coincide with the midpoint of the height. Similar results are observed when the EZ height is set to 700 mm, 600 mm, and 500 mm; specifically, the calculated position of maximum width based on the upside-down drop shape theory aligns with the actual measurements. Furthermore, the analysis of the experimental data indicates that the asymmetry of the drop-shaped EZ corresponds with that of the actual EZ, emphasizing that the primary trend in shape alteration for the drop-shaped EZ is consistent with that of the actual EZ.

In summary, the analysis of variance concerning drop-shaped and ellipsoidal EZs, along with a comparison of their respective alteration trends, provides quantitative evidence supporting a significant finding. Specifically, the research demonstrates that, contrary to the theories underpinning ellipsoidal shapes, the theory related to inverted drop shapes offers a more accurate depiction of the actual morphology of equidistant zones. Furthermore, this theory effectively captures the nuances of their changing characteristics over time. Consequently, this establishes the superiority of the upside-down drop shape theory in representing and understanding these phenomena.

4.1.5 The Drop-Shape MZ

Due to the limitations of the experimental conditions, the shape of the MZ was not investigated in the tests. However, Castro [10] conducted model experiments to monitor the movement limits of granular media using self-designed sensors.

In this analysis, we examine the transformation of the shape of the MZ, utilizing the experimental data provided by Castro. Specifically, we focus on the test results obtained from gravels with an average diameter of 8 mm-ND. Illustrated in Fig. 4.11 [10] are the vertical sections of the MZ at various mass values. The data reveals that the shape of the MZ closely resembles that of an inverted drop, as predicted by the derived empirical equation. Furthermore, the MZ exhibits notable longitudinal asymmetry, indicating that its structural characteristics vary significantly along its length.

Using the results from gravels with an average diameter of 18 mm-ND as a case study, Fig. 4.12 illustrates the relationship between the height and mass of the MZ, as derived from Eq. (4.1), which has constants $k_1 = 20.422$, $k_2 = 2.397$ and $k_3 = 121.937$. Similarly, Fig. 4.13 depicts the correlation between the height and maximum width of the MZ based on Eq. (4.15), with constants $k_1 = 22.248$, $k_2 = 2.534$ and k_3

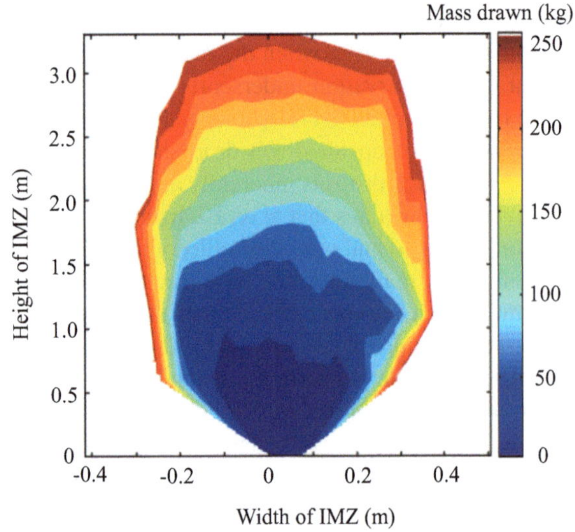

Fig. 4.11 Vertical section of IMZ [10]

= 105.876. An analysis of Figs. 4.12 and 4.13 reveals a strong alignment between the experimental results and the corresponding equations, indicating that the constants are largely congruent. This suggests that the shape of the MZ closely resembles that of a drop-shaped MZ. Additionally, the same analytical approach employed to assess the EZ's test data was utilized to evaluate the variance in the MZ's radius and the evolving nature of its shape, leading to a similar conclusion.

In conclusion, the applicability of the upside-down drop shape theory to the MZ is validated from both qualitative and quantitative perspectives. Qualitatively, this is evident in the shape of the MZ, while quantitatively, it is supported by the relationship between the mass drawn and the height of the MZ.

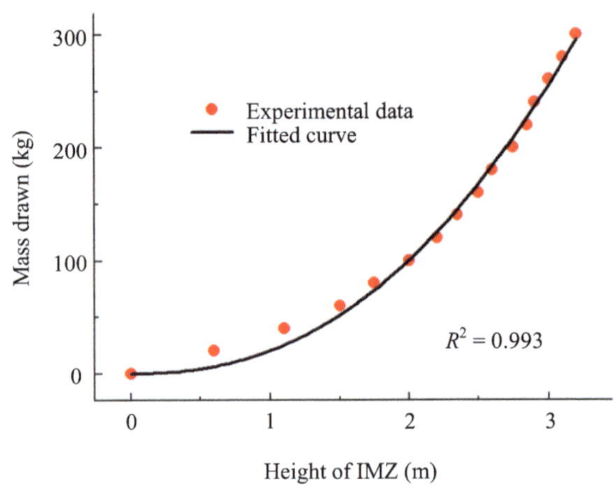

Fig. 4.12 Relationship between the mass drawn and the IMZ's height

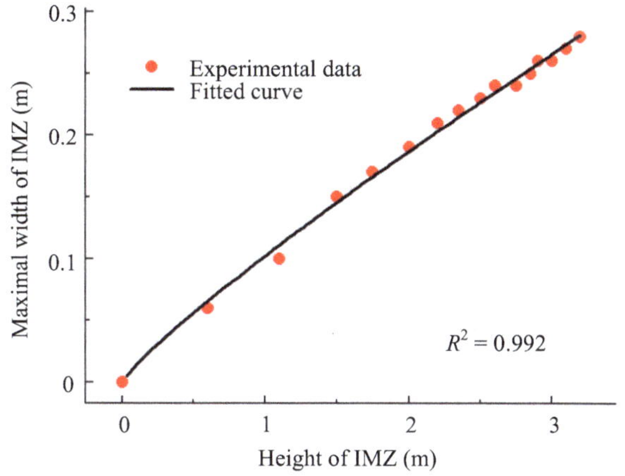

Fig. 4.13 Relationship between maximal width and height of IMZ

4.1.6 *The Forecast of Ore Dilution Rate*

The primary objective of investigating draw problems is to accurately forecast ore loss and dilution. Utilizing the newly developed empirical equation for the drop-shape EZ, as presented in Eq. (4.15), we can derive both the average and instantaneous ore dilution rate equations.

$$D_a = \frac{m_r}{m} = 1 - \left(\frac{z}{H}\right)^{k_2} * \left[k_2 * \ln\left(\frac{H}{z}\right) + 1\right] \tag{4.22}$$

$$D_i = \frac{dm_r}{dm} = 1 - \left(\frac{z}{H}\right)^{k_2} \tag{4.23}$$

Equations (4.22) and (4.23) are the equations of the average ore dilution rate D_a and the instantaneous dilution rate D_i, respectively, in which m and m_r are the total mass drawn and the rock mass drawn, respectively.

Utilizing the model previously discussed, a series of dilution tests were conducted to determine the actual ore dilution rate. These tests aimed to compare the measured dilution rate with the predicted rates derived from ellipsoid theory and newly established empirical equations. In these experiments, rock materials were incorporated into the model and specifically loaded onto the ore layer. The methodology used in this experimentation closely mirrored that of the isolated draw test described earlier, with the primary difference being the measurement of the mass of the extracted rocks. The findings from these experiments indicate that the ore dilution rate predicted by the new empirical equations aligns more closely with the actual measured rate. This suggests that the upside-down drop shape theory exhibits greater predictive capability than the ellipsoid draw theory in forecasting ore dilution rates. Furthermore, the research findings regarding the drop shapes for EZ and MZ have facilitated the derivation and validation of the particle trace equation, as well as the velocity equation pertaining to caved ore and rock.

4.2 Optimization of an Empirical Equation in Drop Shape Theory

The demand for mineral resources is expected to increase significantly due to the rapid advancement of the global economy and the urgent need to transition to a low-carbon economy. Among various extraction methods, caving mining stands out as a widely used technique in metal mining worldwide. This method is distinguished by its large-scale operations, cost-effectiveness, and high efficiency in underground mining environments [25, 27, 31, 32]. The process results in a specific material system comprising caved ore and rock, which forms either through blasting or natural collapse during mining activities. This granular material system is a focal point of research across multiple disciplines, including mining engineering, civil engineering, rock mechanics, physics, materials science, chemistry, and chemical engineering, underscoring its interdisciplinary significance [33]. A comprehensive understanding of the structural characteristics of this caved ore and rock system is essential for grasping the critical physical and mechanical properties associated with the behavior of these particle systems. Despite its importance, current knowledge in this area remains limited, which has hindered the advancement and implementation of innovative draw theories within caving geomechanics. In recent years, the primary emphasis in the field of caving mining has been on enhancing cave flow to optimize economic returns. Unfortunately, this focus has neglected the investigation of fundamental theories underpinning the mining process, particularly the flow characteristics associated with caved ore and rock. Consequently, there is a significant gap in thorough research regarding both the structural characteristics and the mechanisms that govern the formation of the particle flow system within the context of caved ore and rock. This oversight highlights the need for further exploration and analysis to fully understand and advance the field.

Researchers have engaged in both physical and numerical examinations of the structural and mechanical attributes of various granular systems for an extended duration. In the 1960s, Bernal [34] and Scott [35] identified configurations exhibiting short-range order within disordered packings of spherical particles. These configurations significantly influence the macroscopic characteristics of the particles. This finding laid the groundwork for further research into the structure of particle packing systems. Since then, a wealth of literature has emerged, focusing on aspects such as packing density [36], local order degree [37], bridge frameworks [38], topological configurations [39], and the packing of non-spherical particles [40], each examining the fundamental nonuniformity of particle packing systems from various perspectives. Drawing inspiration from the behavior of viscoplastic fluids, Forterre and Pouliquen [41] developed a constitutive model to portray the flow of dense particles.

This model effectively captures the flow dynamics of granular systems under varying boundary conditions and has been extensively applied in the analysis of natural disasters and engineering challenges, such as debris flows and landslides. Furthermore, Yang et al. [42], Fei et al. [43], and Tian et al. [44] conducted comprehensive studies on the physical properties, segregation phenomena, and flow dynamics in chutes, open canals, and silos through various methodologies, including physical experiments, numerical simulations, and theoretical assessments. Wang et al. [45] performed physical experiments designed to replicate the avalanche process within a wedge-shaped hopper, resulting in the development of a shear-thinning viscosity model for particle flow behaviors. Their investigation into the mechanical properties of particle systems yielded significant insights, particularly concerning the relationship between the shear localization region and the two polarization stress fields. This relationship was examined through two-dimensional photoelastic tests, wherein Wang et al. [46, 47] demonstrated that these stress fields align with the direction of maximum shear stress. Moreover, they emphasized the critical role of local anisotropy in the statistical frameworks governing particle materials, positing that a thorough understanding of these local behaviors is essential for grasping the overall mechanics of particle systems. In a different approach, Fonseca et al. [48] employed micro-computed tomography technology to collect comprehensive data regarding the contact interactions among particles. They introduced the contact index, a novel parameter aimed at quantifying the contact area between particles. The results of their study revealed that the contact index demonstrated greater sensitivity compared to other commonly utilized measures, such as the coordination number, branch vector length, and porosity size, when assessing the influence of contact fabric on particle behavior. Furthermore, numerous researchers have conducted quantitative analyses using finite element and discrete element numerical simulations to investigate the complex relationship between macro and micro properties within granular mechanics. These studies seek to clarify the micro-mechanisms that govern the observed macro-physical and macro-mechanical characteristics in particle systems. Significant areas of focus in this research domain have included the evolution of internal force chains [49, 50], anisotropy within contact fabric [51, 52], and the intricate relationships between stress and strain tensors [53]. Additionally, investigations have explored how various particle systems interlock and exhibit arching behavior under loading conditions [54], thereby deepening our understanding of the complexities inherent in particle interactions.

Nevertheless, few investigations have approached the issue of draw from a mesoscopic structural-mechanical perspective, applying the principles of rheology and statistical mechanics mentioned earlier. Research on gravity flow has primarily focused on using either physical or numerical models to evaluate the spatial development of the IEZ, and the IMZ [4, 5, 8, 55–58]. Studies have also explored the secondary fragmentation of caved ore and rock [2, 59, 60], the infiltration of fine particles [27, 61–63], as well as the calculation of ore loss and dilution indices [11, 64], and the enhancement of draw techniques in conjunction with stope structural parameters [11, 65, 66]. Furthermore, several draw theories have been developed, including the ellipsoidal draw theory [20, 67], stochastic medium draw theory [68,

69], and upside-down drop shape (UDDS) theory [28, 70]. It is noteworthy that UDDS theory depicts the configurations of the IEZ and IMZ as resembling inverted water droplets. Although these theories concerning draw have experienced varying degrees of acceptance and practical application, a cohesive understanding of the evolution of the IEZ shape remains elusive. Moreover, the processes contributing to the formation of the IEZ have not been thoroughly investigated.

The present circumstances reveal a significant disparity between the increasing demand for effective underground mining practices and the inadequate theoretical frameworks currently available to support these operations. A particularly pressing issue faced by researchers is the necessity of achieving a deeper understanding of the mesoscopic gravity flow mechanisms associated with caved ore and rock. This challenge is further complicated by variations in void ratios, which can substantially influence the behavior of these materials. To address this complexity, the paper conducts a series of isolated numerical draw tests, systematically examining a range of void ratios and employing a rolling resistance model to analyze the results. The primary objective of this research is to establish a quantitative relationship between void ratios and the shape evolution of the IEZ through the lens of drop shape (DS) theory. By closely examining the structural and mechanical characteristics of both packing and flow systems in the context of caved ore and rock, this study applies principles of statistical mechanics to yield new insights. Furthermore, it investigates the gravity flow mechanisms related to caved ore and rock across varying void ratios, thereby contributing to a more comprehensive understanding of these processes. The findings and concepts articulated herein significantly enhance the existing knowledge surrounding draw operations in underground mining. They also provide essential theoretical and technical guidance aimed at optimizing stope structural parameters, forecasting rates of ore loss and dilution, and promoting the safe and efficient extraction of mineral resources. Additionally, this study introduces an innovative analytical framework for understanding gravity flow, particularly emphasizing the relevance of applying statistical physics parameters within the context of caving mining. The insights gleaned from this research are intended to serve as a significant reference point for future kinetic theoretical investigations involving granular materials, including various forms of caved ore, rock, and associated particulate flows such as landslides, sand, grains, and chemical particles.

4.2.1 Description of the Draw Scheme

In engineering applications, achieving computational accuracy and efficiency with the particle discrete element method presents significant challenges. However, the rolling resistance model effectively accounts for the influence of particle shape without incurring substantial computational costs [71–74]. Consequently, a series of isolated numerical draw tests were performed using rigid balls in conjunction with the rolling resistance model [75] to investigate the flow characteristics of caved ore and rock at varying void ratios.

4.2.1.1 Numerical Draw Model

The authors conducted a series of physical draw experiments using a self-constructed 3D draw model, as illustrated in Fig. 4.1a. In conjunction with these physical tests, they performed numerical draw tests based on the rolling resistance model, as shown in Fig. 4.14. The comparison of results from the physical draw experiments and the numerical tests highlighted both the reliability and advantages of employing the rolling resistance model for numerical simulations of draw scenarios [64]. Furthermore, the study involved identifying the mechanical parameters related to both the wall and the ball, which are comprehensively detailed in Table 4.5. Collectively, these investigative experiments established a robust foundation for examining the gravity flow dynamics of caved ore and rock across various void ratios, thereby providing valuable insights into the subject matter.

A numerical draw simulation was conducted using an isolated draw model with dimensions of 40 × 40 × 80 m^3, as illustrated in Fig. 4.15a. The drawpoint is represented by the red wall at the bottom of the model. The model features a total column height of 80 m, with a draw height of 60 m indicated by the red particles and an overburden height of 20 m represented by the blue particles (Fig. 4.15b).

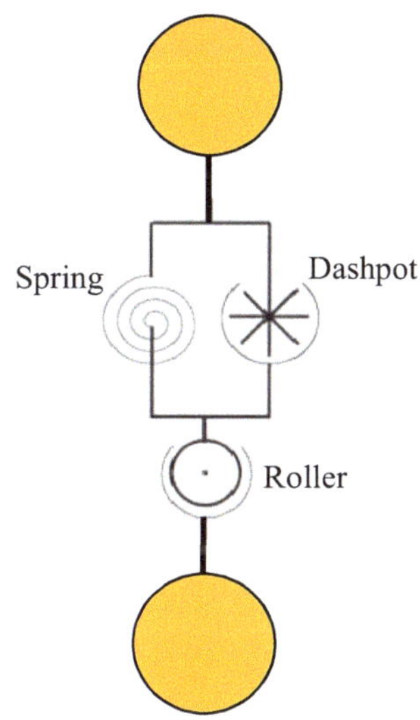

Fig. 4.14 Schematic of the rolling resistance model [64]

Table 4.5 Mesoscopic mechanical parameters of both wall and ball

Wall			Ball				
Normal stiffness/ $(N \cdot m^{-1})$	Shear stiffness/ $(N \cdot m^{-1})$	Friction coefficient	Normal stiffness/ $(N \cdot m^{-1})$	Shear stiffness/ $(N \cdot m^{-1})$	Solid density/ $(kg \cdot m^{-3})$	Friction coefficient	Rolling friction coefficient
5×10^8	5×10^8	0.50	4×10^8	4×10^8	3190	0.55	0.55

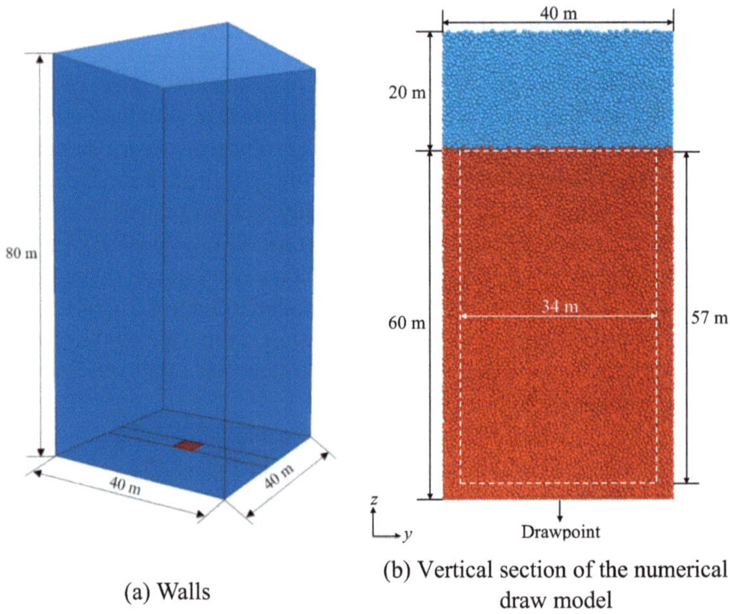

Fig. 4.15 The numerical draw model

4.2.1.2 Schemes and Processes of Numerical Draw Simulation

The particles incorporated in the simulations varied in diameter from 0.4 m to 0.8 m, as illustrated in Table 4.6. The numerical draw simulations encompassed specific processes detailed below:

Table 4.6 Schemes of numerical draw simulation

No	Friction coefficient in the initial packing stage	Rolling friction coefficient in the initial packing stage	Particle size range/m	Condition for the end of the draw
1	0.05	0.05	0.4–0.8	IEZ height reaches 60 m
2	0.10	0.10		
3	0.15	0.15		
4	0.25	0.15		
5	0.30	0.15		
6	0.45	0.15		
7	0.50	0.25		
8	0.50	0.35		
9	0.50	0.45		

(1) Initial packing phase: Following gravity settling, various coefficients of friction and rolling resistance (refer to Table 4.6) were assigned to the balls. This resulted in static packing systems composed of caved ore and rock, characterized by diverse void ratios. Subsequently, structural metrics such as void ratio, bond orientation order, and Minkowski tensor were computed, alongside mechanical metrics including coordination number and stress across different regions of the model.
(2) Draw phase: The mesoscopic mechanical parameters presented in Table 4.5 were assigned to the walls and balls. During the draw phase, the configuration of the IEZ at varying heights and the stresses in different regions were documented. The simulation concluded when the height of the IEZ reached 60 m, while the overburden rock height was maintained at a constant 20 m.

4.2.2 Optimization of the Empirical Equation

In this section, we investigate the qualitative and quantitative evolutions of the IEZ shape, maximum width, and height to optimize the empirical equation within drop shape theory, taking into account various void ratios.

4.2.2.1 The IEZ's Shapes Under Different Conditions of Initial Void Ratio

The shapes of the IEZ were examined at various heights by documenting the identification numbers of particles along with their spatial configurations following gravity settling. As depicted in Fig. 4.16, the IEZ shapes were derived from nine distinct numerical draw tests, specifically showcasing examples at heights of 30 and 60 m. From a qualitative perspective, it was observed that at a height of 30 m, the IEZ shapes were largely unaffected by the initial void ratios and closely resembled UDDSs. Notably, the maximum width of the IEZ remained stable across these tests. In contrast, the shapes observed at a height of 60 m displayed a more dynamic transformation. Here, a gradual transition was noted, starting from a DS for tests 1–3, evolving into approximately ellipsoidal forms for tests 4 and 5, and ultimately leading to UDDSs in tests 6–9 as the initial void ratios increased. Additionally, there was a significant increase in the maximum width of the IEZ at this height. It is crucial to highlight that the initial average void ratios of the packing particle system were computed using only the particles located within the confines of the white polyline box illustrated in Fig. 4.15b. Particles located within a distance equivalent to five times the average particle size (5 $Da = 3$ m) from both the surrounding and bottom boundaries of the draw model were deliberately excluded from consideration. This approach ensured that the statistical analyses related to the structural and mechanical indicators derived from the nine numerical draw tests focused exclusively on the caved ore and rock contained within the designated area outlined by the white polyline box in

Fig. 4.16. The subsequent statistical examination of various structural and mechanical properties across different tests centered on differing numerical draw scenarios, concentrating solely on the caved ore and rock encapsulated within the specific white polyline box indicated in Fig. 4.15b.

From a quantitative perspective, the void ratios at various heights were assessed through numerical tests identified as 1, 5, and 9, as illustrated in Fig. 4.17. In each of these tests, a consistent trend was observed: the void ratios increased with ascending height. This pattern suggests that the local packing structure of the caved ore and the rock density improved as the spatial location approached the drawpoint. Notably, the increase in void ratio was more pronounced in Test 1 compared to Tests 5 and 9. Specifically, the maximum increase recorded was approximately 14%, indicating a significant variation in the density of the packing system in relation to spatial height.

The relationship between various void ratios and the maximum heights and radii of the IEZs was established by analyzing the IEZ heights of 30 m and 60 m in

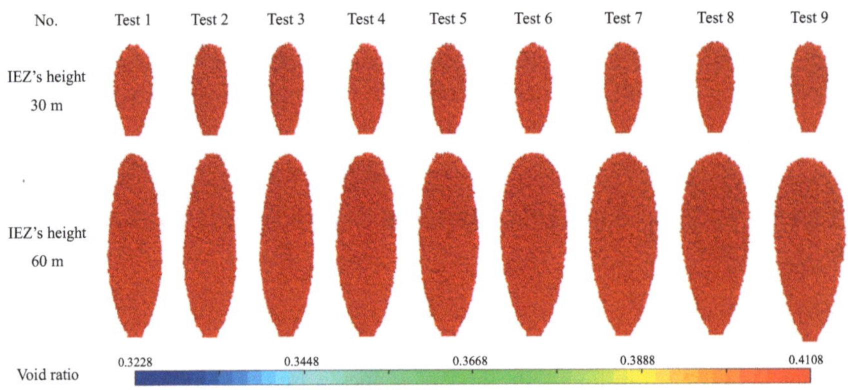

Fig. 4.16 The IEZ's shapes at different draw heights in 9 numerical draw tests

Fig. 4.17 Void ratios at different heights in numerical draw Tests 1, 5 and 9

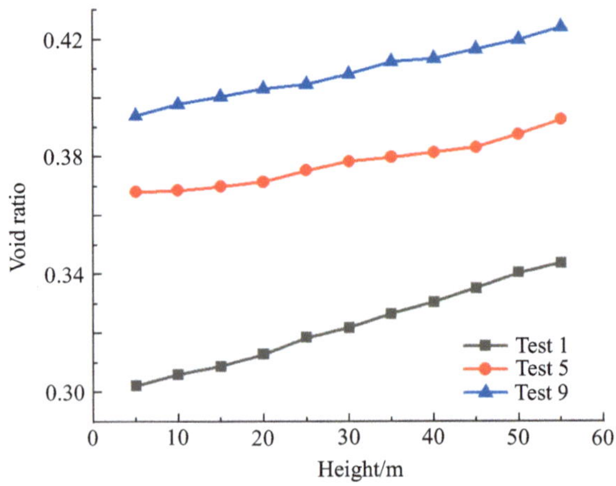

nine numerical draw tests, as illustrated in Fig. 4.18. In Fig. 4.18a, it is evident that for an IEZ height of 30 m, the maximum radii of the IEZs across Tests 1 to 9 exhibited no significant variations with increasing void ratios. Specifically, in Tests 1–3 (represented by the black polyline within the light-blue region in Fig. 4.18a), the maximum radii of the IEZs displayed a minor decrease as the void ratio decreased, with the largest reduction being approximately 0.785 m, equivalent to about 1.3 *Da*, which remains less than the maximum diameter D_{max} of 0.8 m. Conversely, for tests 4 to 9 (illustrated by the black polylines within the light-yellow region in Fig. 4.18a), the maximum radii of the IEZs exhibited a slight increase in conjunction with the rise in void ratios, culminating in a maximum increase of approximately 0.711 m, corresponding to about 1.2 *Da*, and still below D_{max}. In the case of an IEZ height of 60 m, there was no notable alteration in the maximum radius of the IEZs for Tests 1–3, as depicted by the red polyline within the light blue area in Fig. 4.18a. However, an analysis of Tests 4–9 revealed a significant trend in the maximum radii of the IEZs, illustrated by the red polyline within the light-yellow region in Fig. 4.18a. The data indicated a substantial increase in the maximum radii correlated with rising void ratios, culminating in a maximum increase of 3.234 m, approximately equivalent to 5.4 *Da*. This observation suggests a direct relationship between the void ratio and the spatial dimensions of the IEZs. In contrast, the results from Tests 1–3, where the IEZ height remained at 60 m, exhibited no substantial changes in the maximum radii of the IEZs, as represented by the red polyline within the light blue area in Fig. 4.18b. All measurements remained below the midpoint height of 30 m, with a minimum difference recorded at 2.264 m, approximately 3.8 *Da*. This consistency supports the qualitative analysis presented in Fig. 4.16, which concludes that the IEZs generated at a height of 60 m in tests 1–3 can be categorized as DSs. In the case of tests 4–5, the maximum radii of the IEZs, depicted by the red polyline in the pink area of Fig. 4.18b, ranged between 30 and 31 m. The maximum deviation from the midpoint height recorded was 0.784 m, approximately equal to 1.3 *Da*, which is less than the maximum diameter noted. Consequently, the analysis indicates that the IEZs at this height in tests 4 and 5 can be interpreted as having nearly ellipsoidal shapes. The findings from Tests 6–9 revealed a further increase in the maximum radii of the IEZs, as represented by the red polyline in the light-yellow region of Fig. 4.18b. As the void ratio continued to rise, the measurements exceeded 32 m. The minimum difference compared to the midpoint height reached 2.527 m, approximately 4.2 *Da*, leading to the conclusion that the IEZ shapes at the height of 60 m in these tests can be classified as UDDSs. This classification reflects the adjustments in shape influenced by variations in the void ratio.

In conclusion, during the initial draw phase, or for particles located closer to the drawpoint, it was determined that the drawpoint itself, rather than the void ratio, was the primary factor influencing the configuration of the IEZ [76]. This observation specifically pertains to particles positioned at elevations below 30 m in the previously discussed numerical draw trials. Consequently, it can be inferred that the void ratio had a negligible influence on the shape of the IEZ at elevations both below and above 30 m, as long as the vertical height of the maximum IEZ radius remained under 30 m. Furthermore, it was noted that when the average void ratio within the

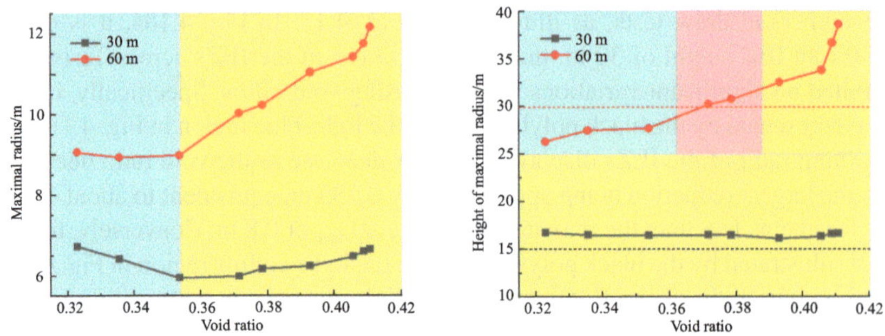

(a) Relationship between void ratio and the IEZ's maximal radius

(b) Relationship between void ratio and height of the IEZ's maximal radius

Fig. 4.18 Relationship between different void ratios and the maximal radii and their heights of the IEZs in 9 numerical draw tests

drawing model fell below approximately 0.36, while keeping the drawpoint constant, there were no significant alterations in the shape of the IEZ. As the height of the IEZ increased, the influence of the drawpoint progressively diminished, with the void ratio ultimately becoming the dominant factor affecting the geometry of the IEZ for particles positioned farther from the drawpoint, specifically those located at heights exceeding 30 m in the aforementioned numerical draw tests. Additionally, it was observed that an increase in the void ratio led to expansions in both the maximum width and height of the IEZ, indicating a rise in longitudinal asymmetry in its shape. A comprehensive examination of these results from a mesoscopic structural-mechanical perspective is provided in Sect. 4.2.3.

4.2.2.2 Quantitative Relationship of Void Ratio, IEZ's Width and Height

Initially, Tests 1, 5, and 9 were utilized as case studies to investigate the correlation between the heights and maximum radii of the IEZs, as depicted in Fig. 4.19. The analysis demonstrated that the relationship between the heights and maximum radii of the IEZs across these three tests adhered to the power function relationship established in the UDDS theory [70], as expressed in Eq. (4.24). Moreover, it is noteworthy that the R^2 values, which represent the correlation coefficients for each test, surpassed the threshold of 0.970, indicating a robust correlation between the analyzed variables:

$$r_{\max} = f_1 * H^{f_2} \tag{4.24}$$

Here, $r_{\max}$ is the maximum radius of the IEZ, H is the IEZ height, and f_1 and f_2 are the fitting constants.

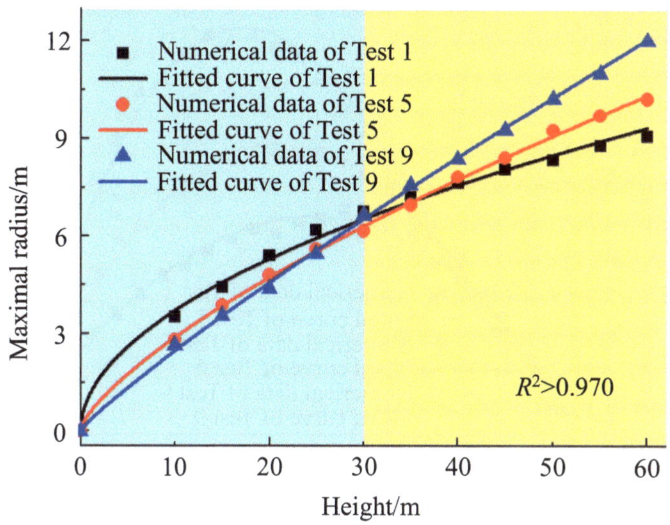

Fig. 4.19 Relationship between heights and maximal radii of the IEZs in numerical draw tests 1, 5 and 9

As illustrated in Fig. 4.19, the maximum width of the IEZ did not exhibit statistically significant variation among the three tests when the IEZ height was below 30 m. Test 1, which demonstrated the lowest void ratio, achieved the greatest maximum width of the IEZ, followed by Test 5 and Test 9 in that order. The maximum width difference among the three tests was 0.710 m, approximately equal to 1.2 $Da < D_{max}$. Conversely, for IEZ heights exceeding 30 m, the variances in the maximum widths of the IEZs progressively increased in conjunction with the void ratio. In this scenario, Test 1, with the lowest void ratio, recorded the minimal maximum width of the IEZ, with Test 5 and Test 9 following. The most significant width difference among the three tests reached 2.951 m, approximately 5.0 *Da*.

Given that the relationship between the height and maximum radius of the IEZ adheres to the power function described in UDDS theory, it is essential to investigate whether the alterations in the shape of the IEZ, influenced by varying void ratios, correspond with the empirical equation for the IEZ shape [70] (Eq. (4.25)) as delineated in UDDS theory. The equation is as follows:

$$y^2 = a * z^b * \ln\left(\frac{H}{z}\right) \tag{4.25}$$

Here. y is the radius of the IEZ at different vertical heights, H is the IEZ height, z is the vertical height, and a and b are the fitting constants.

In Tests 1, 5, and 9, the study analyzed the relationship between vertical height and the IEZ radius at a constant height of $H = 60$ m, as illustrated in Fig. 4.20. The results from these tests exhibited a consistent trend in the changes to the shape of the

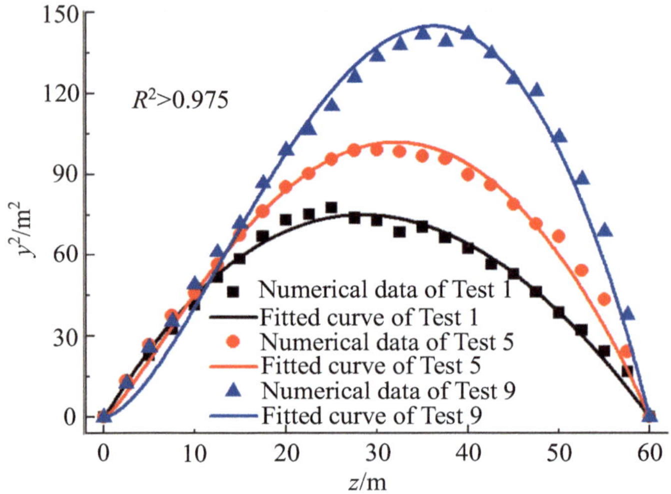

Fig. 4.20 Relationship between vertical heights and IEZs' radii in numerical draw Tests 1, 5 and 9

IEZ, aligning with the relationship defined by the empirical equation presented in Eq. (4.25). Notably, the correlation coefficient R^2 values for each test exceeded 0.975, indicating a robust correlation. This finding underscores that the empirical equation governing the shape of the IEZ is applicable not only to UDDSs but also to DSs and configurations approximating ellipsoidal shapes. Furthermore, the transformations in the shape of the IEZ are significantly influenced by variations in void ratios within the system. This influence can be quantitatively characterized by the fitted coefficients a and b, which provide a more detailed understanding of how different void ratios impact the IEZ shape. The results imply a complex relationship between these parameters, highlighting the need for further investigation into their interactions to enhance our comprehension of the dynamics at play in these systems.

Further research is needed to determine the quantitative relationship involving the void ratio v and the fitted coefficients a and b. Based on nine tests, Fig. 4.21a presents a scatter plot that demonstrates the connection between the void ratio and the fitted coefficients a and b for an IEZ height of 60 m. The light blue region in Fig. 4.21a shows that small void ratios (below 0.354, Tests 1–3) have a negligible effect on the IEZ configuration, which is consistent with the results depicted in Fig. 4.18. In contrast, the light-yellow area in Fig. 4.21a indicates that higher void ratios (exceeding 0.354, tests 4–9) lead to a reduction in coefficient a and an increase in coefficient b as the void ratio rises. A linear fitting graph in Fig. 4.21b illustrates the relationship between the void ratio and the fitted coefficients a and b for Tests 4–9, with correlation coefficient R^2 values exceeding 0.994. By incorporating the linear equation that depicts the relationship between the void ratio v and the fitted coefficients a and b from Fig. 4.21b into the empirical equation representing the IEZ structure (Eq. (4.25)), the following relationship can be established:

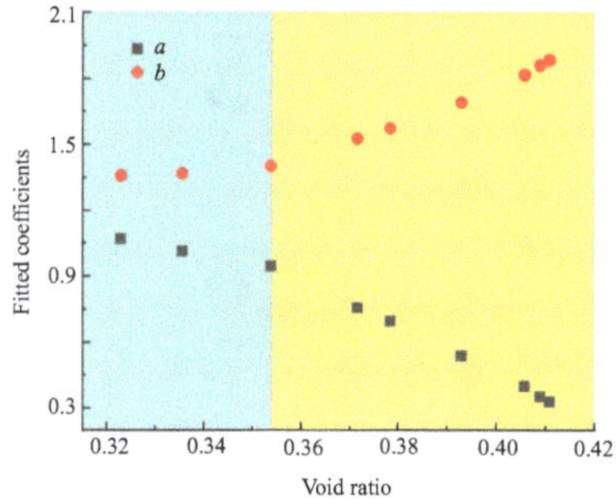

(a) Scatter graph of simulated data

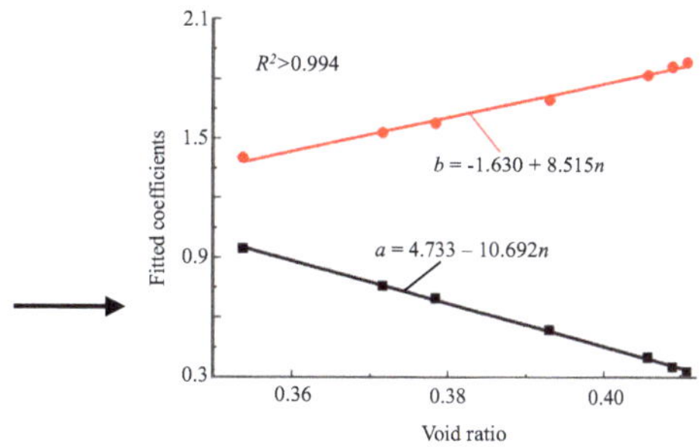

(b) Linear fitting graph of simulated data

Fig. 4.21 Relationship between void ratio and fitted coefficients a and b in 9 numerical draw tests

$$y^2 = (4.733 - 10.692n) * z^{-1.630+8.515n} * \ln\left(\frac{60}{z}\right) \tag{4.26}$$

To assess the accuracy of Eq. (4.26), the actual simulated void ratio data and IEZ radii at various vertical heights were compared with the values derived from Eq. (4.26). Specifically, an analysis was conducted on the IEZ radii at vertical heights of 40 m and 50 m. Figure 4.22 illustrates the simulated IEZ radii data for different void ratios, represented by black and red scatter points. Meanwhile, the light blue and light-yellow regions indicate the IEZ radii calculated by Eq. (4.26) within a 5% error margin. From Fig. 4.22, it is evident that the simulated IEZ radii data for both 40 m and 50 m vertical heights fell within the specified ranges of the computed results, thereby confirming the validity of the empirical equation for the IEZ shape (Eq. (4.26)). Similar findings were observed for additional vertical heights, including 35, 45, and 55 m; however, these results will not be elaborated upon in this context.

According to the findings presented in Sects. 4.2.2.1 and 4.2.2.2, it is possible to expand the UDDS theory [70] to encompass the DS theory. In particular, if the void ratio v is below a designated threshold v_c (which is 0.354 in this study), the maximum width of the IEZ can be found nearer to the drawpoint, with a vertical distance of less than 30 m as noted in this paper. In this case, it is the drawpoint that primarily influences the configuration of the IEZ, leading to the formation of a DS. Conversely, when the void ratio v within the draw model surpasses v_c, the maximum width of the IEZ is observed to be considerably distant from the drawpoint. Consequently, the void ratio increasingly becomes the predominant factor that determines the shape of the IEZ, causing it to transition from an elliptical shape to that of a UDDS. In conclusion, the trend of the IEZ shape transformation can be extrapolated from Eq. (4.25) to Eq. (4.27) as follows:

$$\begin{cases} y^2 = a * z^b * \ln\left(\frac{H}{z}\right), & v \le v_c \\ y^2 = (k_1 + k_2 * v) * z^{(k_3 + k_4 * v)} * \ln\left(\frac{H}{z}\right), & v > v_c \end{cases} \tag{4.27}$$

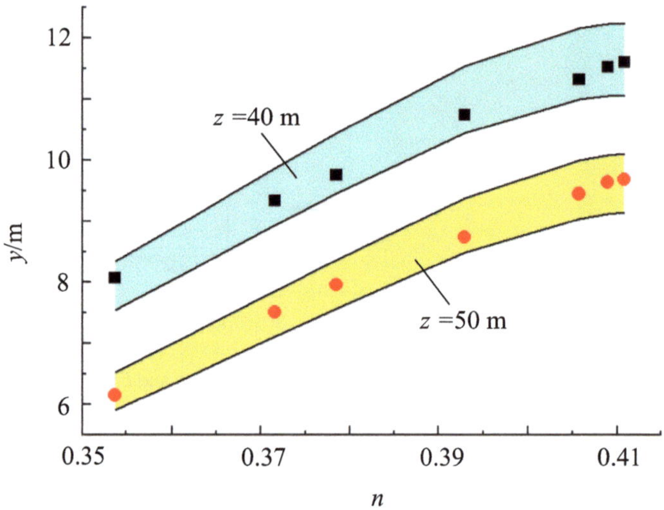

Fig. 4.22 Relationship between void ratio and IEZ's radius at 40 m and 50 m vertical heights

Here, k_1–k_4 are fitted coefficients.

4.2.3 Structural-Mechanical Characteristics Analysis

To investigate the meso-mechanism by which the void ratio influences the changing trend of the IEZ shape, this study analyzes the structural-mechanical characteristics of static packing (Stage 1) and gravity flow (Stage 2) within caved ore and rock systems.

4.2.3.1 Stage 1: Static Packing of Caved Ore and Rock

Utilizing the open-source library Voro++ [77], the particles closest to all points within the interior of the caved ore and rock packing system were identified through radical Voronoi tessellation [78]. This segmentation into distinct cells facilitated a quantitative analysis of the local structure of the system. Subsequently, the structural-mechanical attributes were analyzed using the Minkowski tensor, bond orientation order, coordination number, and stress.

Initially, the Minkowski tensor [79] serves to describe the morphology of the Voronoi cells [34] created within the static packing system of caved ore and rock. For a given cell K, the associated Minkowski tensor $\mathbf{W}_1{}^{0,2}$ is defined as follows:

$$W_1^{0,2}(K) = \frac{1}{3}\int_{\partial K} \mathrm{n} \otimes \mathrm{n}dA \tag{4.28}$$

Here, **n** is the external normal of the surface ∂K of the cell, $\otimes$ represents the direct product of the vectors, and A is the cell surface area.

Based on the Minkowski tensor $\mathbf{W}_1{}^{0,2}$, the degree of the anisotropy of a Voronoi cell can be defined as follows:

$$\beta = \frac{\eta_{\min}}{\eta_{\max}} \tag{4.29}$$

Here, $\eta_{\max}$ and $\eta_{\min}$ are the maximum and minimum eigenvalues of $\mathbf{W}_1{}^{0,2}$, respectively. Among them, $\beta = 1$ corresponds to an isotropic cell, and the smaller the value of β, the greater the anisotropy of the local packing system structure.

The analysis of the caved ore and rock packing systems during Tests 1, 5, and 9 yielded a probability distribution for β that closely adhered to a Gaussian distribution pattern. As illustrated in Fig. 4.23a, the correlation coefficients (R^2 values) from these tests exceeded 0.990, signifying a robust fit to the Gaussian model. The peak values of β recorded in Tests 1, 5, and 9 were 0.59, 0.63, and 0.65, respectively. This progression of peak values suggests that a denser configuration of caved ore and rock packing corresponds to an increased degree of anisotropy within the structure of the packing system. Additionally, the probability distribution pertaining to the coordination number of particles serves as a crucial mechanical characteristic indicator for elucidating the structural anisotropy prevalent in a particle packing system. In Fig. 4.23b, the distributions of average coordination numbers for Tests 1, 5, and 9 are depicted. The results again demonstrated a Gaussian distribution, with correlation coefficients R^2 exceeding 0.984. The peak average coordination numbers observed were approximately 7.0, 5.5, and 5.0 for Tests 1, 5, and 9, respectively. Furthermore, the range of values within the probability distribution for the average coordination number was most extensive in test 1; this was followed by test 5, with test 9 exhibiting the narrowest range. The findings suggest that a more compact structure within the caved ore and rock leads to enhanced anisotropy in the way particle coordination numbers are distributed. Focusing on the typical local packing system measuring $6 \times 6 \times 6$ m^3 (10 $Da \times 10\ Da \times 10\ Da$), which was assessed in Tests 1, 5, and 9, calculations were made for the standard deviation α of the probability distribution concerning the average coordination number, as well as the β value for each particle and its mean value. The outcomes illustrated in Fig. 4.24 reveal that with an increase in the void ratio, there is a corresponding decline in the ratio of blue and green particles within the local packing system of the caved ore and rock. Conversely, the proportion of red and yellow particles shows a gradual rise. Furthermore, the mean value displays an upward trend, while the standard deviation α exhibits a downward trend. According to the analysis depicted in Figs 4.23 to 4.24, there is a noticeable positive correlation between the packing system's structural anisotropy of the caved ore and rock and the anisotropy of the particle coordination number distribution. This suggests that an increase in the void ratio may lead the static packing system of the caved ore and rock to become more isotropic.

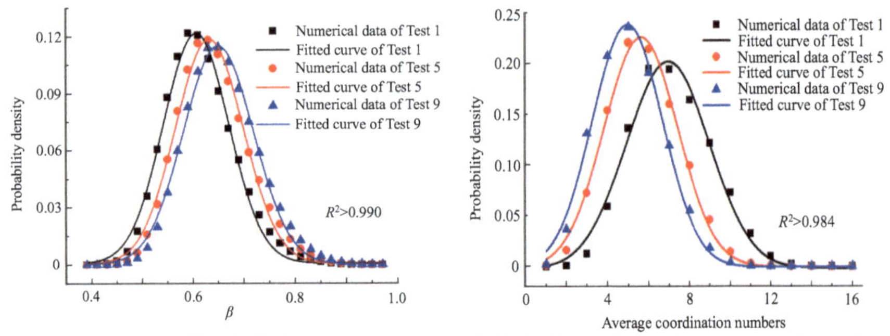

(a) Probability distribution of β (b) Probability distribution of average coordination numbers

Fig. 4.23 Probability distributions of β and average coordination numbers in numerical draw tests 1, 5 and 9

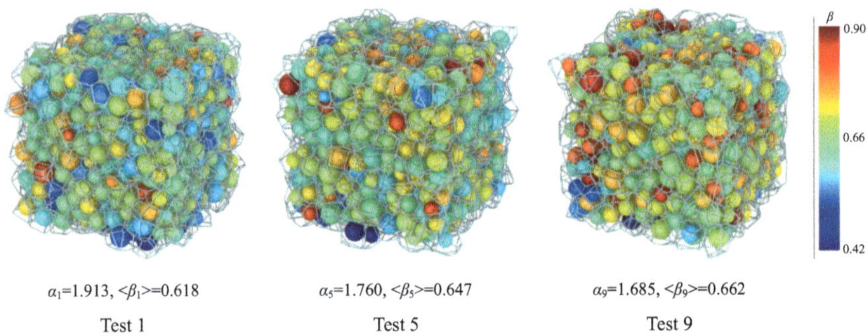

Fig. 4.24 α and <β> of typical local packing system of caved ore and rock in numerical draw tests 1, 5 and 9

The correlation between the structural anisotropy of the caved ore and the rock packing system, as well as the anisotropy of the particle coordination number distribution, was examined using the Pearson correlation coefficient, as described in Eq. (4.30):

$$\rho(X, Y) = \frac{\text{cov}(X, Y)}{\sigma_X \sigma_Y} \tag{4.30}$$

Here, ρ represents the Pearson correlation coefficient, which has a variation range of [−1, 1], where −1 indicates an entirely negative linear correlation, 0 means no linear correlation, and +1 indicates an entirely positive linear correlation. Furthermore, X and Y represent two variables.

The average value <β> and the standard deviation α were computed and normalized for Tests 1–9. To assess the correlations, the normalized values of α and <β> were compared, as shown in Fig. 4.25. A Pearson correlation coefficient ρ of −0.992 was obtained, which signifies a strong negative linear correlation between α

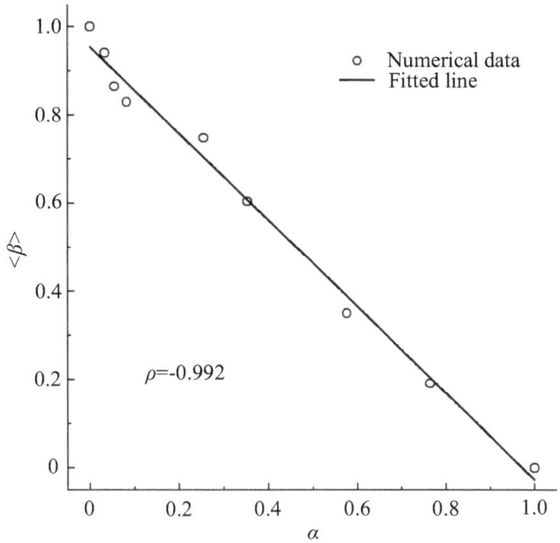

Fig. 4.25 Correlation comparisons between normalized α and $<\beta>$ in numerical draw Tests 1–9

and $<\beta>$. This suggests a significant positive linear correlation between the structural anisotropy of the caved ore and the rock packing system, as well as the anisotropy of the distribution of particle coordination numbers.

Bond orientational order is a well-established measure for assessing local ordering within particle systems. It quantitatively evaluates the regularity and symmetry of particle arrangements at a short-range level [80]. It is defined as follows:

$$q_l(i) = \sqrt{\frac{4\pi}{2l+1}\sum\nolimits_{m=-l}^{l}\left|\frac{1}{n}\sum\nolimits_{j=1}^{n} Y_{lm}(\chi_j, \psi_j)\right|^2} \tag{4.31}$$

Here, j represents one of the n particles adjacent to particle i and $Y_{lm}(\chi_j, \psi_j)$ are the spherical harmonics, where χ_j and ψ_j are the polar and azimuthal angles, respectively, formed by particle j. Different l values represent various forms of rotational symmetry and can capture different ordered forms. Among them, $l = 6$ represents the dominant ordered form of spherical or nearly spherical particles. Furthermore, Q_6 represents the average value of the bond orientational order parameter q_6. The formula $Q_6 = \langle q_6(i)\rangle$ can be used to characterize the order degree of the overall or local structure of the caved ore and rock packing system. Generally speaking, the larger the Q_6 value, the greater the degree of order in the particle packing system.

The average bond orientational order, denoted as Q_6, was assessed at various heights by analyzing the caved ore and rock packing systems during Tests 1, 5, and 9. The results, illustrated in Fig. 4.26a, reveal a consistent downward trend in Q_6 values as the vertical height increases. Specifically, Test 1 recorded the highest Q_6 value, ranging from 0.365 to 0.369, indicating a relatively orderly structure at this height. This was followed by test 5, which exhibited a slightly lower range of Q_6

values between 0.344 and 0.349. In contrast, Test 9 demonstrated the lowest Q_6 values, with a range from 0.332 to 0.338. These observations imply that greater density in the caved ore and rock packing system corresponds with an enhanced degree of structural order. Furthermore, Fig. 4.26b, c provide insights into the average coordination numbers and stresses recorded at different vertical heights across Tests 1, 5, and 9. The data indicate a decline in both average coordination numbers and stress values as vertical height increases in all conducted tests. Among the three, test 1 showcased the highest average coordination numbers, which varied from 6.675 to 7.519, along with stress values that ranged from 0.395 to 1.217. On the other hand, Test 5 displayed lower average coordination numbers, sitting between 5.473 and 6.014, with stresses ranging from 0.323 to 0.753. Finally, the results from Test 9 indicated the lowest average coordination numbers, ranging from 4.828 to 5.315, with corresponding stress values between 0.264 and 0.609. A significant finding from this analysis is the strong positive correlation observed between the structural order of the caved ore and the rock packing system, as well as the mechanical indicators, which include both the average coordination number and stress. Furthermore, it appears that the sensitivity of average bond orientational orders, coordination numbers, and stress values to variations in vertical height and void ratios exhibits different levels of responsiveness. In conclusion, these findings highlight the substantial influence of structural order within the caved ore and rock packing system on the mechanical behavior of the materials involved.

To examine and contrast the stress distribution patterns within the caved ore and rock packing system at various void ratios, the normalized stress contour illustrations for packing systems in Tests 1, 5, and 9 are presented in Fig. 4.27. The average stress measured on the central vertical section of the draw model (as depicted in Fig. 4.15b) serves as a case study. An increase in vertical height correlated with a decline in average stress values across the three tests, with the highest average stress recorded at the base of the draw model, consistent with the trend shown in Fig. 4.26c. Furthermore, Fig. 4.27 provides data on the area proportions situated between two neighboring normalized stress contours. Only the area proportions of the six middle regions were analyzed, while the proportions of the upper and lower areas were omitted. As illustrated in Fig. 4.28a, a denser packing system of caved ore and rock resulted in a more gradual variation in average stress along the vertical axis, while this relationship was inverted in less dense configurations. Additionally, Fig. 4.27 includes statistics regarding the standard deviation of height z for the same normalized stress contour at varying horizontal positions x, focusing exclusively on the five central normalized stress contours and excluding the topmost and bottommost contours. Figure 4.28b demonstrates that a denser configuration of the caved ore and rock packing system produced less fluctuation in average stress along the horizontal direction, with this trend being reversed for less dense systems.

Fig. 4.26 Average bond orientational order Q_6, coordination numbers and stress at different heights in numerical draw Tests 1, 5 and 9

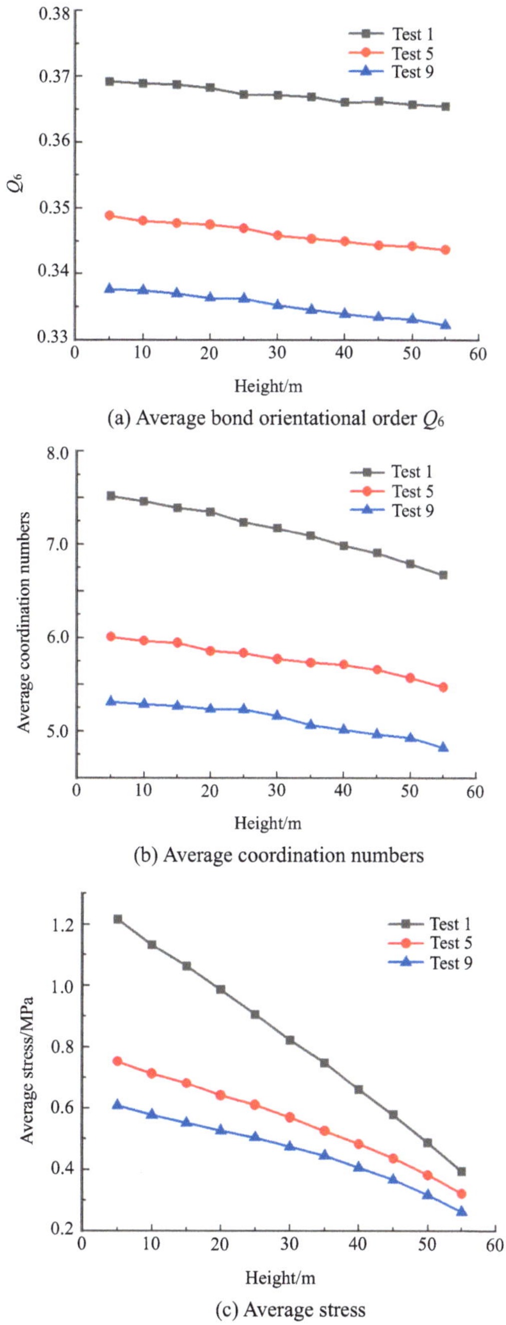

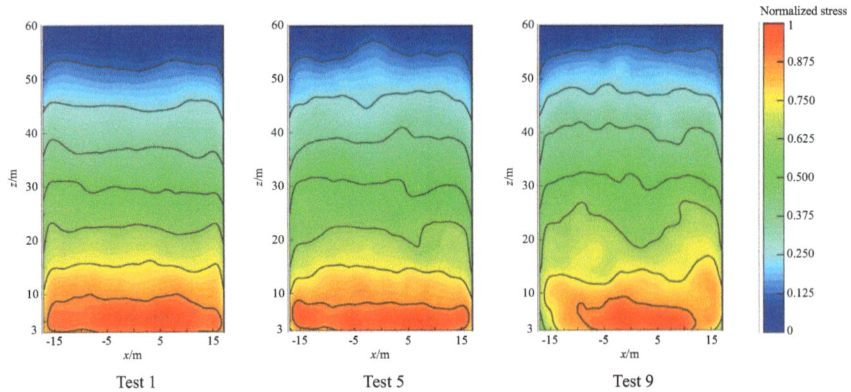

Fig. 4.27 Normalized stress contour plots of static packing system of caved ore and rock in numerical draw Tests 1, 5 and 9

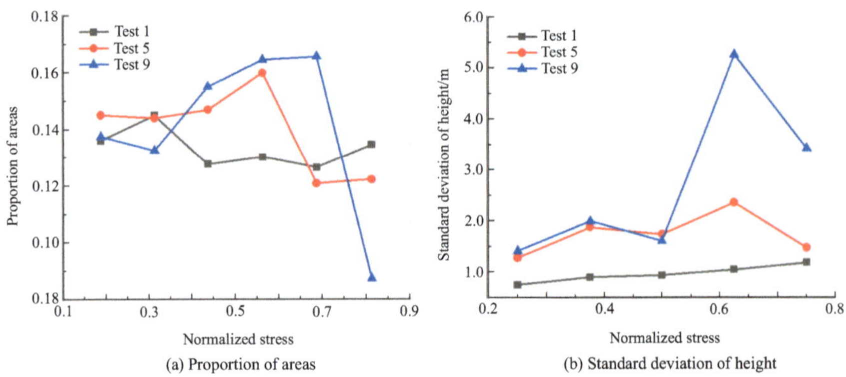

Fig. 4.28 Proportion of areas and standard deviation of height of normalized stress contour in numerical draw Tests 1, 5 and 9

4.2.3.2 Stage 2: Gravity Flow of Caved Ore and Rock

The analysis in Sect. 4.2.3.1 highlights the impact of various void ratios on the structural-mechanical characteristics of the caved ore and rock packing system (Stage 1), which subsequently affects the gravity flow of the system (Stage 2). The distribution and evolution of stress significantly influence the migration and evolution of caved ore and rock. Therefore, it is essential to examine the mechanical mechanisms underlying the changing trends in the shapes of the IMZ and IEZ under different void ratios.

The IMZ is identified as the precursor to the IEZ in mining operations. Analyzing the average stresses within the central vertical section of the draw model, specifically at IMZ heights of 25 and 50 m, provides valuable insights into the stress distribution and flow behavior of the mining system. As illustrated in Fig. 4.29, the normalized stress contour plots for the flow of caved ore and rock particles from Tests 1, 5, and 9 reveal significant findings. Notably, the presence of an isolated stress contour zone (ISCZ) exhibiting uniform normalized stress contours mirrors the shapes of both the IMZ and the IEZ while remaining in proximity to a DS. The analysis indicates that the average stress diminishes progressively from the outer boundaries toward the internal regions of the IMZ. This reduction in stress is attributed to the more loosely structured nature of the caved ore and rock system within the IMZ compared to the surrounding areas. Moreover, the phenomenon of interlocking arching occurring above the IMZ further complicates the stress distribution; it facilitates the transfer of stress to the surrounding regions, consequently leading to higher stress levels outside and adjacent to the IMZ than within it. The examination of stress arch spans at IMZ heights of 25 and 50 m, as shown in Fig. 4.30, highlights noteworthy variations. At the height of 25 m, the differences in the stress arch span across the three tests are minimal, with the maximum discrepancy recorded at approximately 0.807 m, which is roughly equivalent to 1.3 times D_a or D_{max}. In contrast, when the IMZ height is increased to 50 m, a significant disparity in the stress arch span emerges among the tests, revealing a maximum difference of approximately 4.149 m, equivalent to about 6.9 *Da*. This finding indicates that the height of the IMZ has a substantial impact on the stress distribution and the behavior of the stress arches within the system. The findings suggest that a denser system of caved ore and rock leads to the formation of a narrower stress arch span. The analysis underscored a distinct correlation and consistency between the variations in the stress arch span and the alterations in the shapes of the IMZ and the IEZ, as elaborated in Sect. 4.2.2. More specifically, when the height of the IMZ was relatively low—illustrated by the example of 25 m depicted in Fig. 4.29—there was minimal variation observed in the stress arch span positioned above the IMZ, irrespective of differing void ratios. This observation indicates that the maximum width of the IMZ remained relatively unchanged under these conditions. Conversely, when examining a scenario with a comparatively higher IMZ height, such as 50 m as shown in Fig. 4.29, the data revealed that the span of the stress arch above the IMZ increased as the void ratio rose. This trend signifies a clear relationship whereby an increase in the void ratio directly correlates with a greater maximum width of the IMZ. The processes surrounding the formation, reinforcement, and eventual collapse of the stress arch situated above the IMZ contribute to the loosening of the surrounding caved ore and rock within the relevant area, ultimately facilitating the gradual evolution of the IMZ. Consequently, the ISCZ can be regarded as a precursor to the development of the IMZ.

4.3 Conclusions

The examination of the changing trends in the shapes of the EZ and MZ serves as a fundamental aspect for understanding the dynamics of caved ore and rock flow. This understanding is crucial, as it provides essential insights for optimizing the structural parameters of stopes. One of the established theoretical frameworks employed to

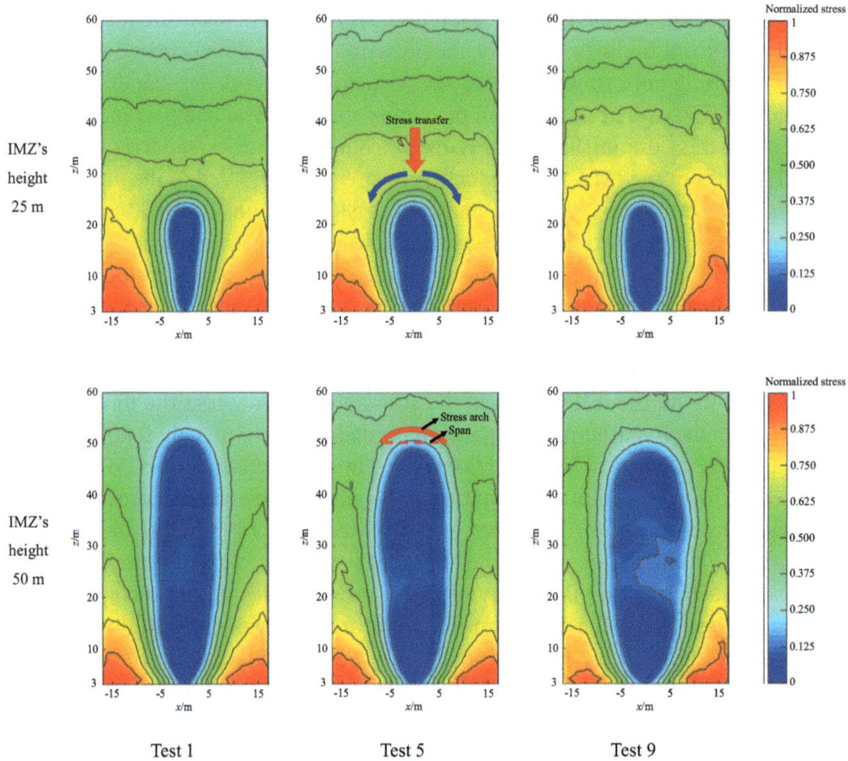

Fig. 4.29 Normalized stress contour plots of particle flow system of caved ore and rock in numerical draw Tests 1, 5 and 9

Fig. 4.30 Spans of stress arch at heights of 25 m and 50 m in numerical draw Tests 1, 5 and 9

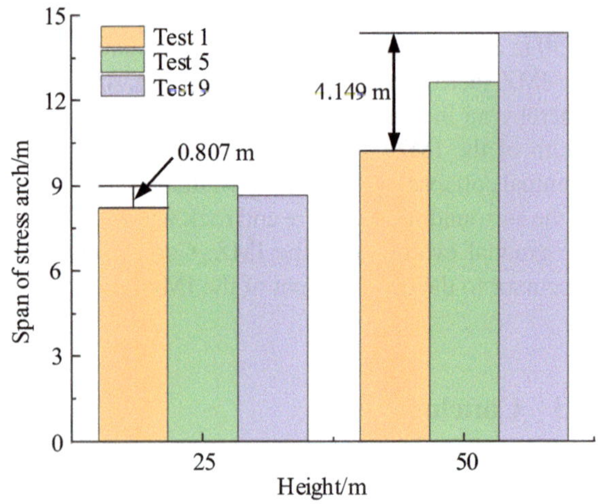

analyze the shapes of the EZ and MZ is the ellipsoid draw theory. Historically, this theory has been significant in addressing various draw-related issues associated with the caving mining method. Throughout this study, the results of isolated draw experiments corroborated previous observations regarding the geometry of the EZ, revealing that it exhibits marked longitudinal asymmetry rather than conforming to the characteristics of a standard ellipsoid. Building upon a thorough quantitative analysis of the experimental data, this paper introduces a novel approach to investigate the shapes of both the EZ and MZ, as well as to explore their evolving trends over time. This innovative method culminates in the derivation of new empirical equations that facilitate the calculation of these shapes, thereby enhancing the accuracy and reliability of the analyses performed in the context of mining operations.

In 1995, Nedderman [31] applied a modified kinematic model to forecast the evolving shape of the flowing zone during the batch discharge from a cylindrical bunker, successfully validating his predictions against Watson's model experiments. The equation describing the non-ellipsoidal shape of the MZ, as derived by Nedderman, bears some resemblance to the empirical equation for EZ/MZ introduced in this paper, which effectively substantiates the validity and precision of both the empirical equation and the theory regarding the inverted drop shape. It is crucial to highlight the distinctions between the two equations mentioned earlier. Firstly, the methodologies employed to derive these equations significantly differ: the equation presented in this paper stems from experimental data analysis coupled with straightforward mathematical derivation, whereas Nedderman's equation arises from a modified kinematic model based on an assumed differential equation governing velocity distribution. Secondly, the constants in the equation within this paper, which represent the impact of various factors on the EZ shape, are easily attainable through a basic isolated draw test, making them comprehensible and acceptable, particularly for mining companies that may lack a robust theoretical understanding of draw. In contrast, Nedderman's equation incorporates constants related solely to density, variations in density, and particle size. Numerous physical and numerical draw experiments conducted by Power [9], Castro [10], Pierce [27], Hancock [2], and other researchers have demonstrated that, in addition to particle size and density, several other significant factors are not currently considered in Nedderman's theories [31].

The main conclusions of this part are as follows:

(1) Based on the analysis of variance and the examination of the changing trends in the shapes of ellipsoids and inverted droplets, it was confirmed that the droplet shape closely corresponds with the actual configurations of the EZ and MZ. Consequently, the theory regarding the inverted droplet shape is considered to be more effective than that of the ellipsoid model.
(2) The relationship between the mass drawn and the height of the EZ or MZ follows a power function, while the mass and radius of the EZ or MZ demonstrate an index function relationship. Furthermore, the height and maximum width of the EZ or MZ are correlated by a power function.
(3) Based on the findings related to the drop shape of the EZ and MZ, a novel approach to represent the results with equations and diagrams has been

suggested: Initially, employ labeled tracers to quantify the mass drawn m at various heights H of the EZ; subsequently, conduct a fitting analysis between m and H to derive the constants k_1 and k_2; ultimately, these constants can be incorporated into the resulting empirical equation to ascertain the geometry of the EZ or MZ. This method, in contrast to traditional techniques, is both precise and efficient, demonstrating marked advantages that can significantly reduce experimental efforts and the processing of data.

The newly developed empirical equations, grounded in the theory of upside-down drop shape, represent a significant advancement over the traditional ellipsoid draw theory. However, these equations have inherent limitations. To enhance their utility and effectiveness in studying the flow mechanisms associated with caved ore and rock, further optimization and refinement are essential. This optimization should be based on a robust framework of empirical data obtained from extensive physical and numerical testing.

In this research, we conducted numerical experiments to examine isolated draw under varying conditions of bulk density, utilizing the rolling resistance model as a foundational approach. Through this investigation, we established a quantitative relationship between the initial void ratio present in the particle system of caved ore and rock and the corresponding shape variations observed in the draw zone. This relationship refines the shape characterization equation within the context of drop shape theory. Building on these findings, we applied statistical mechanics methods to analyze the structural and mechanical characteristics of the static packing system of caved ore and rock prior to discharge, as well as the dynamic flow system during the draw process. This analysis explores the mechanisms of movement and evolution of caved ore and rock, offering insights into how different levels of porosity can affect these interactions from a microscopic perspective.

The main conclusions of this part are as follows:

(1) The void ratio is crucial in determining the structural and mechanical characteristics of particle packing and flow systems. An increase in the void ratio leads to a transition in the configuration of the IEZ, shifting from a DS shape to a roughly ellipsoidal form and subsequently to a UDDS shape. Furthermore, it has been observed that with a rise in the void ratio, both the maximum width and height of the IEZ increase, while the longitudinal asymmetry of the IEZ becomes more pronounced. Additionally, regardless of the shape of the IEZ, its height and maximum radius follow a power function correlation. This research incorporates the void ratio into the existing empirical equation governing the IEZ and proposes an enhanced empirical equation, as presented in Eq. (4.27). The findings suggest that the UDDS theory may be applicable to the DS theory, which has significant implications for mine design.
(2) As the particle packing system of caved ore and rock becomes denser, there is an increase in the degree of order, average coordination number, average stress, and the anisotropic characteristics of both the structure and the distribution of particle coordination numbers. Conversely, an increase in the void ratio leads to

a packing structure of the caved ore and rock that approaches isotropy. Furthermore, a positive linear relationship exists between the anisotropy of the particle packing system and the anisotropic nature of the particle coordination number distribution. Additionally, a denser arrangement of caved ore and rock results in a reduced gradient of average stress variation along the vertical axis and less fluctuation along the horizontal axis.

(3) In the flow system associated with caved ore and rock, the configuration of the ISCZ created by uniform normalized stress contours above the drawpoint exhibited characteristics akin to those of both the IMZ and the IEZ, taking on a shape that can be approximated as a DS. This finding indicates that the stress distribution is not uniform and demonstrates a gradation, with average stress values diminishing as one move from the external environment toward the core of the structure. Throughout the drawing process, a distinct arch of stress and its subsequent transfer were evident, revealing that the stress levels surrounding and external to the IMZ were significantly higher than those found within it. This observation underscores the importance of spatial stress variations in mining operations, as they can influence the integrity and stability of the mined zones. Moreover, a strong correlation and consistency were identified regarding how changes in the spans of stress arches interacted with both the IMZ and IEZ, illustrating the interconnected dynamics that govern these areas. When examining the height of the IMZ, it was found that lower heights yielded minimal variations in the stress arch spans across different void ratios. This suggests that the maximum width of the IMZ remains relatively stable, regardless of the fluctuations in the void ratio under these specific conditions. Conversely, as the IMZ height increased, the span of the stress arch showed a marked increase in correlation with rising void ratios, resulting in a significant enlargement of the IMZ's maximum width. This observation leads us to conclude that the ISCZ may function as an early indicator or precursor to the formation and development of the IMZ, highlighting its potential role in mining operations and stress management strategies.

The study presents several noteworthy findings [85] that warrant consideration:

(1) Figures 4.16 and 4.18 illustrate a significant phenomenon regarding the behavior of caved ore and rock when densely packed, as evidenced by Tests 1–3. It was observed that an increase in the void ratio corresponded with a slight reduction in the maximum width of the IEZ. This observation raises the intriguing possibility that a further decrease in the void ratio could induce a structural transition similar to nucleation or crystallization within the local or overall packing arrangement of the caved ore and rock [81, 82]. Essentially, this implies that the behavior of caved ore and rock may transition from a fluid-like state to one that resembles a solid-like state. To fully comprehend this potential 'phase transition' and its implications, further research is essential. Specifically, it is imperative to investigate how this shift in local or overall structure affects both the shape and maximum width of the IEZ. Such investigations will yield valuable insights

into the dynamics of the particle system and may contribute to advancements in understanding the behavior of caved materials under varying conditions.

(2) The design and operation of caving mines are affected by the dimensions and shape of the IEZ. It is essential to take into account the void ratio threshold, which dictates how the shape of the IEZ transitions from a DS to a roughly ellipsoid form and ultimately to a UDDS. This evaluation encompasses the shape, size distribution, and strength of both the caved rock and the ore. Future research will broaden the range of parameters employed in the correlation analysis, with a focus on integrating appropriate structural parameters for medium- and long-range assessments. This approach aims to enhance the understanding of the mechanical processes that govern gravitational flow.

(3) The shape of the IMZ and IEZ is directly influenced by the stress arching effect associated with gravitational flow. Future analyses will aim to elucidate the evolution, strengthening, and failure processes of the stress arch. Furthermore, techniques from statistical mechanics and machine learning models will be employed to uncover the mechanisms of stress that drive the alterations in the shapes of thc IMZ and IEZ. The findings of this research will serve as a crucial resource for investigating hang-ups in applications related to caving mines [83, 84].

Note. Figures 4.1, 4.2, 4.3, 4.4, 4.5, 4.6, 4.7, 4.8, 4.9, 4.10, 4.11, 4.12, 4.13 and Tables 4.1, 4.2, 4.3, 4.4 in Sect. 4.1 are reprinted from the authors' previously published work [70]:

Jin AB, Sun H, Wu SC, et al. Confirmation of the upside-down drop shape theory in gravity flow and development of a new empirical equation to calculate the shape[J]. International Journal of Rock Mechanics and Mining Sciences, 2017, 92: 91–98.

Figures 4.14, 4.15, 4.16, 4.17, 4.18, 4.19, 4.20, 4.21, 4.22, 4.23, 4.24, 4.25, 4.26, 4.27, 4.28, 4.29, 4.30 and Tables 4.5, 4.6 in Sect. 4.2 are reprinted from the authors' previously published work [85]:

Sun H, Zhao LS, Elmo D, et al. Optimization of an empirical equation in drop shape theory and a structural-mechanical analysis of a caved ore and rock particle flow system[J]. Phys. Fluids, 2025, 37(2): 023,379.

Permission has been obtained from the publisher for the reuse of these materials.

References

1. Qiao DP (2003) Research on particle movement rule and ore drawing theory. Northeastern University, Shenyang
2. Hancock W (2013) Gravity flow of rock in caving mines: numerical modelling of isolated, interactive and non-ideal draw. The University of Queensland, Brisbane
3. Halim AE (2004) 3D large scale physical modelling for studying interactive draw and drawpoint spacing in block caving mines. In: JKMRC international student conference, p 123
4. Kvapil R (1965) Gravity flow of granular material in hoppers and bins. Int J Rock Mech Min Sci 2(1):35–41

5. Kvapil R (1965) Gravity flow of granular material in hoppers and bins in mines—II. Coarse material. Int J Rock Mech Min Sci 2(3):277–292
6. Mccormick RJ (1968) How wide does a drawpoint draw?. Eng Min J 106–116
7. Watson GR (1993) Flow patterns in flat bottomed silos. University of Edinburgh, Edinburgh
8. Laubscher D (1994) Cave mining-the state of the art. J South Afr Inst Min Metall 94:279–293
9. Power G (2004) Modelling granular flow in caving mines: large scale physical models and full-scale experiments. University of Queensland, Brisbane
10. Castro R (2006) Study of the mechanisms of gravity flow for block caving. University of Queensland, Brisbane
11. Castro R, Pineda M (2015) The role of gravity flow in the design and planning of large sublevel stopes. J South Afr Inst Min Metall 115(2):113–118
12. Tao GQ, Yang SJ, Ren FY (2009) Experimental research on granular flow characters of caved ore and rock. Rock Soil Mech 30(10):2950–2954
13. Wang HJ, Ying SH, Wu AX et al (2010) Experimental study of the factors affecting the ore flow mechanism during block caving. J China Univ Min Technol 39(5):693–698
14. Marano G (1980) The interaction between adjacent draw points in free flowing materials and its application to mining. Chamber Mines J 22:25–32
15. Wang CH (1982) Draw. Metallurgical Industry Press, Beijing
16. Kvapil R, Baeza L, Rosenthal J et al (1989) Block caving at El Teniente mine, Chile. Trans Inst Min Metall Min Technol 98:43
17. Litwiniszyn J (1956) Application of the equation of stochastic processes to mechanics of loose bodies. Arch Appl Mech 8(4):393–411
18. Ren FY (1994) Stochastic medium draw theory and its applications. Metallurgical Industry Press, Beijing
19. Chen G (1997) Stochastic modelling of rock fragment flow under gravity. Int J Rock Mech Min Sci 34(4):323–331
20. Brown ET (2003) Block caving geomechanics. Julius Kruttschnitt Mineral Research Centre, University of Queensland, Brisbane
21. Verdugo R, Ubilla J (2004) Geotechnical analysis of gravity flow during block caving. In: Proceedings of MassMin 2004, Santiago, pp 195–200
22. Hancock WR, Weatherley DK (2008) 3D simulations of block caving flow using ESyS-Particle. In: Proceedings of the 1st southern hemisphere international rock mechanics symposium, Australian Centre for Geomechanics, Perth, pp 221–229
23. Hancock WR, Weatherley DK, Chitombo GP (2010) Large-scale simulations of gravity flow in block caving. In: Proceedings of the 2nd international symposium on block and sublevel caving, Perth, pp 553–556
24. Yashar P (2013) Mathematical programming for sequence optimization in block cave mining. University of Alberta
25. Chitombo GP (2010) Caving mining-16 years after Laubscher's 1994 paper 'Cave mining-state of the art'. In: Proceedings of the 2nd international symposium on block and sublevel caving, Perth, pp 45–61
26. Castro R, Gonzales F, Arancibia E (2009) Development of a gravity flow numerical model for the evaluation of drawpoint spacing for block/panel caving. J South Afr Inst Min Metall 109(7):39
27. Pierce M (2010) A model for gravity flow of fragmented rock in block caving mines. The University of Queensland, Brisbane
28. Fröström J (1970) Examination of equivalent model materials for development and design of sublevel caving. Stockholm: Royal Institute of Technology (in Swedish)
29. Rustan A (2000) Gravity flow of broken rock-what is known and unknown. In: Proceedings of MassMin 2000, Brisbane, p 557
30. Kuchta ME (2002) A revised form of the Bergmark-Roos equation for describing the gravity flow of broken rock. Miner Resour Eng 11:349
31. Nedderman RM (1995) The use of the kinematic model to predict the development of the stagnant zone boundary in the batch discharge of a bunker. Chem Eng Sci 50(6):959

32. Brown ET (2007) Block caving geomechanics. Brisbane: Julius Kruttschnitt Mineral Research Centre
33. Binquan K, Yixin C, Jindong L et al (2017) Granular materials flow like complex fluids. Nature 551:360–363
34. Bernal JD, Mason J (1960) Packing of spheres: coordination of randomly packed spheres. Nature 188(4754):910–911
35. Scott GD (1962) Radial distribution of the random close packing of equal spheres. Nature 194(4832):956–957
36. Landauer J, Kuhn M, Nasato DS et al (2020) Particle shape matters-Using 3D printed particles to investigate fundamental particle and packing properties. Powder Technol 361:711–718
37. Seidler GT, Martinez G, Seeley LH et al (2000) Granule-by-granule reconstruction of a sandpile from X-ray microtomography data. Phys Rev E 62(6):8175–8181
38. Cao YX, Chakrabortty B, Barker GC et al (2013) Bridges in three-dimensional granular packings: experiments and simulations. Europhys Lett 102(2):24004
39. Lin MC, Zhou M, Liu JY et al (2022) A topological view on microscopic structural evolution for granular material under loading and unloading path. Comput Geotech 141:104530
40. Shepherd RF, Conrad CJ, Sabuwala T et al (2012) Structural evolution of cuboidal granular media. Soft Matter 8(17):4795–4801
41. Forterre Y, Pouliquen O (2008) Flows of dense granular media. Annu Rev Fluid Mech 40(1):1–24
42. Yang GC, Liu QY, Hu MB et al (2014) Phase transition and bistable phenomenon of granular flows down a chute with successive turnings. Phys Lett A 378:1281
43. Fei ML, Sun QC, Zhong DY et al (2012) Simulations of granular flow along an inclined plane using the Savage-Hutter model. Particuology 10(2):236–241
44. Tian T, Su JL, Zhan JH et al (2018) Discrete and continuum modeling of granular flow in silo discharge. Particuology 36(1):127–138
45. Wang NS, Xu JL, Guo XL et al (2018) Velocity profiles of avalanches during hopper discharge. Fuel 218:350–356
46. Wang YQ, Wang YJ, Zhang J (2020) Connecting shear localization with the long-range correlated polarized stress fields in granular materials. Nat Commun 11:4349
47. Wang YQ, Shang J, Wang YJ et al (2021) Contact force measurements and local anisotropy in ellipses and disks. Phys Rev Res 3:043053
48. Fonseca J, O\'Sullivan C, Coop MR, et al (2013) Quantifying the evolution of soil fabric during shearing using scalar parameters. Géotechnique 63(6):487–499
49. Hou SQ, Wang W, Wang ZY et al (2018) Force chain characteristics and effects of a dense granular flow system in a third body interface during the shear dilatancy process. J Appl Mech Tech Phys 59:153–162
50. Imseeh WH, Alshibli KA (2018) 3D finite element modelling of force transmission and particle fracture of sand. Comput Geotech 94:184–195
51. Guo N, Zhao JD (2013) The signature of shear-induced anisotropy in granular media. Comput Geotech 47:1–15
52. Zhao SW, Evan TM, Zhou XW (2018) Shear-induced anisotropy of granular materials with rolling resistance and particle shape effects. Int J Solids Struct 150:268–281
53. Li X, Yu HS, Li XS (2009) Macro-micro relations in granular mechanics. Int J Solids Struct 46:4331–4341
54. Jenkins MC, Haw MD, Barker GC et al (2011) Does gravity cause load-bearing bridges in colloidal and granular systems? Phys Rev Lett 107(3):038302
55. Power G (2004) Modelling granular flow in caving mines: large scale physical models and full scale experiments. University of Queensland, Brisbane, pp 57–168
56. Castro R, Trueman R, Halim A (2007) A study of isolated draw zones in block caving mines by means of a large 3D physical model. Int J Rock Mech Min Sci 44(6):860–870
57. Song Z, Konietzky H, Herbst M (2020) Drawing mechanism of fractured top coal in longwall top coal caving. Int J Rock Mech Min Sci 130:104329

58. Valdes JR, Leleu SL (2008) Influence of mineral composition on the simple shear response of sands: experimental study. J Geotech Geoenviron Eng 134(12):1820–1824
59. Dorador L (2016) Experimental investigation of the effect of broken ore properties on secondary fragmentation during block caving. The University of British Columbia, Vancouver
60. Sun H, Gao YT, Elmo D et al (2019) A study of gravity flow based on the upside-down drop shape theory and considering rock shape and breakage. Rock Mech Rock Eng 52(3):881–893
61. Bridgwater J, Cooke MH, Scott AM (1978) Inter-particle percolation: equipment development and mean percolation velocities. Trans Inst Chem Eng 56:157–167
62. Hashim M (2011) Particle percolation in block caving mines. University of New South Wales, Australia
63. Castro R, Arancibia L, Gomez R (2022) Quantifying fines migration in block caving through 3D experiments. Int J Rock Mech Min Sci 151:105033
64. Sun H, Jin AB, Davide E et al (2020) A numerical based approach to calculate ore dilution rates using rolling resistance model and upside-down drop shape theory. Rock Mech Rock Eng 53(10):4639–4652
65. Trueman R, Castro R, Halim A (2008) Study of multiple draw-zone interaction in block caving mines by means of a large 3D physical model. Int J Rock Mech Min Sci 45(7):1044–1051
66. Castro R, Gómez R, Pineda M et al (2020) Experimental quantification of vertical stresses during gravity flow in block caving. Int J Rock Mech Min Sci 127:104237
67. Kvapil R, Baeza L, Rosenthal J et al (1989) Block caving at El Teniente mine, Chile. Trans Am Inst Min Metall Eng 98:43–56
68. Jerzy L (1956) Application of equation of stochastic processes to mechanics of loose bodies. Arch Mech Stos 8:393–411
69. Chen G (1997) Stochastic modelling of rock fragment flow under gravity. Int J Rock Mech Min Sci 34:323–331
70. Jin AB, Sun H, Wu SC et al (2017) Confirmation of the upside-down drop shape theory in gravity flow and development of a new empirical equation to calculate the shape. Int J Rock Mech Min Sci 92:91–98
71. Iwashita K, Oda M (1998) Rolling resistance at contacts in simulation of shear band development by DEM. J Eng Mech ASCE 124:285–292
72. Ai J, Chen JF, Rotter JM et al (2011) Assessment of rolling resistance models in discrete element simulations. Powder Technol 206:269–282
73. Wensrich CM, Katterfeld A (2012) Rolling friction as a technique for modelling particle shape in DEM. Powder Technol 217:409–417
74. Jiang MJ, Shen ZF, Wang JF (2015) A novel three-dimensional contact model for granulates incorporating rolling and twisting resistances. Comput Geotech 65:147–163
75. Itasca Consulting Group (2016) PFC3D (particle flow code in 3 dimensions) user's guide in PFC2D version 5.0. Minneapolis: Itasca Consulting Group Inc.
76. Jin AB, Sun H, Ma GW et al (2016) A study on the draw laws of caved ore and rock using the discrete element method. Comput Geotech 80:59–70
77. Rycroft CH (2009) VORO++: a three-dimensional Voronoi cell library in C++. Chaos 19(4):041111
78. Zhao SW, Zhao JD, Guo N (2020) Universality of internal structure characteristics in granular media under shear. Phys Rev E 101(1–1):012906
79. Schröder-Turk GE, Mickel W, Kapfer SC et al (2011) Minkowski tensor shape analysis of cellular, granular and porous structures. Adv Mater 23(22–23):2535–2553
80. Steinhardt PJ, Nelson DR, Ronchetti M (1983) Bondorientational order in liquids and glasses. Phys Rev B 28(2):784–805
81. Saadatfar M, Takeuchi H, Robins V et al (2017) Pore configuration landscape of granular crystallization. Nat Commun 8:15082
82. Cosiovich D, Ozawa M, Berthier L (2018) Local order and crystallization of dense polydisperse hard spheres. J Phys Condensed Matt 30(14):144004
83. Castro R, Gómez R, Hekmat A (2016) Experimental quantification of hang-up for block caving applications. Int J Rock Mech Min Sci 85:1–9

84. Castro R, López S, Gómez R et al (2021) Experimental study of the influence of drawbell geometry on hang-ups in cave mine applications. Rock Mech Rock Eng 54:1–10
85. Sun H, Zhao LS, Elmo D et al (2025) Optimization of an empirical equation in drop shape theory and a structural-mechanical analysis of a caved ore and rock particle flow system. Phys Fluids 37(2):023379

Chapter 5
Flow Characteristics of Caved Ore and Rock

Gravity flow, which pertains to the study of the flow characteristics of caved ore and rock, is a fundamental issue in the investigation of draw in caving mines. This chapter presents a series of research findings, including the flow characteristics of caved ore and rock under both near-field and far-field conditions, the flow characteristics of caved ore and rock with uneven particle size distributions, the evolution characteristics of the movement zone influenced by a single coarse particle, the flow characteristics of caved ore and rock under the influence of multiple coarse particles, and the uniform flow and stress evolution laws of caved ore and rock across multiple drawpoints. These studies were conducted using both physical tests and numerical simulations.

5.1 Research on the Near/Far-Field Flow Characteristics of Caved Ore and Rock Based on Rigid Block Model

5.1.1 A Brief Overview of Near/Far-Field Draw Research

The characteristics of the block caving mining method, especially the natural block caving technique, encompass continuous mining, extraction from beneath the overlying strata, and the management of ground pressure through the filling of mined-out areas with caved overburden. This method is acknowledged for its low cost and high efficiency, rendering it a widely adopted large-scale mining technique in both domestic and international metal mines [1, 2]. Based on the varying heights of the caved rock mass or the draw height, the challenges associated with drawing in underground metal mines can be categorized into two primary types: near-field draw and far-field draw. Generally, draw occurring at a height of less than 100–200 times the average particle size of caved ore and rock is classified as near-field draw, while

H. Sun et al., *Study on Flow and Hang-up Characteristics of Caved Ore and Rock*,
https://doi.org/10.1007/978-981-92-0335-2_5

draw at a height greater than 100–200 times the average particle size is classified as far-field draw [3]. Currently, both domestic and international scholars have undertaken extensive indoor tests, numerical simulations, and theoretical research on the near-field draw problem in metal mines. Various drawing theories have emerged from these efforts, including ellipsoid theory [4], elliptical-like theory [5], stochastic medium theory [6], and upside-down drop shape theory [7, 8]. These theories have made significant contributions to the advancement and refinement of draw theory and technology in the context of metal mining. Indoor draw tests have garnered considerable attention from scholars and consistently play a vital role in draw research. In relation to the near-field draw problem, researchers such as Kvapil [9, 10], Janelid [11], Laubscher et al. [12], Power et al. [13], Castro et al. [14], Tao et al. [15], Wang et al. [16], and Jin et al. [8] have utilized physical models with varying similarity ratios (1:30 and 1:100) to investigate the movement laws of caved ore and rock under different conditions, including particle size, column height, drawpoint size, and vertical stress. Additionally, scholars such as Wang et al. [17], Shao [18], Xu et al. [19], Castro et al. [20], and Sun et al. [21] conducted a series of physical tests based on different combinations of sublevel height, drift spacing, drawpoint spacing, and wall angle to optimize the stope-structure parameters in caving metal mines, including Chengchao Iron Mine, Shouyun Iron Mine, Jinchanggouliang Gold Mine, Agnico Eagle Gold Mine (Canada), and Meishan Iron Mine. With the continuous development of computer technology and draw theory, numerical simulation methods have become increasingly crucial and indispensable in the study of drawing problems. Among these methods, the PFC software, which is based on the particle discrete element method, can analyze the movement laws of granular media from a mesoscopic perspective, making it particularly suitable for draw research [22, 23].

Currently, both domestic and international scholars have made significant advancements in the study of near-field draw issues in metal mines utilizing PFC software. Hashim [24] quantified the shear strain in both the IEZ and the IMZ through indoor draw tests, employing the PFC program to investigate the influence of particle shape, block size distribution, and other factors on the flow characteristics of small particle diameters. Song et al. [25] utilized the PFC software to examine how particle shape affects the configuration of the IEZ. Hu et al. [26] conducted a four-factor, three-level orthogonal numerical simulation, along with a profit factor evaluation function, to explore the flow dynamics of caved ore and rock, identifying the optimal combination of stope structural parameters, specifically focusing on sublevel height, drift spacing, drawpoint spacing, and dilution rate. Sun et al. [27–29] investigated the movement dynamics of caved ore and rock under three types of boundary conditions using the PFC program, successfully visualizing the process of shape changes in both the IEZ and IMZ.

Currently, the heights of ore—rock layers in domestic and foreign mines using the natural caving method, such as the Cadia Valley copper—gold mine in South Africa and the Pulang copper mine in China, have exceeded 200–300 m, which belongs to the typical far-field draw problem in underground metal mines. Castro et al. [30] employed laboratory test methods to study the variation laws of the vertical stress borne by the bottom structure under different draw methods in far-field draw. Rafiee et al. [31] utilized the discrete element numerical simulation method to study the caving ability of ore-rock under the influence of different factors before far-field draw. They found that in-situ stress and hydraulic radius are the primary factors influencing the caving ability in the natural caving method. However, current draw theory, laboratory tests, and numerical simulation studies have yet to establish a unified understanding of the shape evolution laws of the IMZ and the IEZ in the far field. Additionally, research on the migration mechanisms of caved ore and rock remains insufficiently in-depth. This has contributed to the relatively slow advancement of draw theories and methods, thereby restricting the application of caving mining techniques, particularly the natural caving method, in underground metal mines across China. Consequently, this study employs a comprehensive approach that includes physical draw tests, numerical simulations, and theoretical analyses to investigate the flow characteristics of caved ore and rock in the near and far fields of metal mines, based on the rigid block model. This research aims to provide theoretical and technical support for optimizing structural parameters of caving mines and ensuring the safe and efficient extraction of mineral resources.

5.1.2 Research on the Near-Field Draw Tests

5.1.2.1 Construction of the Near-Field Draw Model

This study first compares the results of physical and numerical tests of draw under near-field conditions to verify the reliability of the near-field draw simulation. It then analyzes the flow characteristics of caved ore and rock in these conditions, thereby laying the foundation for the study of far-field draw.

(1) Design of the near-field draw model

In this near-field indoor test, a large-scale three-dimensional physical testing platform was employed, capable of accommodating various similarity ratios, as illustrated in Fig. 4.1a. This platform is constructed from PVC (Polyvinyl Chloride) and acrylic glass. The model's maximum height is 2.5 m, and its volume is adequate for housing particles measuring 0.8 m × 0.8 m × 2.0 m. The test design utilizes a single bottom draw configuration with a similarity ratio of 1:25, simulating an actual draw height of 50 m. The drawpoint is centrally located at the bottom of the model, measuring 0.12 m × 0.12 m.

The near-field numerical draw model serves as an in-situ representation (Fig. 4.15a), geometrically analogous to the physical test model, with the numerical model being 25 times larger than its physical counterpart. The red wall at the center of the model's base functions as the draw structure and can be removed prior to the draw simulation to establish the drawpoint.

(2) Materials for the near-field draw test

Given that the focus of this study is on the flow characteristics of caved ore and rock, limestone particles, which are commonly used in construction due to their low density and availability, were selected as the granular materials and tracer particles for the physical tests (Fig. 5.1a).

In previous studies utilizing particle discrete element software such as PFC and EDEM to investigate draw issues, spherical particles or clumps composed of multiple spherical particles were commonly employed to simulate real ore and rock (Fig. 5.1b). However, the overly smooth surfaces of spherical particles impede the generation of sufficient interlocking forces between them, rendering it impossible to quantitatively and accurately characterize the shape and change patterns of the IEZ and the IMZ. If the reliability of the draw simulation results cannot be effectively enhanced, conducting further quantitative analyses of the migration and evolution laws of the caved ore and rock, as well as their mechanical mechanisms, will prove challenging in future research. Consequently, in these numerical tests, a rigid block model [32] was employed using PFC software to construct three different aspect ratios of irregular blocks, as illustrated in Fig. 5.1c. This methodology aims to accurately simulate the actual shapes of caved ore and rock, which include triangular, symmetrical, and slender types. Moreover, in addition to enhancing the accuracy of simulating these shapes, utilizing the Rigid Block Model for draw problem research offers a notable advantage in terms of computational efficiency compared to the clump model. In PFC, each rigid block or spherical particle is treated as an independent entity, resulting in only one contact with other entities. In contrast, a clump composed of *N* spherical particles results in *N* contacts with other elements. Consequently, the substantial reduction in the number of contacts between particles or blocks effectively enhances the efficiency of draw simulations.

The particle gradation curves derived from both physical and numerical tests for the draw process are presented in Fig. 5.2. In the physical test, the particle sizes range from 3 to 45 mm, with an average particle size of 19.2 mm. The distribution of various particle sizes of the crushed limestone loose materials is illustrated in Fig. 5.3. In the numerical test, the particle size distribution corresponds closely to that of the physical test, maintaining a similarity ratio of 1:25. Additionally, the three differently shaped rigid blocks, as depicted in Fig. 5.1c, employed in the numerical test are equally represented, with each shape constituting one-third of the total number of blocks.

(3) Test process of near-field draw

In this physical test, the filling height of the granular material was established at 2.0 m, with a layer of tracer particles positioned at intervals of 0.1 m in the vertical direction within the model (see Fig. 5.4). The shape of the IEZ was delineated using the marker particle method [8]. Throughout the test, granular material was replenished at the top of the model to maintain a constant height of the ore-rock layer during the draw process. Ore extraction was ceased when the draw height reached 2.0 m. Due to the difficulties associated with directly observing the movement of caved ore and rock within the three-dimensional physical model, the shape of the IMZ was neither measured nor delineated in this physical draw test.

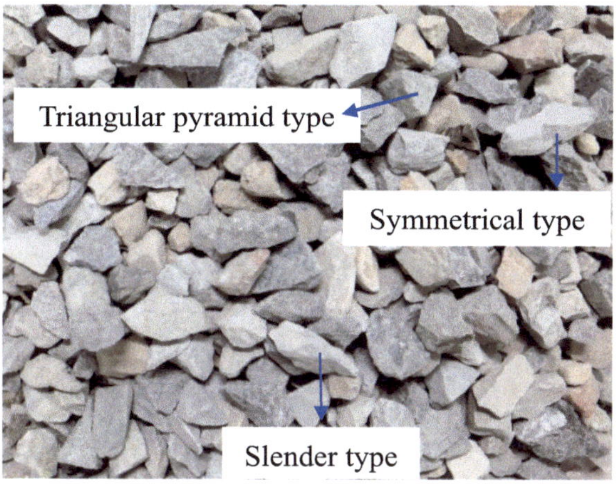

(a) Particle shapes in physical test

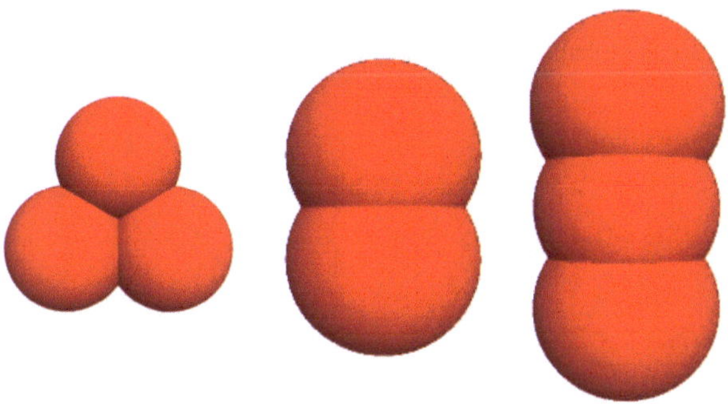

(b) Particle shapes used in previous numerical simulations

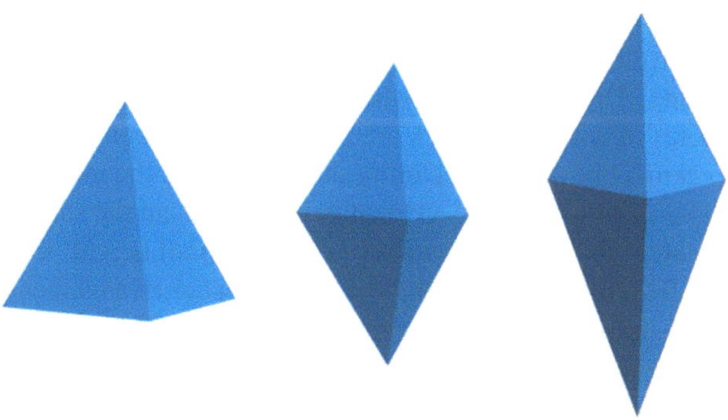

(c) Particle shapes used in these numerical simulations

Fig. 5.1 The 3D particle shapes used in physical and numerical draw tests

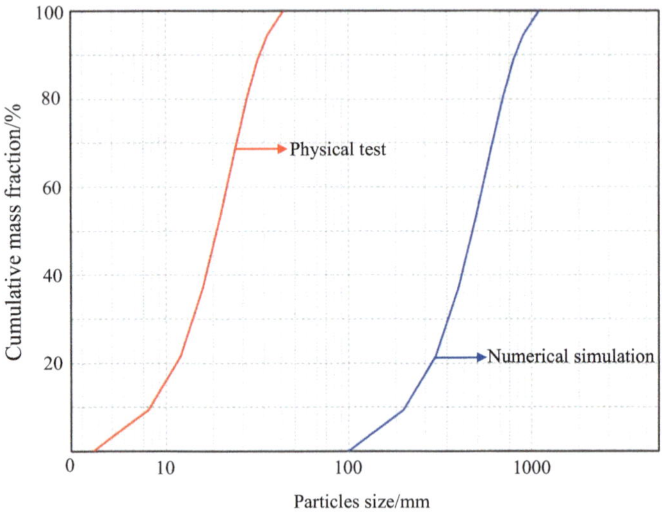

Fig. 5.2 Particle size distribution curves in physical and numerical draw tests

(a) 3~8 mm (b) 8~16 mm (c) 16~25 mm (d) 25~45 mm

Fig. 5.3 Limestone particles with different sizes after sieving

The process of the near-field draw numerical test aligns with the methodology of physical testing. Drawing ceases when the height reaches 50 m. Initially, the spatial coordinates of each rigid block are recorded by the compiling program prior to the commencement of the draw. During the simulation, the initial positions of all blocks drawn at a specific moment are captured to represent the shape of the IEZ. In subsequent research, the focus is primarily on the overall height, width, and shape evolution of the IEZ, rather than on the shape and spatial arrangement of individual rigid blocks being drawn. To enhance computational efficiency, the IEZ derived from inversion is represented as spherical particles with volumes equivalent to those of the irregular rigid blocks. And the IMZ produced in this simulation consists of irregular rigid blocks. Additionally, the meso-mechanical parameter values for the wall and rigid blocks in the draw numerical test are presented in Table 4.5. If the normal and shear stiffness of the wall and the rigid block are too low, excessive overlapping may occur between the blocks and between the blocks and the wall. Conversely, if the stiffness is too high, it can adversely affect the timestep during calculations,

Fig. 5.4 The layout of labeled markers

significantly reducing computational efficiency. Additionally, the stiffness of the wall should be slightly greater than that of the block to prevent the error of the 'block passing through the wall'. The density of the rigid block is consistent with that of the limestone aggregates used in the physical test, which is 2620 kg·m^{-3}. By comparing the results of the physical test of the natural repose angle [33] with the numerical test results, it has been determined that the friction coefficients of both the wall and the rigid block in this draw simulation are set to 0.5.

5.1.2.2 Analysis of the Results of the Near-Field Draw Test

The shape composed of the spatial positions occupied by the caved ore and rock in the initial particle packing system is referred to as the IEZ, while the spatial configuration composed of all the mobile caved ore and rock in the particle packing system is termed the IMZ [34]. The total mass drawn and the positional information of the tracer particles drawn are statistically recorded. Utilizing the interpolation method, the IEZ at various heights, derived from the physical tests as illustrated in Fig. 5.5a, is delineated. Figures 5.5b and c represent the IEZ and IMZ at a height of 50 m, respectively, as obtained from the numerical tests. From Fig. 5.5, it is evident that the shapes of the IEZ and IMZ obtained from both the physical and numerical tests resemble inverted droplet shapes [8], which qualitatively supports the applicability of the rigid block model in the numerical draw study.

In the statistical physics and numerical tests, the maximum radii (defined as half of the maximum width) of the IEZ and IMZ at various heights were measured. Figure 5.6 illustrates the fitting curves that represent the relationship between the heights of the IEZ and IMZ and their corresponding maximum radii.

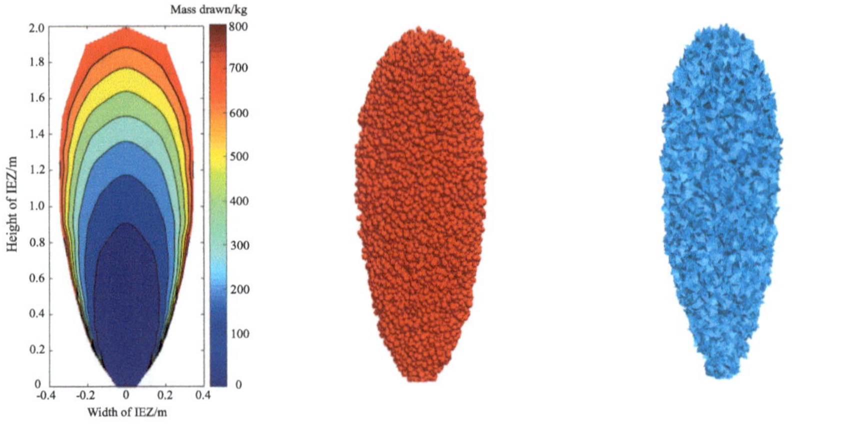

Fig. 5.5 The longitudinal profiles of the IEZ's and IMZ's shapes in physical and numerical draw tests

Fig. 5.6 Comparison of relationship between the height and maximal radius of IEZ/IMZ in near-field physical and numerical draw tests

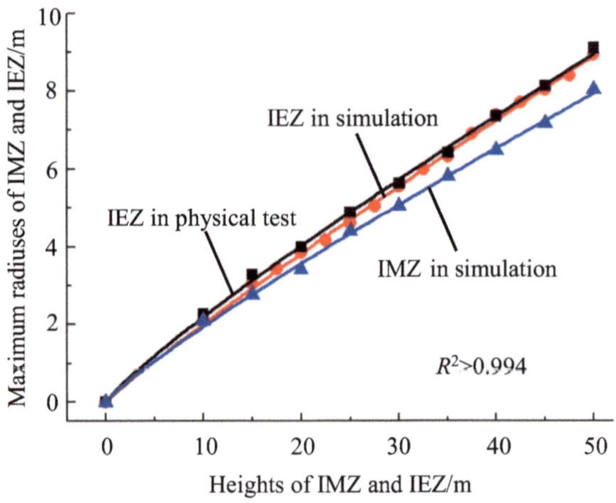

Figure 5.6 illustrates that, under near-field conditions, the heights of both the IEZ and IMZ, as derived from physical and numerical tests, conform to the power function relationship outlined by the upside-down drop shape theory [8] (Eq. (5.1)). Furthermore, the correlation coefficient R^2 for all three curves exceeds 0.994:

$$R_{\max} = a * H^b \quad (5.1)$$

Here, $R_{\max}$ represents the maximum radius of the IEZ/IMZ, H represents the height of the IEZ/IMZ, and a and b are fitting constants.

As illustrated in Fig. 5.6, the maximum radius of the IEZ obtained from numerical tests is generally consistent with that derived from physical tests at the same height. This observation quantitatively validates the reliability of the rigid block model used in the numerical draw study. Moreover, at equivalent heights, the maximum radius of the IMZ is notably smaller than that of the IEZ. This difference can be attributed to the fact that the IMZ is consistently considered a precursor to the IEZ [35]. In essence, the shape change patterns of the IMZ and the IEZ are correlated, with the IMZ forming earlier than the IEZ at the same height. For example, in this numerical test, when the height of the IMZ reached 50 m, the height of the IEZ was only 23.7 m. As the draw process advances, the particle system structure of the caved ore and rock becomes increasingly loose, resulting in the maximum width of the IMZ being smaller than that of the IEZ at the same height.

5.1.3 Research on the Far-Field Draw Numerical Test

Based on the physical and numerical tests conducted on near-field draw, further numerical investigations into far-field draw were performed using a rigid block model to analyze the flow characteristics of caved ore and rock under far-field conditions.

5.1.3.1 Design of Far-Field Draw Numerical Test

Using the same three shapes of rigid blocks (Fig. 5.1c), gradation curves (Fig. 5.2), and meso-mechanical parameters (Table 4.5) as employed in the near-field draw test, a numerical draw model with dimensions of 60 m × 60 m × 105 m was constructed, as illustrated in Fig. 5.7. The model height is approximately 220 times the average particle size of the rigid blocks (0.48 m), thereby satisfying the requirements for far-field draw [3]. Nine measurement spheres, each with a diameter of 6 m, were strategically positioned within the model to monitor variations in horizontal and vertical stresses on the blocks at different spatial locations throughout the draw process. Specifically, measurement domains 1, 4, and 7 were situated 20 m, 55 m, and 90 m above the bottom drawpoint, respectively, ensuring that the axis of the draw model intersects the centers of these three measurement spheres. The spacing between adjacent measurement domains at the same height was maintained at 12 m. Given the consistency of the morphological changes in the IMZ and the IEZ, as well as the constraints on calculation time, this far-field draw numerical test was simulated only until the height of the IMZ reached 105 m. The morphological change patterns of the entire IMZ were analyzed, and the flow characteristics and mechanical mechanisms of the caved ore and rock under far-field conditions were examined based on the stress monitoring results.

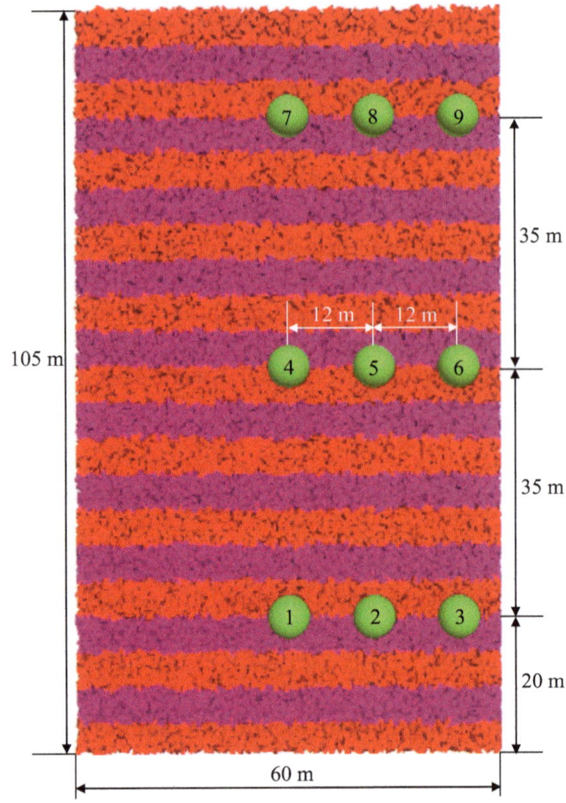

Fig. 5.7 The longitudinal profile of far-field numerical draw model and layout of stress measurement regions

5.1.3.2 Analysis of the Results of the Far-Field Drawing Quantity Test

The far-field draw numerical tests were conducted to determine the maximum radius of the IMZ at various heights. The fitting curve that relates the height of the IMZ to its maximum radius, as depicted in Fig. 5.8, was obtained alongside the morphology of the IMZ at different heights, specifically at 35 m, 65 m, and 95 m. From Fig. 5.8, it is evident that the morphology of the IMZ under far-field conditions conforms to the shape of an inverted water drop. Furthermore, the relationship between the height of the IMZ and its maximum radius aligns with the power function described by the inverted water-drop theory (Eq. (5.1)), demonstrating a high correlation coefficient of $R^2 = 0.996$. During the initial draw stage, when the top of the IMZ is relatively close to the drawpoint (less than 25 m), the drawpoint exerts a significant influence on the evolving shape of the IMZ [27], and its maximum width increases rapidly with height. In subsequent draw stages, particularly during far-field draw, the morphological evolution is primarily governed by the intrinsic physical and mechanical properties of the caved ore and rock, as well as the stress state of the bulk assembly. Under these conditions, the maximum width of the IMZ increases approximately linearly

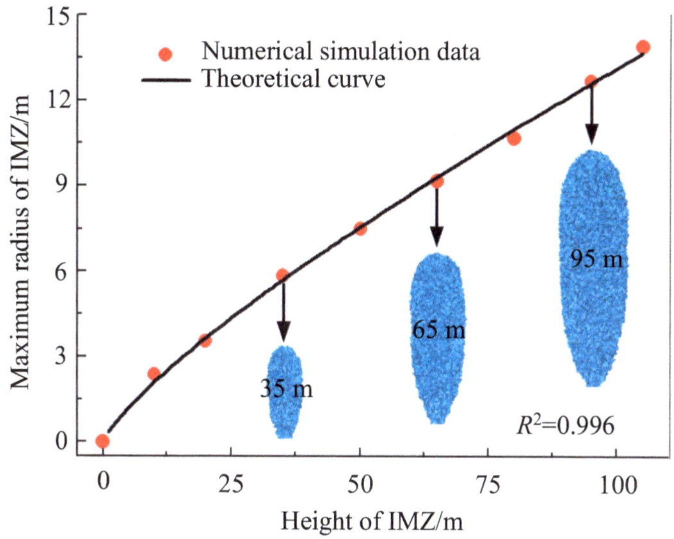

Fig. 5.8 Comparison between the data of far-field numerical draw test and theoretic curve for the relationship between the height and maximal radius of IMZ

with height. Consequently, for large-scale block-cave operations involving substantial draw heights (greater than 100 m), the approximate linear relationship between height and maximum width of the IMZ under far-field conditions greatly facilitates the design and optimization of stope structural parameters, such as drawpoint spacing.

The changes in horizontal and vertical stress experienced by the rigid blocks across nine different spatial positions, as illustrated in Fig. 5.7, were statistically analyzed throughout the entire drawing process. Specifically, Fig. 5.9 depicts the variations in vertical stress within the 1st, 4th, 5th, 6th, and 7th measurement domains. During the initial stage of draw, the curves for the 1st, 4th, and 7th measurement domains in Fig. 5.9 indicate that rigid blocks located in higher overlying rock layers experienced greater vertical stress. In contrast, the curves for the 4th, 5th, and 6th measurement domains reveal that rigid blocks at the same elevation exhibited no significant differences in vertical stress. Furthermore, when the loose range of caved ore and ore did not influence the corresponding measurement domains, their vertical stress remained relatively unchanged (as shown in curve No. 6); conversely, the vertical stress at the relevant spatial positions exhibited a significant decrease.

Figure 5.10 illustrates the changes in horizontal stress within the 1st, 4th, and 7th measurement domains at various heights along the model's central axis. As depicted in Fig. 5.10: (1) In the initial stage of draw, prior to the loose range of the caved ore and rock reaching the height of the 1st measurement domain, the horizontal stress on the caved ore and rock at higher elevations within the model is greater. (2) When the loose range of the caved ore and rock has not yet reached the heights of the 4th or 7th measurement domains, there is no significant change in horizontal stress. (3) As

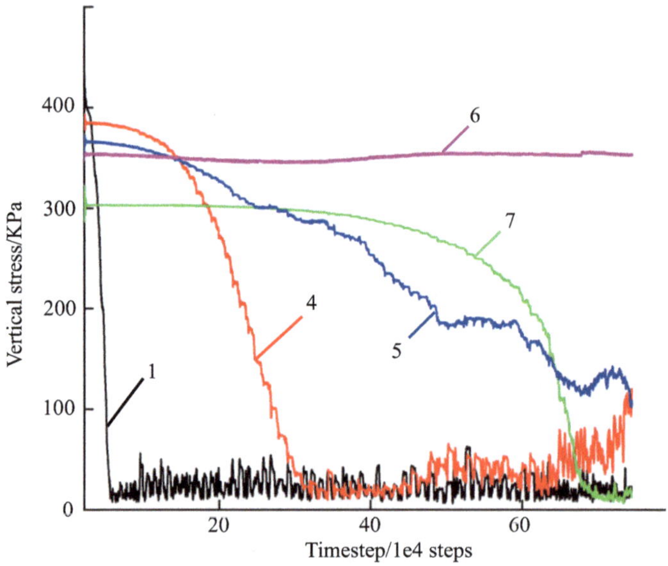

Fig. 5.9 Variations of vertical stresses within measurement regions Nos. 1, 4, 5, 6 and 7

the loose range of the caved ore and rock progressively reaches the heights of the 1st, 4th, and 7th measurement domains, the horizontal stress exhibits a sharp increase, as indicated by the green circles in Fig. 5.10. (4) Once the loose range of the caved ore and rock surpasses the heights of the 1st, 4th, and 7th measurement domains, indicating that the caved ore and rock at those heights are in a relatively loose state, the horizontal stress demonstrates a sharp decrease.

The lateral pressure coefficient reflects the interrelationship between horizontal and vertical stress at various spatial positions within the particle system of caved ore and rock. Given the relatively small vertical stress within the IMZ, which enhances clarity, this analysis considers the reciprocal of the lateral pressure coefficient, representing the ratio of vertical stress to horizontal stress. Figure 5.11 illustrates the changing process of this reciprocal across the 4th, 5th, and 6th measurement domains. As shown in Fig. 5.11, prior to the IMZ reaching the height of the 4th measurement domain, the ratio of vertical stress to horizontal stress decreased significantly. Subsequently, as the position of the 4th measurement domain entered the IMZ range, the ratio exhibited a fluctuating trend. When the IMZ gradually extended to the 5th measurement domain, the ratio of vertical stress to horizontal stress increased steadily; however, once the 5th measurement domain entered the IMZ range, the ratio showed a significant downward trend. Furthermore, the 6th measurement domain remained in the non-loosened area throughout the draw process, with the ratio of vertical stress to horizontal stress exhibiting a gradual increase.

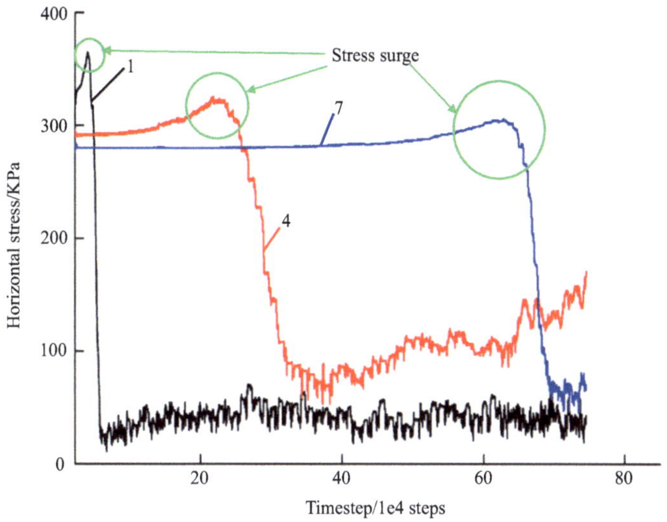

Fig. 5.10 Variations of horizontal stresses within measurement regions Nos. 4, 5 and 6

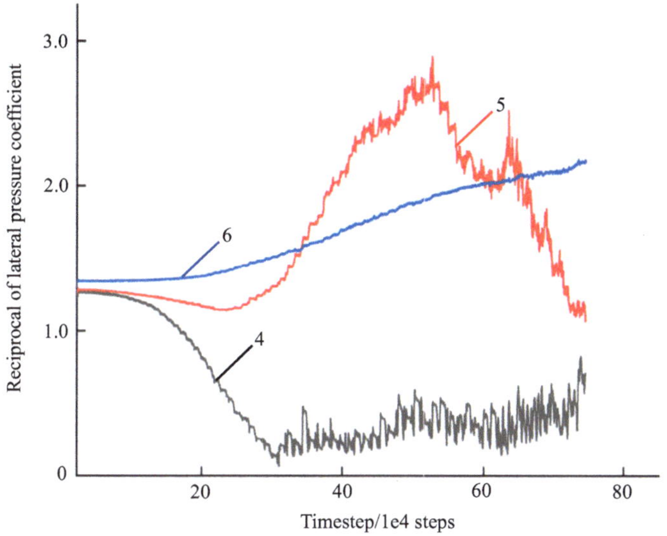

Fig. 5.11 Variations of reciprocal of the lateral pressure within measurement regions Nos. 4, 5 and 6

5.1.4 Discussion on Stress Arch and Stress Transfer Within the Particle Flow System of Caved Ore and Rock

During the draw process, a distinct arch structure emerges within the particle flow system of caved ore and rock [36]. This phenomenon indicates that the particle system can establish a stable configuration through mutual support. The caved ore and rock forming the arch structure can withstand specific stresses while maintaining stability. Consequently, this arch structure is commonly referred to as a stress arch. The formation, evolution, and collapse of the stress arch significantly influence the movement of the caved ore and rock [3, 35]. Based on an analysis of the shape changes and stress evolution patterns of the IMZ, the following discussion explores the mechanical mechanisms from the perspective of the stress arch and stress transfer (Fig. 5.12).

(1) In the context of isolated draw under infinite boundary conditions, the vertical stress induced by the self-weight of the overlying rock is initially uniform across various positions at the bottom of the model before the draw begins. As the draw process initiates, the drawpoint's presence causes a redistribution of vertical stress at the model's bottom, shifting towards the areas adjacent to the drawpoint. This shift gradually intensifies, resulting in the formation of several stress arches within a specific height range above the drawpoint. As illustrated in Fig. 5.12, these stress arches create a scenario where the vertical stress at point A does not equate to the total self-weight of the overlying rock; instead, it corresponds to the vertical stress generated by the self-weight of the caved ore and rock situated between two adjacent stress arches (between points A and B). As the stress arches above point A continue to develop and intensify, the vertical stress at point A exhibits a continuous decrease, as depicted in Fig. 5.9. Concurrently, the vertical stress above point A is transmitted through the stress arches (the contact force chain between aggregates) to both sides, leading to a continuous increase in horizontal stress at point A. This increase becomes particularly pronounced when the top of the IMZ approaches point A, resulting in the stress surge phenomenon illustrated in Fig. 5.10. The ongoing variations in the ratio of vertical stress to horizontal stress at different positions, as shown in Fig. 5.11, provide evidence for the existence of stress transfer phenomena within the particle flow system of caved ore and rock.
(2) When the stress at point A in Fig. 5.12 exceeds the bearing limit of the stress arch at that location, the critical contact between the caved ore and the rock forming the stress arch fails. This failure results in the collapse of the stress arch, allowing the caved ore and rock at point A to enter the loosened area, which subsequently continues to expand upwards. As the loosened area expands, the span of the stress arches above the IMZ increases, leading to a continuous increase in the maximum width between the stress arches that make up the stress arch. Following the collapse of the stress arch, the caved ore and rock will shift, gradually forming an inverted droplet-shaped IMZ. Kvapil and Janelid et al. [9–11] initially conducted a physical draw test using sand as the medium and

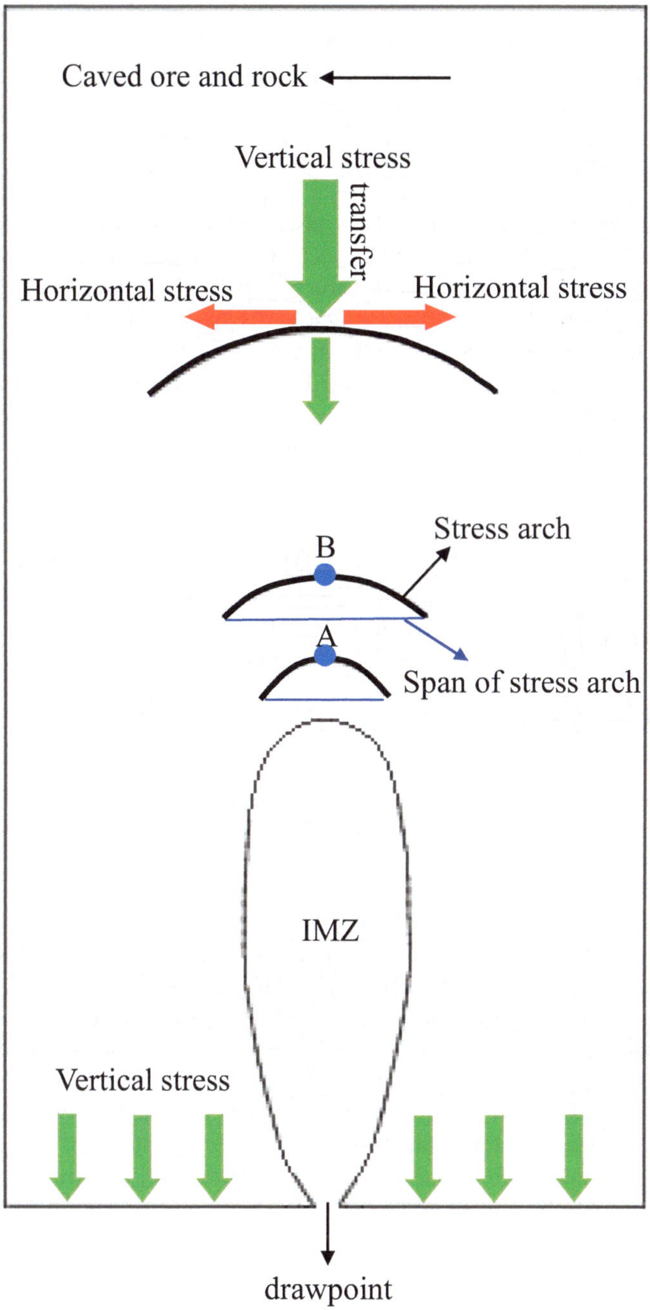

Fig. 5.12 Schematic of stress arch and stress transfer within particle flow system of caved ore and rock

found that the shape of the IEZ and the IMZ resembled an ellipsoid. However, in contrast to sand, the caved ore and rock exhibit significant non-uniformity in particle size and shape, resulting in a greater internal locking force between the particles and a more pronounced stress arch effect during the gravity flow process. Consequently, the shape of the formed IMZ is more akin to an upside-down drop shape than to an ellipsoid.

Subsequent researchers will employ physical tests and numerical simulations to further investigate the evolution of mechanical indicators, such as contact force chains, stress states, and arch-shaped structures within the particle flow system of caved ore and rock. They will utilize principles of rheology and statistical mechanics to analyze the characteristics of structural non-uniformity and the dynamic behaviors of particles in the particle flow system from both macro and mesoscopic perspectives. From a mechanical standpoint, these studies aim to elucidate the formation mechanisms of the structures within the particle flow system of caved ore and rock.

5.2 Research on the Flow Characteristics of Caved Ore and Rock with Non-Uniformity Particle Size Distribution

5.2.1 A Brief Overview of Particle Non-Uniformity Research

In block caving mines, the caved ore and rock generated by blasting or natural caving consist of numerous irregularly shaped, variably sized, and intricately arranged particles. Their size distribution demonstrates significant non-uniformity. Traditional draw theories and the majority of existing draw tests have employed single or uniform-sized particle aggregates as the test medium, neglecting the non-uniformity in size gradation and spatial distribution (packing state) of the actual caving ore and rock. This oversight leads to unreasonable designs of stope structure parameters and inaccurate predictions of the ore dilution index [37]. Consequently, it is crucial to investigate the effects of non-uniform size distribution on the flow characteristics of caved ore and rock.

For an extended period, scholars both domestically and internationally have conducted numerous indoor draw tests and numerical simulations using monodisperse or uniformly sized caved ore and rock. Peters [38] constructed the largest two-dimensional physical draw model of its time, investigating how the size of the drawpoint affects the morphology of the IEZ through draw tests. Power [13] and Castro [14] employed gravel as the medium and executed systematic three-dimensional physical draw tests to examine the morphological changes of the IEZ and the flow dynamics of caved ore and rock under three-dimensional conditions. Wu et al. [39] and Sun et al. [29, 33] developed uniform-sized particle models using the PFC3D program, exploring the morphology of the IEZ while optimizing structural parameters within the non-pillar sublevel caving, under varying end-wall conditions and different particle media. Chen et al. [40] performed numerical tests to investigate the evolution characteristics of the internal force chain within a dispersed medium system, utilizing a flexible isolation layer beneath the dispersed medium in the PFC2D program. They integrated concepts from both contact mechanics and statistical mechanics to quantitatively analyze the evolution characteristics of the length, quantity, strength, direction, and straightness coefficient of the internal force chain during the multi-drawpoint draw process.

Current studies on the non-uniformity of particle size primarily focus on highway rockfill materials and dust particle size [41–43]. Zheng et al. [44] and Zhu et al. [45] utilized the R–R (Rosin–Rammler) distribution function to investigate the particle size distribution laws in dust and liquid droplets within the spray tower. In the draw field, Li et al. [46] examined three factors affecting the non-uniformity of caved ore and rock: the particle size dispersion coefficient, the spatial distribution dispersion coefficient, and the pore distribution dispersion coefficient. They studied the non-uniformity of caved ore and rock and derived definitions, expressions, and quantitative relationships among these three factors and the non-uniformity of the granular medium using fuzzy comprehensive evaluation decision-making. Li et al. [47] utilized Monte Carlo simulations to predict the spatial distribution of the joint surface in a large nickel ore stope, based on measured spatial geometric parameters of the joint surface. They also estimated the equivalent size of the caved ore and rock. Castro et al. [30] conducted indoor tests involving multiple drawpoints to investigate the variation of vertical stress experienced by the bottom structure under consistent particle gradation conditions and varying draw methods. Their findings revealed that the range of vertical stress variation during the draw process was between 0.3 and 2.8 times the initial vertical stress, which closely correlated with the draw method and spatial position within the stope. Wang et al. [16] conducted a physical draw test with a similarity ratio of 1:100. They crushed the ore sample and divided it into two groups based on particle size composition. This study preliminarily explored the influence of coarse particle content on the morphology of the IEZ, the hang-up, and the processes and results of ore-rock interface evolution during the draw process. Similarly, Zhang et al. [48] employed a combination of physical simulation tests and theoretical analysis to investigate the impact of size gradation on the initial coal-draw volume, particle transport trajectory, and the evolution characteristics of the coal-rock interface. Utilizing the mechanics theory of the dispersed medium, they derived the dynamic evolution equations for the initial coal-rock interface on both the support side and the caving area side. Additionally, they determined the correction coefficient values for the theoretical equations of the coal-rock interface under various top coal size gradation conditions based on the findings from the draw test. Ren et al. [49] conducted an investigation into the flow characteristics of tailings under ore discharge conditions, specifically examining the influence of the height of particles and tailings. They utilized high-speed photography technology to visualize the tailings movement process and analyze the governing dynamics of tailings flow. This research serves as a valuable reference for addressing similar tailings dry discharge collapse scenarios in draw operations. Li [50], utilizing the LBM-DEM coupling framework within an isolated draw model, defined the maximum and minimum particle sizes as 1.25 m and 0.5 m, respectively. This study simulated the impact of variations in particle sizes of caved ore and rock on the draw process. Furthermore, Liu et al. [51] established a two-dimensional model to simulate the collision motion of non-uniformly scattered particles, effectively depicting the flow dynamics of these particles. Sun et al. [52] introduced a method for generating irregular clusters using the PFC program and investigated the variations in the shape of the IEZ and the particle sizes of caved ore and rock, particularly under conditions that account for secondary fragmentation.

Currently, most physical draw tests and numerical simulations in the existing literature primarily focus on the differences in particle size among caved ore and rock. However, they inadequately consider the impact of the non-uniform spatial distribution of caved ore and rock on the draw results. This study employs a comprehensive approach that integrates physical tests, numerical simulations, and theoretical analysis to investigate the flow characteristics of caved ore and rock from two key perspectives: particle size gradation and the non-uniformity of the spatial distribution of caved ore and rock. The findings of this research significantly enhance the current draw theory and provide both theoretical and technical support for optimizing structural parameters in underground mines, predicting loss indicators, and ensuring the safe and efficient extraction of mineral resources.

5.2.2 *Physical Draw Test Study*

Firstly, this study will investigate the influence of the size distribution of caved ore and rock on the morphological changes of the IEZ through physical tests. The findings will provide reference data for subsequent numerical draw tests utilizing the mesoscopic discrete element method.

5.2.2.1 Physical Test Model and Materials

The material testing (Fig. 4.1a) employed a three-dimensional physical model of a bottom drawpoint, with a total height of 2.5 m and a similarity ratio of 1:25. To improve visibility during the loading and unloading processes of the caved ore and rock, the front section of the model was constructed using detachable, semi-transparent organic glass plates, while the remainder of the structure was made from lightweight yet durable PVC. To better simulate the actual conditions of stope sidewalls, the inner surfaces of the model were intentionally roughened around the edges to enhance texture. The model was designed to accommodate particles measuring $0.8 \times 0.8 \times 2.0\ m^3$, with the drawpoint centrally positioned at the bottom, sized at $0.12 \times 0.12\ m^2$, facilitating physical testing under infinite boundary conditions.

Due to the substantial quantity of granular medium required for this physical test, and based on various physical and numerical studies [13, 15, 33] indicating that the type of granular medium does not significantly influence the draw process

Fig. 5.13 Limestone particles of different sizes and labeled particles

and outcomes within a specific density range, we adopted a practical approach. To preserve the integrity of the test results while minimizing the effort involved, we selected lower-density building limestone particles (illustrated in Fig. 5.13a–d) and marking particles (shown in Fig. 5.13e) as substitutes for actual caved ore and rock. Prior to conducting the test, the granular medium was sorted by size, and the marking particles were dyed and labeled. It is important to note that particles smaller than 3 mm, which resemble powder, do not accurately reflect the shape, interlocking forces, and other physical and mechanical characteristics of most real caved ore and rock, and are challenging to recover completely; therefore, we excluded the 0–3 mm granular medium during the sieving process.

5.2.2.2 Design of the Physical Draw Test

The physical draw test was structured around the widely utilized Rosin–Rammler distribution function (refer to Eq. (5.2)), which is prevalent in studies of granular media. As illustrated in Fig. 5.14, three categories of isolated bottom draw tests were conducted: the fine particle category, the medium particle category, and the coarse particle category, each featuring distinct particle size gradations of granular media. Adhering to the principle of variable control, the similarity ratio for all tests was established at 1:25. The coefficient of uniformity [46] Cu for the particle size gradation of the granular media remained constant at 2.60, with average particle sizes D_{50} recorded at 11.5 mm, 14.9 mm, and 19.2 mm, respectively.

$$F(D) = 1 - e^{-\left(\frac{D}{D_{50}}\right)^{C_u}} \tag{5.2}$$

Here, $F(D)$ denotes the total mass fraction of particles that have a diameter smaller than D, D is the particle size. D_{50} indicates the mean diameter, also in millimeters, while C_u refers to the coefficient of homogeneity.

The three tests conducted were identical in nature, differing solely in the gradation of the granular media sizes. All other testing conditions and methodologies remained consistent. In this specific test, the morphology of the IEZ was defined using the marking particle technique, as illustrated in Fig. 5.13e. To ensure a uniform spatial distribution of various sizes of caved ore and rock within the model, the central four-point method [53] was utilized for filling the granular media. Furthermore, by adding

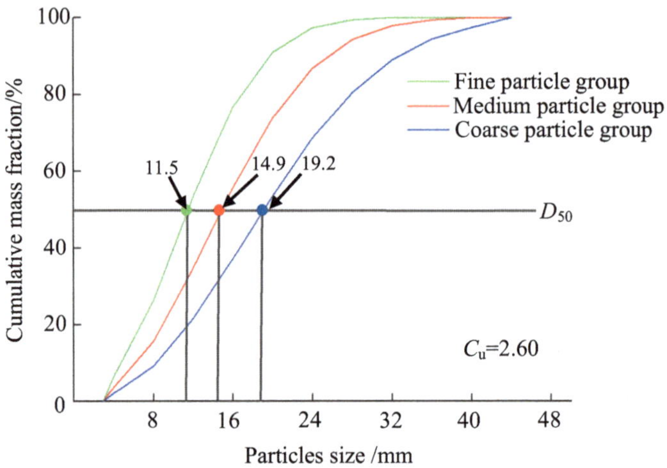

Fig. 5.14 Particle size distribution curves in physical draw tests

granular media at the top of the model, the overall particle filling height and the final draw height for each test were maintained at 2.0 m, simulating the actual 50 m height of the rock layer and the draw height. Additionally, to closely replicate the real draw process, the test aimed to uphold consistent parameters, such as the time intervals between draws and the quantity extracted in each instance, thereby minimizing the influence of human factors on the physical test outcomes related to the draw.

5.2.2.3 Analysis of the Physical Test Results

Through a statistical evaluation of the tracer particles drawn alongside the overall mass drawn, lincar interpolation was executed utilizing the averaging calculation feature in Matlab to determine the maximum width of the IEZ across various mass drawn. The resulting three longitudinal cross-sectional diagrams, derived from physical tests, are illustrated in Fig. 5.15. Additionally, the maximum radius of the IEZ at varying heights was computed, leading to the fitting curves that correlate height with maximum radius, as depicted in Fig. 5.16. Analysis of Figs. 5.15 and 5.16 reveals that the shapes of the IEZ from the three tests, corresponding to different particle size gradations, consistently exhibit upside-down drop shapes. Furthermore, both the height and maximum radius of these bodies conform to the power function relationship outlined by the upside-down drop shape theory [8] (Eq. (5.3)), with a correlation coefficient R^2 exceeding 0.996:

$$r_{\max} = a \times H^b \tag{5.3}$$

Here, $r_{\max}$ denotes the largest radius of the IEZ, expressed in meters; H indicates the height of IEZ, likewise measured in meters; a and b serve as the fitting parameters.

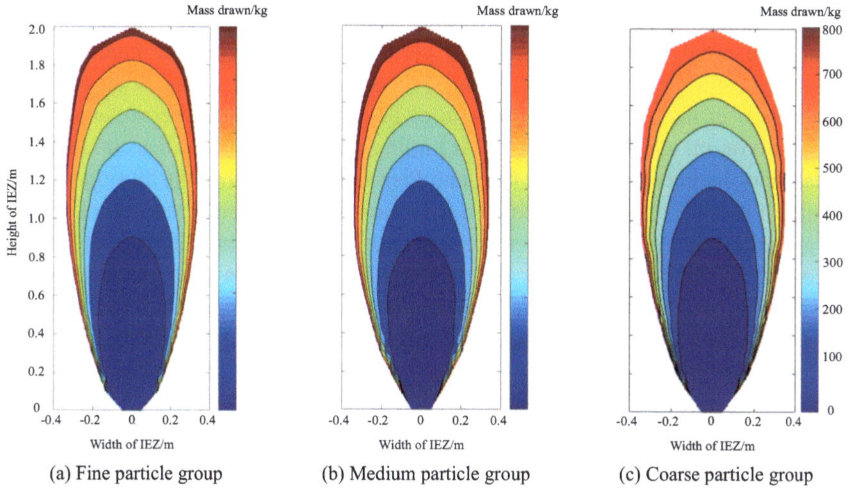

Fig. 5.15 Longitudinal profiles of the IEZ's shapes in three physical draw tests

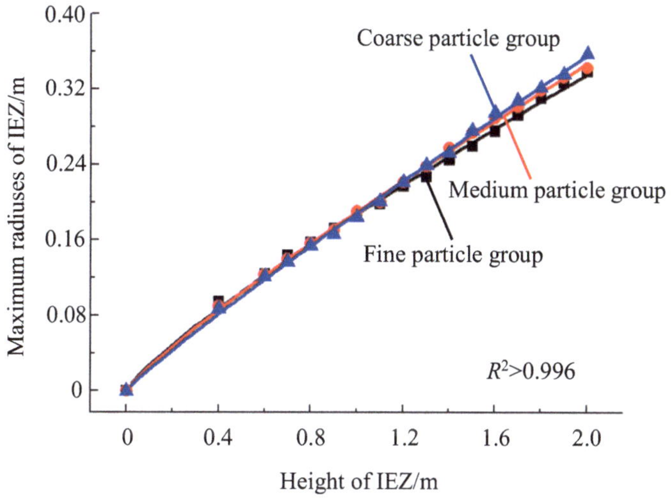

Fig. 5.16 Fitting curves between the height and maximal radius of IEZ in three physical draw tests

Additionally, as illustrated in Figs. 5.15 and 5.16, when the draw height is minimal, the maximum widths of the IEZ from the three tests are nearly identical, with the drawpoint being the primary factor influencing this outcome. Conversely, at greater draw heights, which are further from the drawpoint, the maximum width of the IEZ from the coarse particle group (represented by the blue curve) is marginally larger, followed by the medium particle group (red curve), while the fine particle group (black curve) exhibits the smallest radius. Given that the testing methods, procedures, granular material types, and non-uniformity coefficients across the three tests remain constant, the only variable is the average particle size of the granular material. Thus, within the examined parameters, it can be concluded that at a specific draw height, an increase in the average particle size of the granular material correlates with a larger maximum width of the IEZ. This finding aligns with the studies conducted by researchers such as Castro et al. [35] and Pierce et al. [3] From the aforementioned results, it can be reasonably deduced that variations in the average block size of the granular material across the three tests not only result in differing interlocked forces among particles [35], but also affect the configuration state quantities of the granular material packing system (void ratio and coordination number). Specifically, a larger average particle size leads to a higher void ratio in the packing system and a lower coordination number, facilitating better flow of the granular material. The subsequent sections will employ numerical testing methods to quantitatively analyze the flow characteristics of the caved ore and rock under varying packing system conditions (Spatial distribution non-uniformity of different particles), focusing on the morphological changes of the IEZ and the void ratio, average coordination number, and their probability distributions within the particles.

5.2.3 *Study on the Numerical Draw Test*

In both domestic and international block caving mines, the non-uniform size distribution of caved ore and rock not only reflects the inherent differences in particle sizes but also involves the non-uniformity in the spatial distribution (packing state) of the various sizes of caved ore and rock. To investigate how different size gradations of granular media influence the morphological changes of the IEZ, we conduct physical draw tests, complemented by numerical simulations that examine draw behavior across various particle packing systems of the caved ore and rock.

5.2.3.1 Design of Numerical Test Scheme

It is well established that natural block caving mining operations experience phenomena such as natural fragmentation, primary fragmentation, and secondary fragmentation during the draw process [54]. Consequently, the particle size distribution within the mine exhibits a decreasing trend from the top to the bottom, as illustrated in Fig. 2.1 [52]. In block caving mines characterized by high ore strength and low rock strength, as well as in situations involving tailings storage and land occupation within collapse areas, the concept proposed by Ren et al. [49] of discharging dry tail sand to the active surface collapse area under draw conditions results in a particle size distribution phenomenon where particles are smaller at the top and larger at the bottom within a specific stope height. Furthermore, due to the impact and compaction effects of blasting on the existing caved ore and rock, coupled with the prolonged compaction resulting from the accumulation of caved ore and rock, varying packing densities or void ratios of caved ore and rock will develop within the stope.

To investigate the flow behavior of caved ore rock under various particle size spatial distributions, four distinct numerical tests were conducted, as illustrated in Fig. 5.17. The particle size distribution for these tests was aligned with the red curve in Fig. 5.14, maintaining a similarity ratio of 1:25 compared to the physical draw test. Consequently, the particle size distribution for the numerical tests is detailed in Table 5.1, yielding an average particle size of $D_{50} = 0.0149\text{ m} \times 25 = 0.3725\text{ m}$. Test 1 mirrored the physical draw test for the medium particle group, where particles of varying sizes accumulated naturally and were evenly distributed in height (Fig. 5.17a), serving to validate the numerical draw model. In contrast, Test 2 featured densely packed particles of different sizes, uniformly distributed in height (Fig. 5.17b). Tests 3 and 4 employed two natural packing methods, with particle sizes decreasing from top to bottom (Fig. 5.17c) and increasing from top to bottom (Fig. 5.17d), respectively.

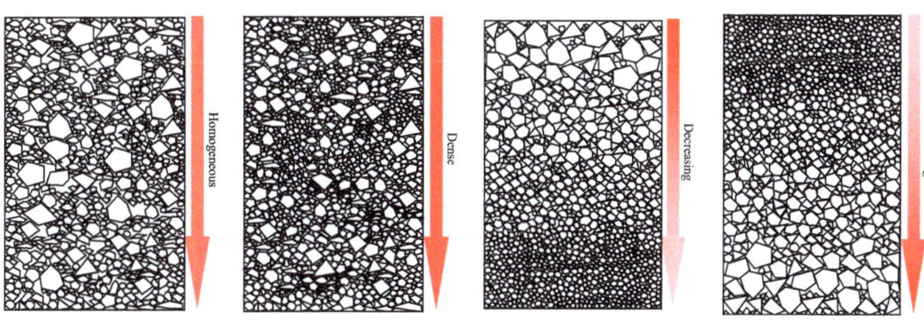

(a) Numerical draw test 1 (b) Numerical draw test 2 (c) Numerical draw test 3 (d) Numerical draw test 4

Fig. 5.17 Schematic of four numerical draw tests under different spatial distributions of caved ore and rock

Table 5.1 Particle size distributions in four numerical draw tests

Particle size/m	Cumulative mass fraction/%
0.20	3.53
0.30	15.75
0.40	34.74
0.50	55.75
0.60	74.00
0.70	86.86
0.80	94.34
0.90	97.93
1.00	100.00

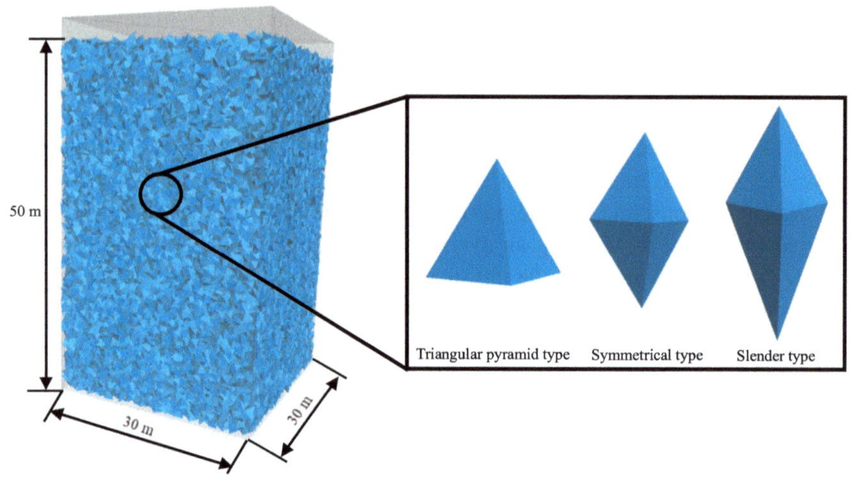

Fig. 5.18 Model used in numerical draw tests

5.2.3.2 The Process of Numerical Test

To mitigate the impact of boundary effects during the draw operation, the dimensions of the numerical draw test model (illustrated in Fig. 5.18) are established at $30 \times 30 \times 50$ m^3. The surfaces of spherical particles are excessively smooth, lacking adequate interlocked force among them [55]. Consequently, this simulation employs the rigid block model within the PFC software [32] to create three varieties of irregular rigid blocks: triangular pyramids, symmetrical shapes, and slender forms. These three block types effectively represent the common shapes of actual caved ore and rock. The quantity of rigid blocks of varying sizes utilized in the four numerical test sets is based on the particle size distribution detailed in Table 5.1.

Before initiating each draw test series, the initial arrangement of particles is evaluated, and calculations are performed to determine the void ratio and average coordination number at various heights. During the draw phase, the shape of the IEZ is dynamically reconstructed at different intervals using the FISH program, enabling the measurement of its height, maximum width, and additional characteristics. All other simulation steps conform to the physical testing procedures for drawing outlined in Sect. 5.2.2.2. The extraction of ore ceases once the draw height reaches 50 m. The mesoscopic parameter values for the wall and rigid block across the four numerical draw tests are detailed in Table 4.5.

5.2.3.3 Analysis of Numerical Test Results

Following the completion of four numerical draw tests, a comparative study was conducted to investigate the patterns of morphological changes within the IEZ. This analysis also took into account the void ratio, average coordination number, and the probability distribution across various particle packing systems, with a specific focus on the flow behavior of the caved ore and rock.

(1) Analysis of the morphological change laws of the IEZ

To enhance computational efficiency, this numerical test utilized spherical particles that matched the volume of the rigid block to dynamically reconstruct the spatial configuration of the IEZ. The morphologies of IEZ at a height of 50 m, derived from four distinct numerical tests, are illustrated in Figs. 5.19 and 5.20, alongside the fitting curves correlating height with the maximum radius. A comparison between the morphology from Test 1 and the physical test of the medium-sized particle group (depicted in Figs. 5.19a and 5.15b), as well as the quantitative relationship between height and maximum radius (represented by the black curve in Fig. 5.20 and the red curve in Fig. 5.16), reveals that the outcomes of the numerical and physical tests are largely aligned under identical conditions. This consistency validates the numerical draw model and its associated mesoscopic parameters. Furthermore, an analysis of the four numerical tests concerning draw across various particle packing systems indicates, as shown in Fig. 5.20, that the heights of the IEZ conform to the power function relationship outlined in Eq. (5.2), with a correlation coefficient R^2 exceeding 0.996. For smaller heights (e.g., 20 m), Test 2, which utilizes the natural particle packing system, exhibits the widest IEZ. In contrast, the other three tests conducted under random packing demonstrate similar widths. Conversely, at greater heights (e.g., 50 m), Test 2 yields the narrowest IEZ, followed by Test 4, while Tests 1 and 3 display the widest dimensions. Notably, as illustrated in Fig. 5.20, Tests 1, 3, and 4, which employed the natural particle packing system, produced IEZs resembling an upside-down drop shape with minimal variation. In contrast, Test 2, conducted under dense packing conditions, resulted in an upright drop shape characterized by a 'lower large and upper small' configuration.

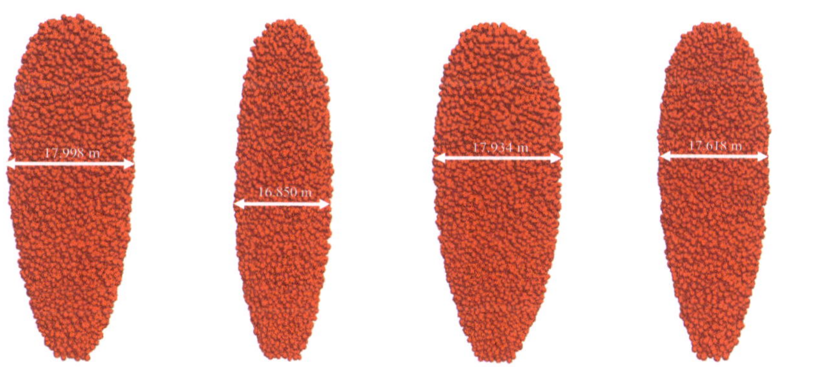

(a) Numerical draw test 1 (b) Numerical draw test 2 (c) Numerical draw test 3 (d) Numerical draw test 4

Fig. 5.19 The shapes of IEZs with the height of 50 m in four numerical draw tests

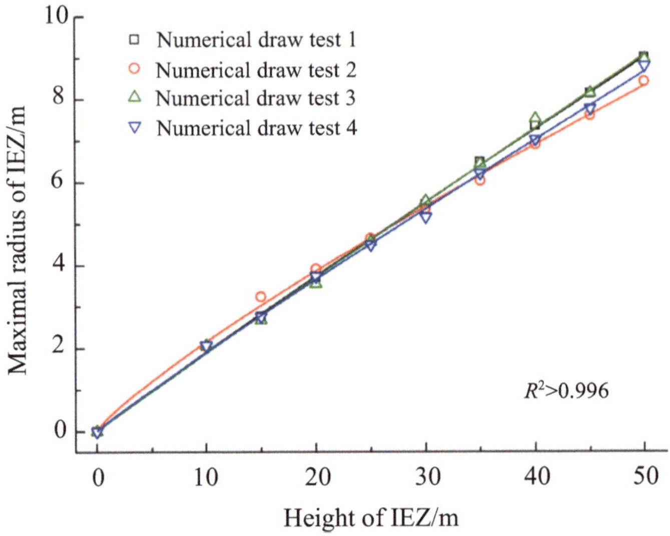

Fig. 5.20 Fitting curves between the height and maximal radius of IEZ in four numerical draw tests

According to the qualitative analysis presented in Fig. 5.19, a quantitative assessment of the spatial morphology variations of the IEZ under different packing techniques was conducted, while disregarding the impact of the drawpoint on morphology when the draw height is relatively low [33]. The maximum width of the IEZ at heights ranging from 30 to 50 m, derived from four numerical draw tests, was statistically evaluated (see Fig. 5.21). It is important to clarify that the 'half-height line' illustrated in Fig. 5.21 denotes the horizontal line at half the height of the IEZ (for instance, at a height of 50 m, the 'half-height' is 25 m), which serves to analyze and describe the longitudinal symmetry of the morphology of the IEZ. As depicted in Fig. 5.21, the maximum widths from Tests 1, 3, and 4 occur at heights exceeding their respective half-height lines. This observation indicates that these widths are located in the upper half of the IEZ morphology, which resembles an upside-down drop shape characterized by a narrower base and a broader top. In contrast, the maximum widths from Test 2 are situated below their corresponding half-height lines, placing them in the lower half of the morphology, which resembles an upright drop shape with a broader base and a narrower top. In summary, the morphology of the IEZ derived from the natural block caving method utilizing a natural particle packing system adopts an upside-down drop shape, while that derived from the dense particle packing system under the block caving method appears upright. Consequently, in large-scale natural block caving operations with draw heights of 200–300 m (for instance, a mineral layer height of 200–300 m), if the loose particle packing system is excessively dense under identical drawpoint spacing, the adjacent upright water droplet-shaped IEZ is less likely to intersect, particularly in areas distant from the drawpoint. This phenomenon results in a higher ore loss rate for the ore extracted from the dense

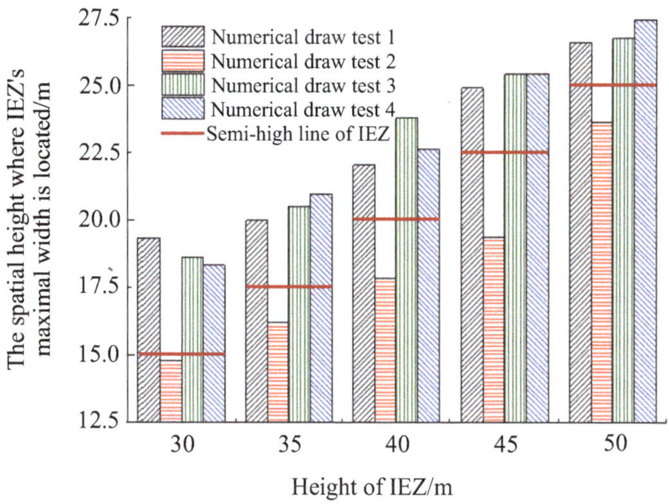

Fig. 5.21 The spatial height where the maximal width of IEZ of different height is located in four numerical draw tests

particle packing system under the block caving method or necessitates a reduction in the aforementioned stope structural parameters under loose particle packing conditions.

(2) Analysis of void ratio and coordination number

To investigate the variations among different particle packing techniques, a statistical analysis was performed on the void ratios, average coordination numbers, and their probability distributions at various heights within the initial packing systems of the four groups prior to the draw (see Fig. 5.22).

Figure 5.22a illustrates that across the four numerical tests, the void ratios of the initial particle arrangements increased with height, indicating that particle packing became denser from the top down due to gravitational effects. Furthermore, the void ratios and their trends in the natural packing systems from Tests 1, 3, and 4 were largely similar and significantly higher than those observed in the dense packing system of Test 2.

Figure 5.22b illustrates that, across the four numerical tests, the mean coordination numbers of the particles' initial packing systems decreased as height increased. This observation indicates a significant inverse relationship between the void ratio and the mean coordination number of the particle packing system. Moreover, the mean coordination numbers for the natural packing systems in Tests 1, 3, and 4 were relatively consistent and substantially lower than those recorded in the dense packing system of Test 2.

The distribution of coordination numbers among particles serves as a key indicator of the structural non-uniformity within a particle packing system [56]. As illustrated in Fig. 5.22c, an analysis of four numerical tests reveals that the average coordination

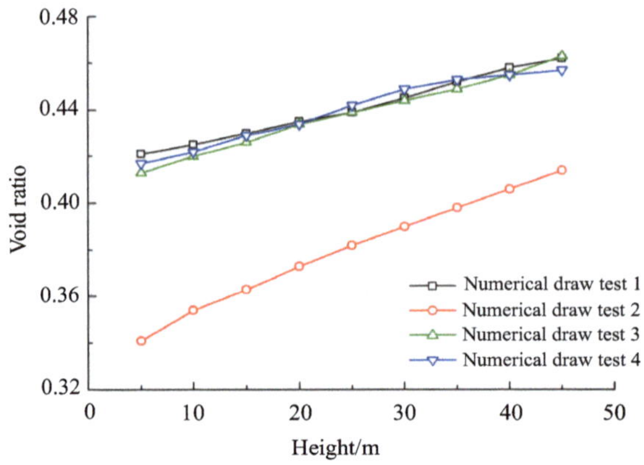

(a) Broken line chart of void ratio of different spatial heights

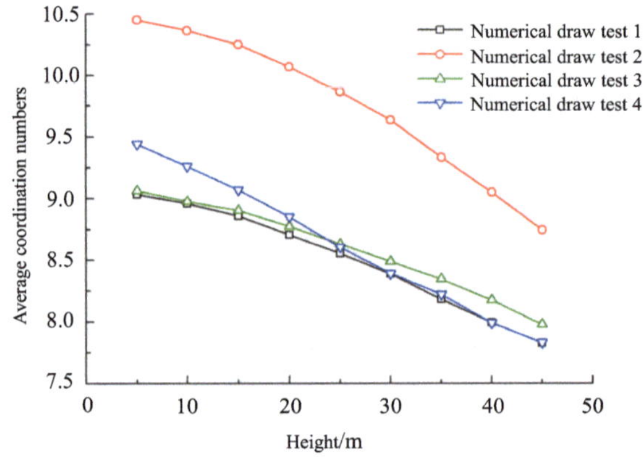

(b) Broken line chart of particle average coordination numbers of different spatial heights

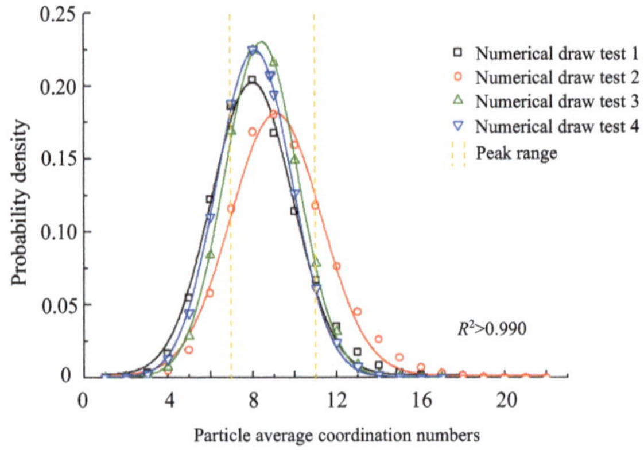

(c) Probability distribution fitting curves of particle average coordination numbers

Fig. 5.22 The void ratio, average coordination number and probability distribution of different heights in four initial particle packing systems

number distribution aligns with a Gaussian profile, achieving a correlation coefficient R^2 exceeding 0.99. The average coordination number peaks between approximately [7, 11]. Notably, in Test 2 (represented by the red fitted curve), this distribution shifts rightward compared to the other tests, indicating an increase in the average coordination number as shown in Fig. 5.22b. Additionally, the peak range for the average coordination number in Test 2 is broader, suggesting a heightened degree of anisotropy in the particle packing system, which further implies that the anisotropic characteristics of the packing system are closely linked to variations in the spatial configuration of the IEZ.

In summary, the analysis presented in Fig. 5.17a, c, and d suggests that varying non-uniform spatial distributions of particles, while maintaining consistent gradation conditions, do not significantly affect the void ratio, average coordination number, or their variations within the overall natural particle packing system. Furthermore, the density of the particle packing system is critical in determining the flow properties of the caved ore and rock.

(3) Discussion on the relationship between particle gradation, packing height, and void ratio spatial variation of the particle packing system of caved ore and rock

The findings from four numerical draw tests conducted under various spatial configurations indicate that, within the examined parameters, the irregular gradation of caved ore and rock does not significantly influence the void ratio at different heights within the packing system. It is established that, assuming other variables remain constant, larger particle sizes correspond to higher average void ratios; similarly, particles positioned at greater heights (with less overlying rock) exhibit higher average void ratios. Consequently, for the two spatial distribution patterns of caved ore and rock illustrated in Fig. 5.17a, b, where different gradation of caved ore and rock are uniformly blended (maintaining a consistent average particle size throughout), the height of the caved ore and rock emerges as the primary determinant of the void ratio, indicating that the void ratio increases with greater spatial height. In contrast, the distribution pattern depicted in Fig. 5.17c demonstrates that, influenced by both caved ore and rock size and spatial position, the caved ore and rock located at lower positions (with more overlying rock) yield a reduced average void ratio. Notably, in the configuration shown in Fig. 5.17d, despite the interplay of caved ore and rock size and spatial position, it remains evident that particles at lower heights have a smaller average void ratio. This suggests that, within the studied range (particle sizes from 0.2 to 1.0 m and a packing system height of 50 m), the height of the overlying ore-rock is the key factor affecting the spatial variation of the void ratio in the packing volume. Furthermore, it can be inferred that the spatial variation patterns of the void ratio in the particle packing system of caved ore and rock may differ across various ranges of particle sizes and packing heights.

Research indicates that in the context of underground mining, the caved ore and rock generated from blasting or natural collapses constitute a distinct macro-scale particle material system [57, 58]. This system exhibits notable features, including the 'stress arch effect' and the 'granary effect' [59, 60]. Future studies will apply principles from statistical physics to quantitatively analyze the mechanical interactions between the caved ore and rock, with the aim of uncovering the relationships among particle size, packing height, and the structural properties of the particle packing system of caved ore and rock.

(4) Discussion on the relationship between the structure of the particle packing system and the shape of the IEZ

This study's physical and numerical tests reveal a significant correlation between the anisotropy of the particle packing system of caved ore and rock structures and the morphological evolutions of the IEZ. These factors ultimately influence the design parameters of stope structures and the indicators of ore loss. Previous research indicates that when the void ratio between particles reaches a critical threshold, the local system forms a mechanically stable and disordered solid with rigidity. For a three-dimensional frictionless spherical particle system, this critical void ratio is 0.36; however, for non-spherical particles with friction, this critical value increases [61]. Consequently, in a densely packed particle system within a localized area, the particles during the draw process are no longer independent and can be considered as 'clusters,' as illustrated in Fig. 5.23. The non-uniformity of this local structure results in variations in the dynamics between the particles, thereby affecting the flow characteristics of the caved ore and rock. Furthermore, if the particle packing system of caved ore and rock is excessively dense, it may lead to 'hang-up' phenomena, which are detrimental to normal production and operational efficiency in the mine.

Subsequent authors will utilize physical tests, numerical simulations, and theoretical analyses to conduct in-depth studies on the critical values of void ratio and coordination number within the void spaces between caved ore and rock, as well as their relationships with particle shapes, sizes, and other influencing factors. They aim to elucidate the relationship between the structure of the particle system of caved ore and rock and the characteristics of dynamic non-uniformity. Furthermore, this

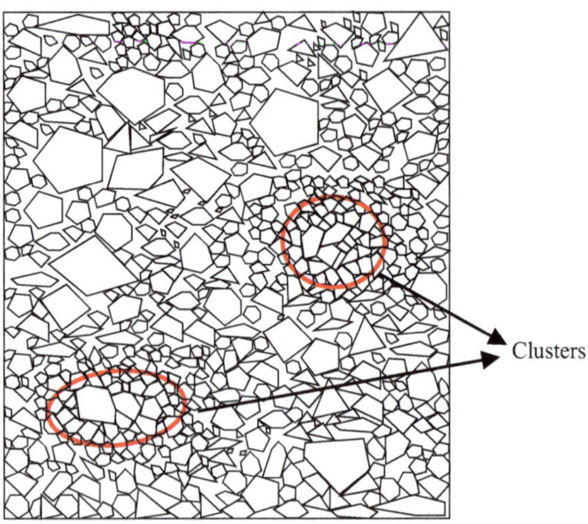

Fig. 5.23 Schematic of local dense packing of caved ore and rock

research will uncover the mechanical mechanisms governing the transport and evolution of caved ore and rock, ultimately leading to the development of a unified characterization model for the morphology of IEZ based on the structural characteristics of the particle system of caved ore and rock.

5.3 Experimental Study on Shape Evolution Characteristics of IMZ Under Influence of Single Coarse Particle

5.3.1 A Brief Introduction to Draw Research Under the Influence of Coarse Particles

The study of the flow characteristics of caved ore and rock is primarily conducted through physical tests and numerical simulation methods, aiming to explore the morphological changes in the IEZ and the IMZ, as well as their impacts on the structural parameters of the stope and indicators of ore dilution [34]. Quantitatively characterizing the morphological evolution of the IMZ is not only crucial for predicting the morphological changes of the IEZ (as IMZ is considered a precursor to IEZ) [35], but it also serves as a key indicator for assessing the flow characteristics of caved ore and rock. Currently, several mature draw theories have been established, including ellipsoid theory [4], ellipsoid-like theory [5], upside-down drop shape theory [8, 33], and stochastic medium theory [6, 62]. However, due to the challenges associated with data monitoring, collection, and analysis, draw physics and numerical tests often simplify various aspects of the test materials, such as geometric and physical mechanical properties (e.g., granular shape, grading range, bulk density, and strength), model similarity ratios (e.g., 1:30, 1:50, and 1:100), draw processes (e.g., draw methods, single draw volume, and draw cut-off conditions), and boundary conditions (e.g., infinite boundaries, semi-infinite boundaries, complex boundaries, and boundary roughness). These simplifications can potentially lead to discrepancies in particle flow behavior and final draw results compared to actual draw conditions. Brunton [63] found, based on the tracer particle method and in-situ draw tests, that the actual flow characteristics of caved ore and rock under near-field conditions significantly differ from the ellipsoid draw theory. The primary reason for this discrepancy is that the coarse particles in the stope substantially influence, or even dominate, the flow behavior of caved ore and rock. While existing draw theories possess distinct advantages and disadvantages, and have achieved varying degrees of recognition and practical application, they have effectively facilitated the development and refinement of draw theories and technologies. Nonetheless, there remains a notable gap between these theories and actual draw results, particularly regarding the impact of coarse particles of different sizes and quantities in real draw scenarios, which has not been adequately addressed. Consequently, it is imperative to investigate the morphological evolution characteristics of the IMZ and the flow characteristics of caved ore and rock under the influence of varying coarse particle conditions.

The disintegrated caved ore and rock represent a typical macro-scale granular material system [57, 58]. The development of research related to particle materials and advancements in testing techniques are closely intertwined. In recent years, measurement technologies for the flow characteristics of granular materials have rapidly advanced. These technologies primarily include digital image methods, laser speckle methods, acoustic techniques, dynamic analysis techniques, and non-contact measurement techniques such as X-ray tomography scanning [64]. Slominski et al. [65] and Katsuragi et al. [66] utilized particle image velocimetry (PIV) and speckle visibility spectroscopy (SVS), respectively, to measure the strain and rate during the flow process of a two-dimensional sand pile. Liu et al. [67] and Chen et al. [68] employed photoelasticity and digital image correlation (DIC) techniques, respectively, to investigate the evolution and failure mechanisms of internal force chains within the particle system. Zhang et al. [69] applied the time-of-flight method to assess the variations in shear wave and longitudinal wave velocities of glass bead particles, discovering that both exhibited power-law scaling in response to pressure changes. Additionally, relevant scholars have conducted systematic studies on the structural and motion characteristics of non-spherical particle systems, including dry balls [70], wet ball systems [71], ellipsoids [72], rod-shaped particles [73], and tetrahedrons [74], utilizing X-ray computed tomography (CT). In the field of draw, Zhang et al. [48] and Ren et al. [49] conducted quantitative studies on the transport trajectory of loose top coal and the flow characteristics of small particles using high-speed photography technology. The limitations of traditional indoor testing methods have become increasingly apparent regarding their ability to study the flow characteristics of caved ore and rock, particularly in terms of breadth and depth. Therefore, it is essential to fully integrate the advantages of non-contact measurement technology with traditional draw testing methods.

Currently, scholars both domestically and internationally have conducted extensive research on the flow characteristics of caved ore and rock, including the evolutionary characteristics of IMZ. Castro and Truema et al. [14, 75] employed an improved extensometer to detect the envelope surface of loose areas in large three-dimensional physical draw models, quantitatively characterizing the morphological changes of IMZ under varying particle size distribution conditions, as well as the interaction rules between adjacent IMZ. Melo and Vivanco et al. [76–78] optimized the particle vacancy diffusion coefficient based on kinematic models and derived the shape equations of IMZ under both two-dimensional and three-dimensional conditions. Chen et al. [79] established a physical test model based on the principle of similarity to label particles, thereby characterizing the morphology of IMZ and IEZ, and investigating the isolated draw in a flexible isolation layer. They obtained the evolution laws of IMZ, IEZ, and cavities, discovering that prior to the peak of ore subsidence, the IMZ were approximately ellipsoidal in shape; post-settlement, the overall shape of the IMZ resembled a trumpet, featuring an exponential curve in the upper part and an approximate ellipsoid in the lower part. Sun et al. [52, 55] utilized discrete element software PFC to customize irregular cluster models and rigid block models, respectively, to explore the flow characteristics of caved ore and rock, taking into account secondary fragmentation and near/far-field conditions. They found that the morphological changes of IMZ under these various conditions conform to the upside-down drop shape theory. During the initial draw stage, the maximum width of IMZ increases rapidly in a power function form with height; in the subsequent draw process, the maximum width of the IMZ increases approximately linearly with increasing height.

During the underground mining process, natural collapses or blasting events generate several large-sized caved ore and rock (Fig. 5.24). These locally densely packed particles form large rigid 'Clusters' [80] (Fig. 5.23), which significantly influence the movement characteristics of the caved ore and rock, the design of stope structural parameters, and the recovery indicators of the ore. Furthermore, although multiple coarse particles are present in an actual stope, the complex and variable interactions among these particles necessitate the examination of a single coarse particle as a starting point. This approach allows for the analysis of how coarse particles affect the movement evolution of surrounding caved ore and rock, thereby providing a reference for subsequent research on varying numbers of coarse particles.

Fig. 5.24 On-site coarse particles

In the context of the actual stope, existing physical tests, numerical simulations, and theoretical studies on draw have inadequately accounted for the impact of coarse particles on the draw process and its outcomes, resulting in overly idealized results. Therefore, this study conducts indoor tests utilizing an independently designed non-contact monitoring and analysis system for caved ore and rock. The aim is to investigate the morphological evolution characteristics of the IMZ and the movement patterns of caved ore and rock influenced by single coarse particles of varying size ratios. The findings of this research serve as a valuable supplement to current discharge theory and provide insightful references for the design of stope structural parameters and the optimization of blasting parameters in block caving mines.

5.3.2 Experimental Study on Draw Considering the Influence of Different Coarse Particles

5.3.2.1 Test System and Materials

This study employs a self-designed non-contact monitoring and analysis system for the particle system of caved ore and rock (see Fig. 5.25). The system is composed of two main components: the caved ore and rock image acquisition subsystem and the caved ore and rock image processing subsystem.

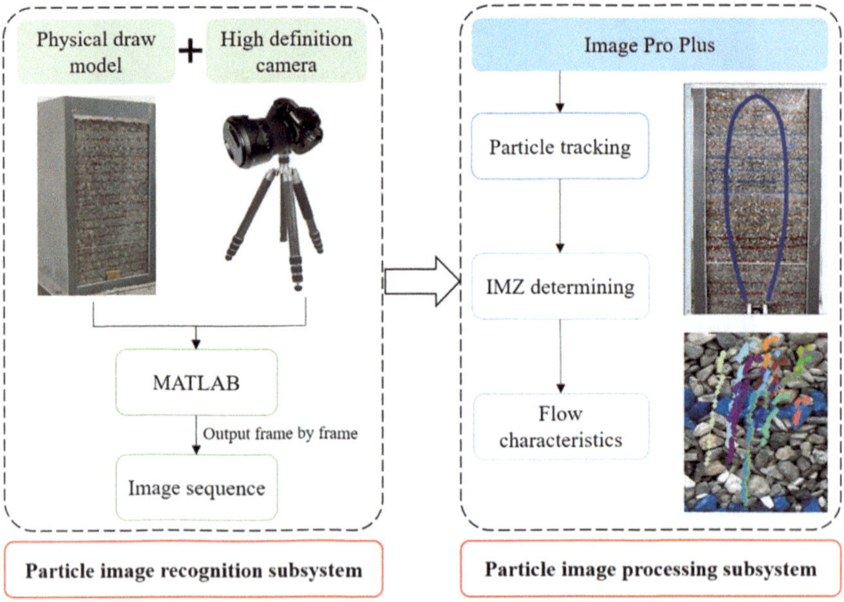

Fig. 5.25 Non-contact draw monitoring and analysis system for particle system of caved ore and rock

The image acquisition subsystem for caved ore and rock is primarily based on a visualized draw model and high-definition cameras to monitor the movement and evolution of caved ore and rock. It employs MATLAB software to output image sequences frame by frame for subsequent processing and analysis. Specifically, (1) In this test, a three-dimensional physical draw model with a similarity ratio of 1:100 was utilized. The internal dimensions of the model were 500 × 500 × 1200 mm^3, with the drawpoint at the end measuring 30 × 30 mm^2. The front end of the model was constructed from 12 mm thick transparent organic glass to facilitate observation of the flow process of caved ore and rock; the rear wall consisted of 14 detachable organic glass plates measuring 50 × 8 mm^2, allowing for the filling of caved ore and rock. The remaining side walls were made of PVC material. Additionally, to better resemble a real mining face wall, the inner walls on both sides of the model were roughened. (2) The camera employed in this test was a Canon 90D, capable of capturing 25 frames per second at a resolution of 6960 × 4640, with an effective pixel count of 325 million. (3) Based on the varying durations of each test group, MATLAB software was used to select appropriate intervals for outputting the monitored movement process of caved ore and rock, frame by frame, as an image sequence.

The caved ore and rock image processing subsystem primarily utilizes the image recognition software IPP to dynamically capture, track, and analyze the overall transport characteristics of various caved ores and rocks during the draw process. IPP, developed by Media Cybernetics in the United States, is an image processing and analysis software capable of measuring and tracking objects based on their contours and colors. It features a comprehensive image optimization function that preprocesses images to highlight the relevant measured information. Currently, IPP is widely employed in fields such as medicine [81] and materials science [82]. In this draw test, IPP software was utilized to dynamically capture different caved ores and rocks, identify particle motion and stable areas, and automatically delineate the shape of the IMZ. Concurrently, trajectory tracking was performed on the various caved ores and rocks to obtain motion characteristics, including particle displacement and velocity, which facilitated the study of the flow characteristics of caved ores and rocks.

The material used in this draw test is light-colored limestone, characterized by a particle density of 2620 kg/m^3. The fine particle size ranges from 3 to 8 mm, with a bulk density of 2050 kg/m^3. For the coarse particle test material, a dispersion of the same type of limestone with a particle size range of 5–30 mm was selected based on the ratio of coarse to fine particle size d_c/d_f. To minimize the impact of dust on the shooting effect, all particles were thoroughly cleaned and dried prior to the test. Additionally, to facilitate the dynamic capture of coarse particles and the surrounding fine particles, the particles were dyed.

5.3.2.2 Testing Program

This study is structured into three main components: (1) an examination of the reliability and advantages of IPP software in conducting indoor draw tests; (2) an

Table 5.2 Draw test scheme considering the influence of different coarse particles

No	Research content	Ratio of particle size of coarse and fine particles d_c/d_f	Particle size range/mm	Location of the coarse particles
I	Software reliability testing	1.0	3–8	-
II	The influence of a single coarse particle on the shape of the IMZ and the flow characteristics of the surrounding fine particles	2.0	3–8	The center right side of the end face glass of the model 115 mm
III		3.0		
IV		4.0		
V		5.0		
VI		5.4		
VII		6.0		

investigation into the morphological alterations of the IMZ induced by coarse particles under single coarse particle conditions; and (3) an analysis of the changes in the flow characteristics of surrounding fine particles resulting from the presence of coarse particles under single coarse particle conditions. The specific testing scheme is detailed in Table 5.2. The fine particles utilized in all experiments range from 3 to 8 mm, while the coarse particle size d_c varies from 1.0 to 6.0 times the fine particle size d_f. Preliminary tests indicated that, compared to other locations, the relative motion between particles on either side of the IMZ is significantly more vigorous, making this area particularly suitable for observing the impact of coarse particles on both the IMZ and the surrounding fine particles. Consequently, selecting a position that remains at the boundary of the IMZ for an extended period can elucidate the influence mode and effects of coarse particles on the morphology of the IMZ. Therefore, as illustrated in Fig. 5.26a, the coarse particles are positioned closely within the glass.

In this test, the filling height of the caved ore and rock was set at 1000 mm. To accurately identify the caved ore and rock, and to enhance the precision of the IMZ mapping, a layer of tracer particles was applied every 50 mm in the vertical direction, tightly adhering to the inner side of the glass, as illustrated in Fig. 5.26b. During the test, approximately 15 g of particles were extracted each time. The test concluded when the coarse particles were released from the model.

5.3.3 *Reliability Verification of IPP Software in Draw Experiments*

The test utilizes IPP software to visualize the shape of the IMZ and gather information on the movement of various caved ore and rock types. However, prior to this, it is essential to verify the reliability of the IPP software in conducting draw tests. There are two methods for delineating the shape of the IMZ. The first method is

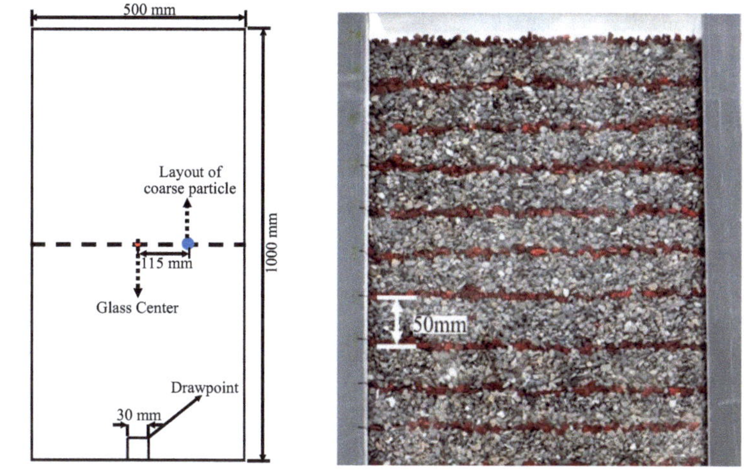

(a) Schematic of layout of coarse particle (b) Layout of labeled markers

Fig. 5.26 Layout of coarse particles and labeled markers

the traditional manual delineation approach, which connects the boundaries of the IMZ based on the movement of tracer particles within the glass. The results of this delineation are presented in Fig. 5.27a. The second method proposed in this study is the IPP-based automatic delineation of the IMZ. This method dynamically captures caved ore and rock using a particle recognition function, obtains cumulative displacement and positional information, and identifies particles that meet the criteria for determining the IMZ, thereby accurately delineating its shape. According to the similarity ratio, particles with a cumulative displacement greater than 10 mm in the model were classified as part of the IMZ in this test [24], with the delineation results illustrated in Fig. 5.27b.

This study compares and analyzes the IMZ shape obtained through the above two methods from both qualitative and quantitative perspectives. Additionally, it verifies the reliability and superiority of the IPP software in conducting draw tests.

From a qualitative perspective, as illustrated in Fig. 5.27a, b, the IMZ shapes obtained through both delimitation methods conform to the upside-down drop shape [9], exhibiting significant vertical asymmetry. However, slight discrepancies are observed in the details between the two methods; specifically, the manually delimited IMZ displays fluctuations in shape, while the automatically delimited IMZ, processed by IPP, presents smoother contours. From a quantitative standpoint, statistical analysis of the maximum radius of IMZ at varying heights yielded fitting curves representing the relationship between IMZ height and maximum radius for both delimitation methods (Fig. 5.28). As depicted in Fig. 5.28, the curves for height and maximum radius obtained from both methods adhere to the power function relationship outlined in the theory of inverted drop shapes [9] (Eq. (5.4)), with correlation coefficients R^2 of 0.983 and 0.985, respectively. In conclusion, the reliability of the IPP software in conducting draw tests has been validated from both qualitative

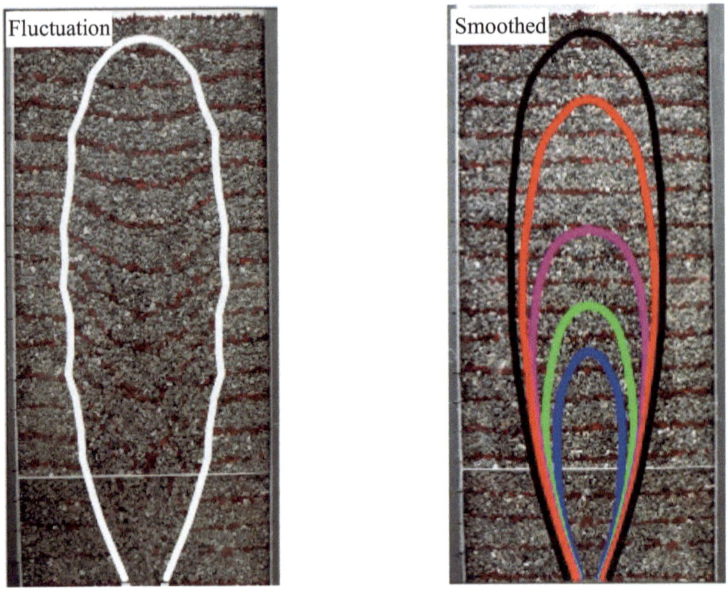

(a) Manual delineation of IMZ (b) IPP-based automatic delineation of IMZ

Fig. 5.27 Front view of IMZ shape obtained by two delimitate methods

and quantitative perspectives. Furthermore, in comparison to manual delimitation methods, the IPP software facilitates more efficient and convenient automatic delimitation of IMZ through particle recognition, demonstrating significant advantages and providing robust technical support for subsequent draw tests and data processing.

$$r = a \times H^b \tag{5.4}$$

Here, r represents the maximum radius of the IMZ, measured in millimeters; H represents the height of the IMZ, also measured in millimeters; a and b are fitting constants.

5.3.4 *The Influence of a Single Coarse Particle on the Shape of IMZ*

Based on the results of the seven draw tests, this study analyzes the influence of coarse particles with varying particle size ratios on the shape of the IMZ from both qualitative and quantitative perspectives.

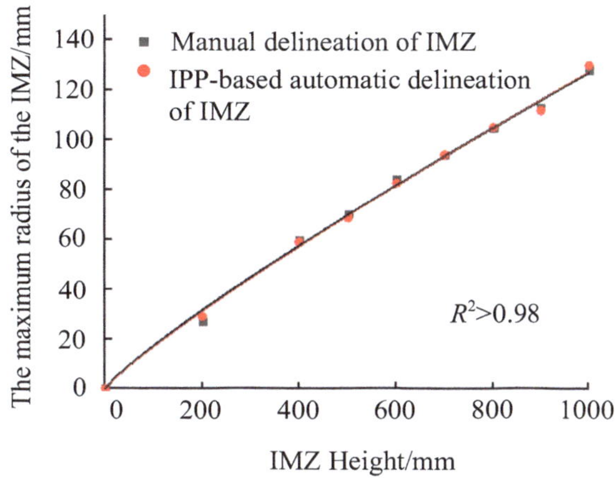

Fig. 5.28 Fitting curves between the height and maximal radius of IMZ in two delimitation methods

5.3.4.1 A Qualitative Study on the Influence of Coarse Particle on the Shape of IMZ

From a qualitative perspective, the impact of coarse particles on the IMZ can be assessed based on their shape as they interact with the IMZ. Consequently, the local shape of the IMZ in contact with coarse particles during various tests was delineated using IPP techniques. The results of the particle size ratio delineation d_c/d_f for coarse and fine particles, which falls within the range of 4.0–6.0, are illustrated in Fig. 5.29. As shown in Fig. 5.29, it can be concluded that only Fig. 5.29d displays a noticeable discontinuous bending around the coarse particles. This observation indicates that coarse particles with a particle size ratio of d_c/d_f equal to 6 can exert a localized influence on the shape of the IMZ.

Test VII investigates the impact of coarse particles on the evolution of the IMZ shape during processing. Utilizing the dynamic recognition function of IPP, we obtained the IMZ shape and the movement status of coarse particles at various heights, as illustrated in Fig. 5.30. To emphasize the location of coarse particles in this test, tracer particles were strategically placed with the upper and lower sections colored

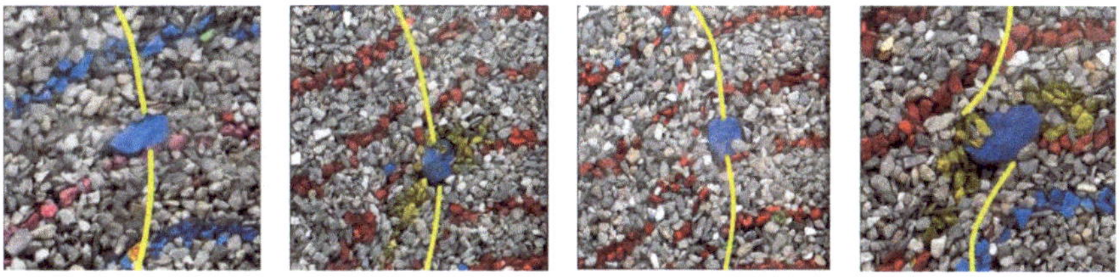

(a) d_c/d_f=4 (b) d_c/d_f=5 (c) d_c/d_f=5.4 (d) d_c/d_f=6

Fig. 5.29 Shape changes of IMZ under the influence of d_c/d_f with different coarse and fine particle size ratio

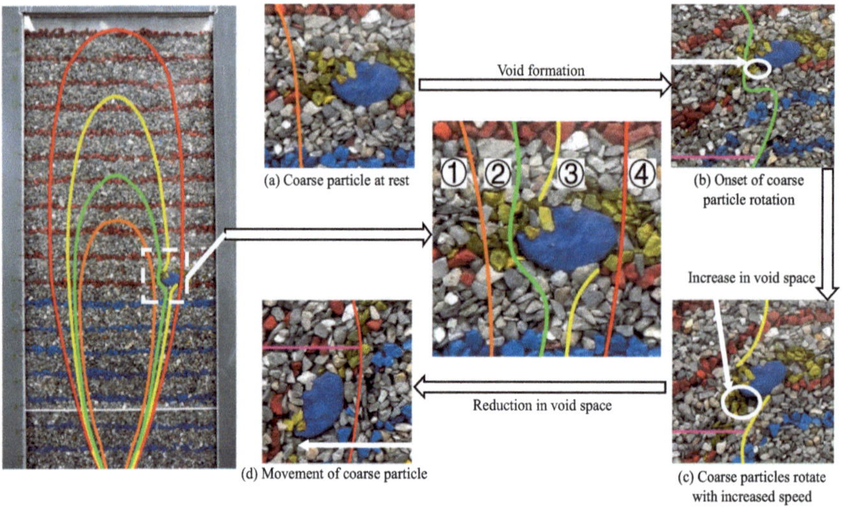

Fig. 5.30 Different stages of IMZ shape change under the influence of coarse particles with the ratio of coarse to fine particle of 6

red and blue, respectively, using the coarse particles as a dividing point. Fine particles surrounding the coarse particles were colored yellow to facilitate observation of their movement and for improved image post-processing.

According to the analysis presented in Fig. 5.30, the evolution process of the IMZ shape under the influence of a coarse particle can be categorized into four distinct stages.

The first stage (represented by the orange curve): When the height of the IMZ is below 700 mm, the coarse particle has not yet made contact with the IMZ, leading to the development of the IMZ in an upside-down drop shape.

The second stage (represented by the green curve): Upon reaching a height of 700 mm, the IMZ begins to interact with the coarse particle, which differ from the surrounding fine particles that start flowing towards the drawpoint. The coarse particles remain stationary, merely rotating in their original positions, which creates gaps beneath them. Consequently, the IMZ continues to evolve around these gaps, resulting in the formation of local depressions.

The third stage (represented by the yellow curve): When the height of the IMZ reaches 800 mm, the single coarse particle continues to rotate. Due to the persistent differences in the flow state of the surrounding fine particles, there is a significant change in their upper and lower void ratios. The gap below the single coarse particle gradually increases, facilitating the loosening of fine particles, which results in the boundary of the IMZ below the single coarse particle shifting outward. Conversely, the void ratio above the single coarse particle decreases, leading to an increased squeezing effect between particles and a slowdown in particle movement. Consequently, the boundary of the IMZ above the single coarse particle bends inward. Unlike the phenomenon observed in the second stage, where fine particles bypassed the single coarse particle, the loosening process in this stage is interrupted by the coarse particle, resulting in a distinct discontinuous bending in the upper and lower sections.

The fourth stage (represented by the red curve): When the height of the IMZ reaches 900 mm, the single coarse particle begins to move towards the drawpoint after sufficient rotation, entering the range of the IMZ. At this stage, the gap below the coarse particle starts to decrease, significantly reducing the hindering effect on the particles above. Consequently, the shape of the IMZ reverts to an upside-down drop shape.

5.3.4.2 Quantitative Study on the Influence of Single Coarse Particle on the Shape of IMZ

Staged changes do not quantitatively reflect specific shape alterations in the IMZ. Building on qualitative research regarding the impact of a single coarse particle on the shape of the IMZ, this study quantitatively characterizes the deformation of the IMZ. As illustrated in Fig. 5.26, the coarse particle is positioned solely on the right side of the model during the experiment. The deformation of the IMZ shape is quantitatively assessed by the difference r_d between the maximum radius r_{rmax} of the right-side IMZ and the maximum radius r_{lmax} of the left-side IMZ. The maximum values of r_d, influenced by the particle size ratio d_c/d_f of coarse to fine particles within the range of 1.0–6.0, are presented in Fig. 5.31.

From Fig. 5.31, it can be observed that when the ratio of coarse to fine particle size d_c/d_f ranges from 1 to 5.4, the maximum deviation of the maximum radius of the IMZ

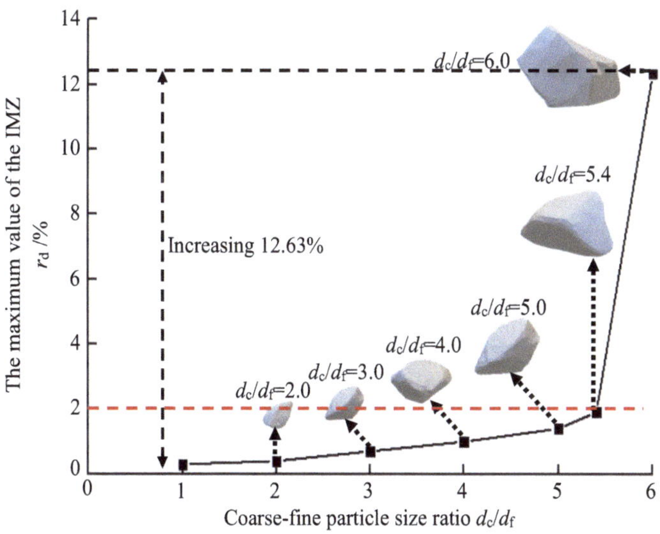

Fig. 5.31 Maximum of r_d under the influence of different coarse particle sizes

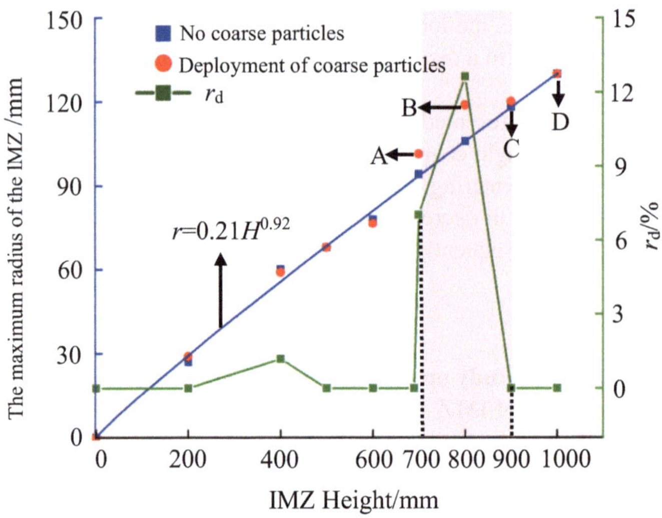

Fig. 5.32 Relationship between the maximum radius of IMZ on both sides and r_d with height under the influence of coarse particle

on both sides exhibits a gradual increasing trend. However, this deviation consistently remains below 2%, which can be considered a normal testing error and thus can be disregarded, indicating that it does not induce any significant alteration in the shape of the IMZ. Notably, the maximum r_d value of the IMZ in Test VII reached 12.63%, quantitatively demonstrating that, under the condition of a single coarse particle, a particle size ratio of coarse to fine particles with a d_c/d_f of 6 or greater can significantly affect the shape of the IMZ. Therefore, in Test VII, the evolution of the IMZ's shape when interacting with a coarse particle was quantitatively characterized by the maximum radius difference r_d between the two sides of the IMZ. The relationship between the maximum radius of both sides of the IMZ and the difference r_d as a function of height is illustrated in Fig. 5.32.

From Fig. 5.32, it can be concluded that r_{lmax} conforms to the power function relationship, but r_{rmax} does not conform to the power function relationship between 700 and 900 mm. At the same time, observing the development trend of r_d, it can also be found that r_d shows significant changes in the same interval. Analyze the local trend of changes in r_d, taking into account the different stages of IMZ changes and the positional relationship between coarse particle and the maximum radius (pink area in Fig. 5.32). The variation trend of the maximum radius on both sides of the IMZ with a height below point A (700 mm) is consistent, as the IMZ is still small and not affected by coarse particle. When the height of the IMZ reaches point A, corresponding to the second stage, r_{rmax} corresponds to a height below the coarse particle. As the packing system below the coarse particle begins to loosen, the maximum radius on one side of the coarse particle begins to increase, and r_d increases from 0 to 7.03%. When the height of the IMZ reaches point B (800 mm), corresponding to the third stage, the r_{rmax} corresponding height becomes closer to the coarse particle due to the development of the IMZ, and is more affected by the course particle. The r_d continues to increase to 12.63%, reaching its maximum value, that is, the local deformation of the IMZ reaches its maximum. When the height of the IMZ reaches point C (900 mm), corresponding to the four stages, r_{rmax} reaches a height above the coarse particle, and r_d gradually approaches 0.

5.3.5 *The Influence of a Single Coarse Particle on the Movement Characteristics of the Surrounding Fine Particles*

The previous section has demonstrated that a single coarse particle can induce local alterations in the shape of the IMZ. Investigating the movement and evolution patterns of fine particles in regions where these changes occur can facilitate a deeper understanding of the various stages of local shape transformations in the IMZ. Consequently, it is essential to examine the evolution patterns of the flow characteristics of fine particles in the vicinity of a single coarse particle. By characterizing the changes in local displacement and velocity around the coarse particle in Test VII, this study explores the influence of the coarse particle on the flow characteristics of the surrounding fine particles.

5.3.5.1 Quantitative Study on the Effect of a Single Coarse Particle on the Shape of IMZ

Based on the regions where the shape of the IMZ varies, a local area centered on the coarse particle, measuring 75 mm in length and 100 mm in width, has been selected. The analysis of the local changes in the IMZ indicates that the influence of a single coarse particle on the surrounding fine particles is heterogeneous. Consequently, the local area is divided into two sections, Z_{ru} and Z_{rd}, with the coarse particle serving as the boundary. For comparative analysis, two corresponding areas, Z_{lu} and Z_{ld}, are selected symmetrically on the left side, as illustrated in Fig. 5.33.

Select all stages affected by the IMZ to calculate the displacement time, and utilize IPP software to track the trajectory of particles within the selected area. This approach allows for the determination of the cumulative displacement of all particles across the four designated areas. The statistical results illustrate the displacement distribution in each area, as depicted in Fig. 5.34.

From Fig. 5.34a, it can be concluded that the displacement probability distributions of the Z_{lu} and Z_{lu} regions are basically consistent, both following a power-law distribution (Eq. (5.5)), and the correlation coefficient R^2 is greater than 0.97.

$$P = cL^d \tag{5.5}$$

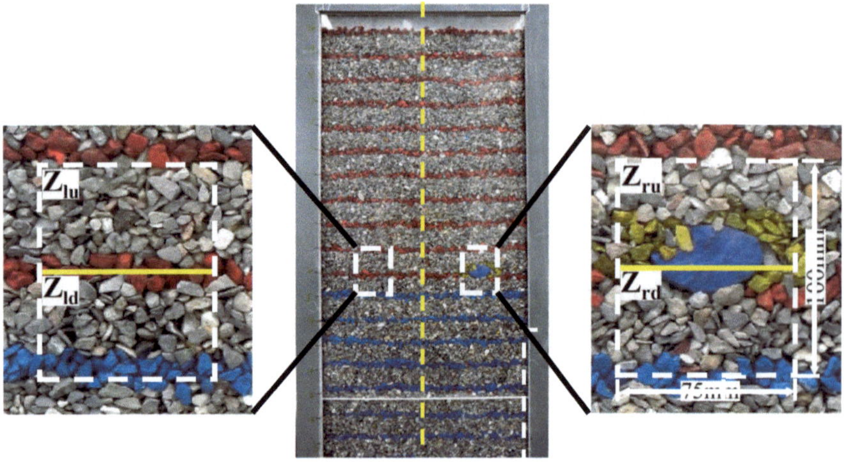

Fig. 5.33 Coarse particle local analysis area

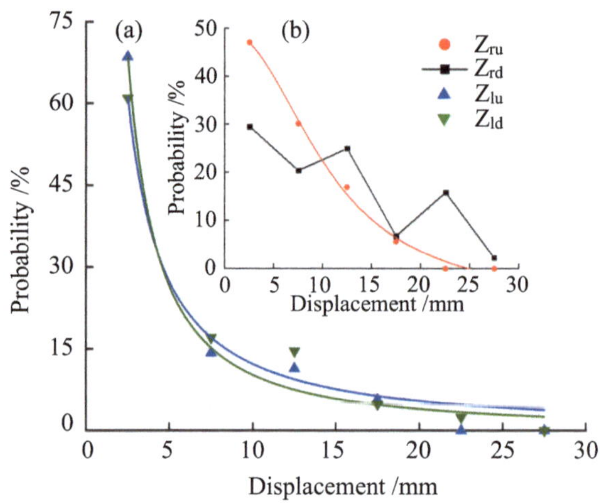

Fig. 5.34 Comparison of displacement probability distribution in different regions **a** probability distribution of displacement in the Z_{lu} and Z_{lu} areas; **b** probability distribution of displacement in the Z_{ru} and Z_{rd} areas

Here, *P* represents probability, expressed as a percentage; *L* represents displacement, measured in millimeters; *c* and *d* are fitting constants.

From Fig. 5.34b, it can be observed that, in contrast to the scenario where the distributions of Z_{lu} and Z_{ld} are identical, the presence of a single coarse particle leads to significant differences in the displacement distributions of Z_{ru} and Z_{rd}. Specifically, the displacement of the upper particles transition from a power-law distribution to a logistic distribution (Eq. (5.6)), with a correlation coefficient R^2 exceeding 0.98. Conversely, the displacement of the lower particles shifts from a power-law distribution to an irregular distribution characterized by greater fluctuations.

$$P = \frac{56.84}{1 + (L/10.44)^{2.18}} - 7.44 \tag{5.6}$$

Compared with Z_{lu} and Z_{ld}, the cumulative probability of particles with displacement greater than 10 mm in the Z_{ru} region (i.e. becoming IMZ) decreased by 10.22%, and the cumulative probability of particles with displacement greater than 10 mm in the Z_{rd} region increased by 28.02%. By comparing the displacement changes of particles in the Z_{ru} and Z_{rd} regions, it can be seen that the influence of the single coarse particle on the displacement of fine particles in the lower region is greater than that in the upper region, indicating that the deformation degree of the IMZ is greater below the coarse particle than above. In response to the phenomenon of varying deformations of the IMZ shape in the upper and lower regions of coarse particles, as illustrated in Fig. 5.34, different interpretations regarding the influence of time on fine particles in these regions can be applied. The relationship curve between average displacement and time for each region can be derived using the IPP method, as demonstrated in Fig. 5.35.

Figure 5.35 illustrates that the timing of changes in Z_{ru} and Z_{rd} differs, indicating that the onset of influence from coarse particles on these two regions does not occur simultaneously. Fine particles in the Z_{ru} region are impacted by a single coarse particle prior to those in the Z_{rd} region. Consequently, the degree of deformation in the IMZ of the Z_{ru} region exceeds that of the Z_{rd} region. This phenomenon can be further elucidated by examining the distinct stages of IMZ changes depicted in Fig. 5.30, where the time points at which the average displacement alters correspond to various stages of IMZ shape modifications. At time t_1, which aligns with the second

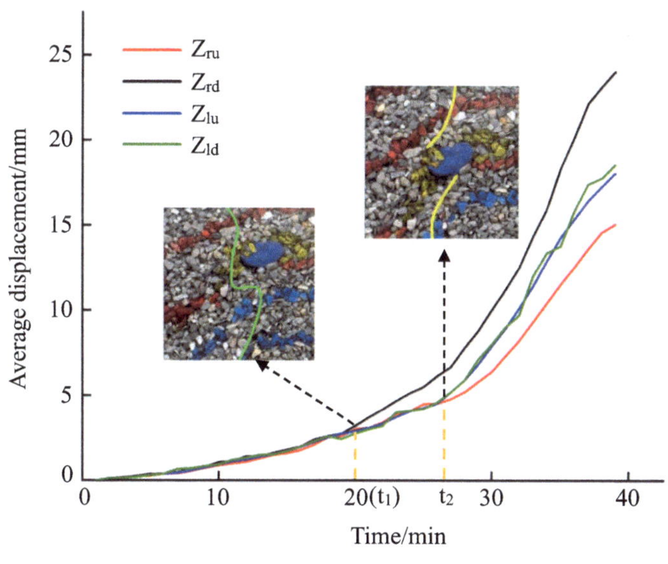

Fig. 5.35 Relationship between average displacement of each area and time

stage, it is evident that the deformation of the IMZ is primarily concentrated beneath the coarse particle. This observation suggests that fine particles in the Z_{ru} region experience accelerated downward displacement due to the presence of the coarse particle, while fine particles in the upper local area remain relatively unaffected. It is not until the third stage of IMZ change (time t_2) that significant deformation is observed in the local area above the coarse particle, indicating that the Z_{rd} region is beginning to be influenced.

5.3.5.2 Quantitative Study on the Effect of a Single Coarse Particle on the Shape of IMZ

In practical engineering, greater emphasis is placed on the speed at which caved ore and rock reach the drawpoint, rather than on changes in particle displacement. Therefore, by monitoring the average velocity of fine particles reaching the drawpoint in the Z_{ru}, Z_{rd}, Z_{lu}, and Z_{ld} regions, this study investigates the impact of a single coarse particle on thc time required for surrounding fine particles to reach the drawpoint. To characterize the proportion of particles influenced by the coarse particle in each region, the probability distribution of velocity was calculated for each area. The statistical results, including the average velocity and the probability distribution of velocity in each region, are presented in Fig. 5.36.

From Fig. 5.36, it can be concluded that the average velocities in the Z_{ru} and Z_{rd} regions are both greater than those in the Z_{lu} and Z_{ld} regions. This indicates that the presence of a single coarse particle promotes the movement of fine particles

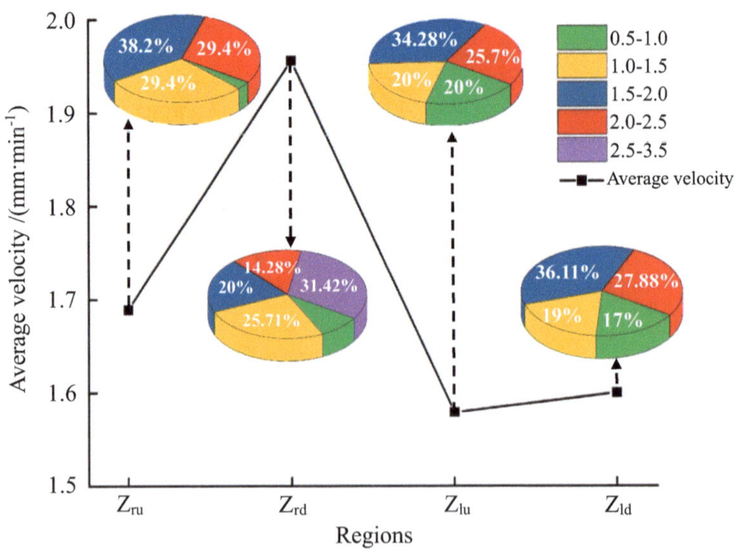

Fig. 5.36 Average velocity and velocity probability distribution of particles in different regions

towards the drawpoint. Compared to Z_{lu}, the average velocity of Z_{ru} particles shows a probability decrease only in the range of [0.5, 1], while exhibiting an increasing trend in all other ranges. Unlike the IMZ boundaries, where coarse particles hinder fine particles in the Z_{ru} region, coarse particles actually accelerate the flow of fine particles towards the drawpoint. Notably, a cumulative 17.02% of fine particles reach the drawpoint ahead of schedule. In comparison to Z_{ld}, the average velocity of particles in Z_{rd} shows a significant increase in the range of [2.5, 3.5], with a cumulative probability increase from 0 to 31.42%. This indicates that 31.42% of particles arrived at the drawpoint ahead of schedule.

The phenomenon of coarse particles promoting the flow of fine particles, as illustrated in Fig. 5.36, can be attributed to the changes in the local void ratio surrounding the coarse particles. By utilizing IPP software to monitor the variations in void ratio during the movement of coarse particles, we obtained the variation curves of void ratio over time for each region, as depicted in Fig. 5.37. From this figure, it is evident that the void ratios in the Z_{lu} and Z_{ld} regions exhibit a consistent trend of change. During the transition from a static to a loose state, the void ratio initially increases before stabilizing. The influence of the single coarse particle results in significant differences in the void ratios of the Z_{ru} and Z_{rd} regions prior to the movement of the coarse particle. This observation aligns with the previously discussed results of accelerated particle movement in the Z_{ru} region and slow particle movement in the Z_{rd} region. It can be inferred that the reason coarse particles not only hinder particle displacement in the Z_{ru} region but also increase the average velocity of particles in this area is due to the differing time periods of interest. In the time interval [0, 40], coarse particles impede movement in the Z_{ru} region. However, when the coarse particle is surrounded by the IMZ during the interval [40, 140], they begin to flow toward the drawpoint. The void ratios of Z_{ru} and Z_{rd} converge and develop steadily. Compared to Z_{lu} and Z_{ld}, the void ratios of Z_{ru} and Z_{rd} are higher, indicating that coarse particles create more space for fine particles to flow. By analyzing the duration of these two influences, we conclude that coarse particles ultimately promote the movement of the surrounding fine particles toward the drawpoint.

5.4 Flow Characteristics of Caved Ore and Rock Under the Influence of Multiple Coarse Particles: Insight from Interparticle Interaction

The caved ore and rock generated by blasting or natural caving during underground mining constitute a complex granular material system. This system not only possesses the fundamental characteristics of typical granular materials, including disorder, multiscale behavior, energy dissipation, self-organization, and nonlinear response [57], but also demonstrates intricate packing patterns due to significant variations in the intrinsic properties of caved ore and rock, such as particle size, shape, density, and strength. Additionally, external factors, including draw methods and boundary

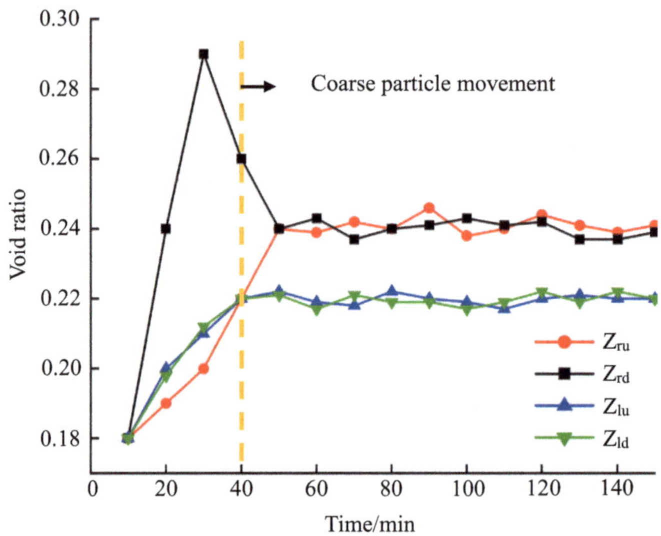

Fig. 5.37 Evolutions of void ratio of areas around coarse particle

conditions, further influence macroscopic statistical laws, such as flow and secondary fragmentation characteristics [35]. Brunton [63] found, based on the tracer particle method and in-situ draw tests, that coarse particles in the stope can significantly affect and even dominate the flow behavior of caved ore and rock. However, considering the operability and efficiency of data monitoring, collection, and analysis processes, coarse particles are often simplified or neglected in most physical and numerical draw tests, resulting in idealized flow laws of caved ore and rock and draw results that deviate from actual draw outcomes to some extent. Therefore, it is necessary to study the flow characteristics and influencing mechanisms of caved ore and rock containing different coarse particles.

The study of the flow characteristics of caved ore and rock primarily involves physical tests and numerical simulations to investigate the laws governing the shape changes of the IEZ and the IMZ, as well as their impact on the structural parameters of the stope and ore loss indicators [2]. The IMZ is not only considered a precursor to the IEZ [52], but also serves as a crucial indicator for assessing the flow characteristics of caved ore and rock. In recent years, researchers have conducted a series of studies focused on the flow characteristics of caved ore and rock, as well as the shape evolution of the IMZ, utilizing both physical tests and numerical simulations. In physical test research, Kvapil [9, 10], Janelid [11], Laubscher [12], Peters [38], Power [13], and Castro [14, 20, 83] frequently utilize sand or gravel as test media to establish various dimensions (2D/3D) and different similarity ratios (e.g., 1:30, 1:50, and 1:100) for physical draw models. They investigate the effects of factors such as particle gradation, drawpoint size and spacing, and vertical stress on the gravity flow and hang-up of caved ore and rock. Furthermore, they establish or enhance theories including ellipsoid theory [84], stochastic medium theory [85], kinematic model [76–78], and upside-down drop shape theory [7, 8]. The physical draw test method is favored by many scholars and holds an indispensable and significant position in draw research. However, traditional measurement methods, such as the tracer particle method, present challenges in accurately tracking the spatial position and movement of caved ore and rock within the model in real time. Additionally, monitoring the magnitude, distribution, and evolution of the contact force between caved ore and rock proves to be difficult. These limitations are the primary reasons why most numerical draw tests continue to focus on describing macroscopic test phenomena and conducting qualitative analyses of test results. Numerical draw tests primarily construct draw models using methods such as the empirical method, finite difference method, finite element method, discrete element method, and cellular automata theory, enabling the examination of various draw problems [3, 86, 87]. Among these, the discrete element method is currently regarded as the most suitable and widely utilized numerical simulation technique in the field of drawing. Hashim [24] employed the PFC program to investigate the influence of particle shape and size distribution on the flow characteristics of fine particles by quantifying shear strain in the IMZ. Similarly, Song et al. [25] utilized the PFC software to explore the impact of particle shape on the morphology of the IEZ. Furthermore, Sun et al. [33, 52, 55, 88], leveraging mesoscopic particle flow theory and the PFC program, examined the movement and fracture mechanisms of caved ore and rock under three types of boundary and near/far field conditions. They visualized the morphological changes in the IEZ and IMZ, and derived an empirical formula for the ore depletion rate. A current challenge and developmental trend in numerical draw research are to fully harness the advantages of discrete element numerical simulation to elucidate the evolution mechanisms of caved ore and rock movement from a microscopic perspective while establishing a continuous medium draw model, all while ensuring calculation efficiency and the reliability of simulation results.

During the actual draw process, the flow of caved ore and rock, influenced by gravity, constitutes a common yet complex non-equilibrium system. The flow characteristics of caved ore and rock are essential manifestations of inter-particle interactions, directly impacting the design of stope parameters and the ore recovery rate. The photoelasticity measurement technique [89] allows for the non-contact measurement of contact forces between particles, making it widely applicable in physical tests to investigate the stress states and force chain structures within compressive and shear particle systems [90–94]. In numerical simulation studies, Imseeh [95] and Hou et al. [96] captured the evolution of force chains in particle systems through finite element and discrete element numerical simulations, respectively, utilizing load distribution rate curves and force chain patterns to describe the carrying behavior of these chains. For coarse particles within the particle system, also referred to as intruders, Díaz-Melián et al. [97, 98] examined the interactions between single and multiple coarse particles with walls when low-velocity projectiles penetrate granular beds. Utilizing the discrete element method, they analyzed the temporal evolution of force chain distributions and the system's energy. Dhiman et al. [99] investigated the effects of static pressure, the velocity of moving intruders, and the friction coefficient on the attractive or repulsive forces between intruders in a confined granular medium, elucidating the origins of long-range attraction or repulsion. Merceron et al. [100] explored the cooperative effects induced by intruders navigating through a granular medium, using stainless-steel disks as the test material, and discovered that the correlation between grain movements diminishes with increasing distance from the intruders over larger length scales.

Existing theories and experimental studies on draw have not adequately addressed the influence of coarse particles on the flow characteristics of caved ore and rock. Research on interparticle interactions has been limited to the statistical analysis of macroscopic mechanical parameters and the description of the macroscopic mechanical properties of particle systems. Consequently, there is a lack of a comprehensive understanding of the correlation mechanisms between the structure and mechanical characteristics of particle flow systems. As a result, the current characterization methods for the formation and evolution of IMZ and IEZ in the particle flow systems of caved ore and rock are diverse yet imprecise, lacking the robustness necessary to guide the design of practical draw plans and production management, which can lead to confusion. This study employs a physical testing method utilizing the image recognition software IPP and a numerical testing approach based on the rolling resistance model to quantitatively investigate the separation of coarse particles, the evolution of IMZ shapes under varying quantities, relative positions, and spacings of coarse particles, and analyzes the influence mechanisms of multiple coarse particles on the flow characteristics of caved ore and rock from the perspective of interparticle interactions. The findings of this research serve as a valuable supplement and enhancement to existing research concepts and methodologies regarding draw, providing useful references for the foundational theoretical research of particulate materials such as sand, rockfill, grains, and chemical products.

5.4.1 *Physical Experimental Study of the Effect of Coarse Particle Numbers on the IMZ Shape*

Based on the self-designed non-contact draw monitoring and analysis system for particle systems of caved ore and rock, a series of end-draw tests were conducted to investigate the evolution of the IMZ shape under the influence of varying coarse particle quantities.

5.4.1.1 Introduction of Test System

The draw monitoring and analysis system (see Fig. 5.25) employed in this draw test comprises two components: the caved ore and rock image acquisition subsystem and the caved ore and rock image processing subsystem [101]. The caved ore and rock image acquisition subsystem primarily monitors the gravity flow process of caved ore and rock using a physical draw model and a high-definition camera. Specifically: (1) This test utilizes a three-dimensional end-drawing physical model with a similarity ratio of 1:100, with dimensions of 500 mm × 500 mm × 1200 mm and an end-drawing opening size of 30 mm × 30 mm. The model is constructed from PVC and high-transparency plexiglass; (2) The camera employed in this test is a Canon 90D, featuring an acquisition resolution of 6960 × 4640 and an acquisition frequency of 25 frames per second; (3) Based on the duration of each test set, suitable intervals are selected to output the monitored caved ore and rock flow process frame by frame as an image sequence. The caved ore and rock image processing subsystem predominantly utilizes image recognition software IPP [82, 102] to conduct dynamic capture, trajectory tracking, and migration characteristic analysis of caved ore and rock during the draw process. To facilitate the dynamic capture of coarse particles and their associated fine particles, staining treatment is necessary for these particles.

5.4.1.2 Schemes of Physical Draw Test

The author's previous research indicates that, in the case of single coarse particles, a coarse-to-fine particle size ratio d_c/d_f exceeding 6.0 significantly influences the shape of the IMZ [101]. Consequently, coarse particles were positioned near the inner side of the plexiglass, as illustrated in Fig. 5.38a. In this experiment, the filling height of the caved ore and rock was set at 1000 mm, utilizing limestone as the test material, which has a solid density of 2620 kg/m^3. The fine particle size range was established at 3–8 mm, with a packing density of 2050 kg/m^3, while the coarse particle size ranged from 5 to 30 mm. To enhance the accuracy of caved ore and rock identification, as well as IMZ determination, a layer of marker particles was placed every 50 mm in the vertical direction, adjacent to the inner side of the glass (Fig. 5.38b). Throughout the test, the output per single discharge was consistently maintained at approximately 15 g. As detailed in Table 5.3, this physical test involved single-drawpoint draw tests with 1, 2, and 4 coarse particles, corresponding to coarse-to-fine particle size ratios d_c/d_f of 6.0, 5.4, and 4.0, respectively.

5.4.1.3 Analysis of the Influence of Coarse Particle Number on the Evolution of IMZ's Shape

Based on the results of three physical draw tests, we conducted a qualitative and quantitative analysis of the influence of coarse particle number on the evolution of the shape of the (IMZ. Utilizing IPP software, we generated contour lines representing the IMZ shape across the three tests. Figure 5.39 illustrates the local particle movements influenced by varying numbers of coarse particles during their motion. From Fig. 5.39, it is evident that the presence of coarse particles—specifically with counts of 1, 2, and 4—exerts a significant impact on the shape of the IMZ.

Taking Test 3 as an example, as shown in Fig. 5.40a, in the area not affected by the four coarse particles, the shape of the IMZ conforms to the upside-down drop

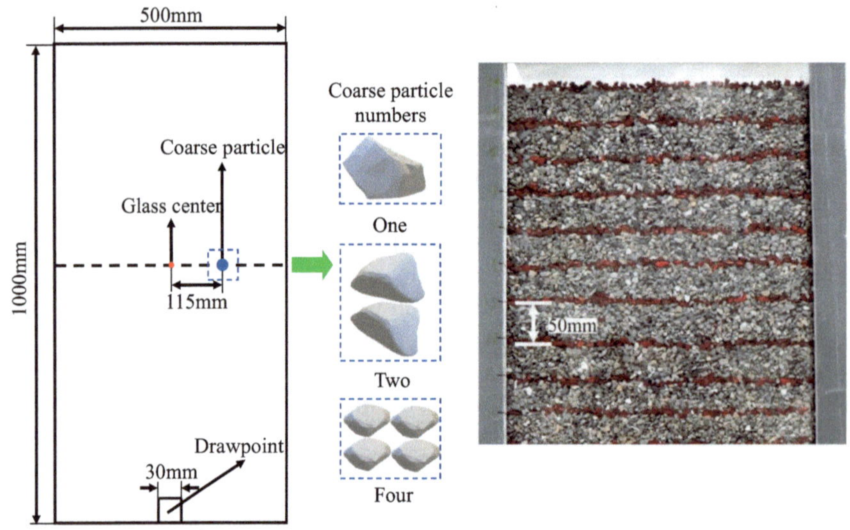

(a) Layout of coarse particles (b) Layout of labelled markers

Fig. 5.38 Layout of coarse particles and labelled markers

Table 5.3 Schemes of physical draw test

No	Coarse particle numbers	Size ratio of coarse and fine particles d_c/d_f	Coarse particle position
1	1	6.0	115 mm away from the central axis
2	2	5.4	
3	4	4.0	

(a) Test 1 (b) Test 2 (c) Test 3

Fig. 5.39 Local particle movements under the influence of different numbers of coarse particles

shape, but the shape of the IMZ in the local area of the coarse particles has undergone variation: (1) As depicted in Fig. 5.40b, when the coarse particles gradually enter the movement zone, the two coarse particles adjacent to the central axis first undergo coordinated movement, creating a void alongside the right coarse particle. The IMZ continues to develop around this void, resulting in a contour line that exhibits a local depression. (2) As shown in Fig. 5.40c, with the progression of the IMZ, the two coarse particles on the left begin to separate and move towards the drawpoint, while the two coarse particles on the right also begin to loosen. Consequently, the void between the coarse particles becomes significantly larger, and the contour line of the IMZ demonstrates a distinct discontinuous bend at the boundary of the coarse particles. (3) As illustrated in Fig. 5.40d, when the height of the IMZ reaches approximately 1000 mm, the local shape of the IMZ has not yet reverted to the normal upside-down drop shape; instead, its contour line shows a tendency to contract inward.

Qualitative insights into how the number of coarse particles shapes the IMZ were first established; their evolution was then captured quantitatively. Invoking the upside-down drop shape model, the IMZ height and its maximum radius obey a power-law relation [8]:

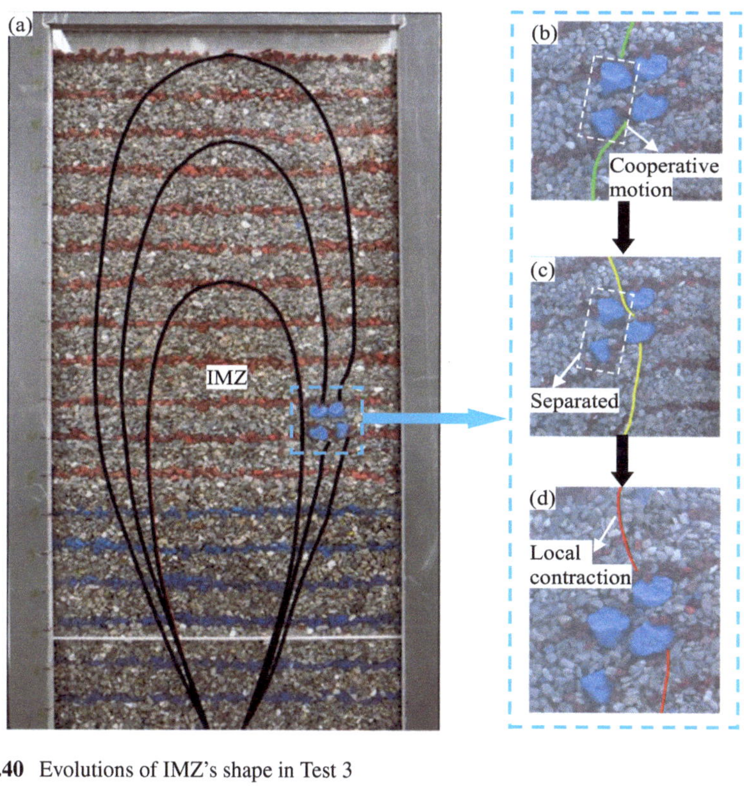

Fig. 5.40 Evolutions of IMZ's shape in Test 3

$$r_{max} = a \times H^b \tag{5.7}$$

Here, r_{max} represents the maximum radius of the IMZ, measured in millimeters; H represents the height of the IMZ, also measured in millimeters; a and b are fitting constants.

Figure 5.41 illustrates the relationship among the height of the IMZ, its maximum radius, and r_d in Test 3. Here, r_d represents the difference between the maximum radius of the IMZ on the side devoid of coarse particles and that on the side containing coarse particles. An analysis of Fig. 5.41 reveals several key points: (1) On the side without coarse particles, the height of the IMZ and the maximum radius exhibit a power function relationship as described by Eq. (5.7), with a correlation coefficient R^2 of 0.996. (2) For IMZ heights below point M (700 mm), no significant difference in maximum radius is observed between the two sides; however, at point N (750 mm), r_d experiences a sudden increase to 11.541%. (3) At point P (800 mm), the two columns of coarse particles begin to separate (see Fig. 5.40c), leading to a diminished influence of the four coarse particles on the local particles, resulting in a subsequent decrease in r_d. (4) Upon reaching point Q (900 mm), the interaction between the two coarse particles on the right side and their local particles impedes the development of the IMZ, which is manifested as a contraction of the contour line on this side of the IMZ, resulting in r_d being −17.632%. Furthermore, at a draw height of 1000 mm, the maximum radius of the IMZ remains influenced by the two coarse particles on the right side, preventing r_d from approaching 0 (see Fig. 5.40d).

Figure 5.42 shows the evolution of the increment of maximum IMZ radius on the side with coarse particles r_d in Tests 1–3. Based on the analysis of Fig. 5.42,

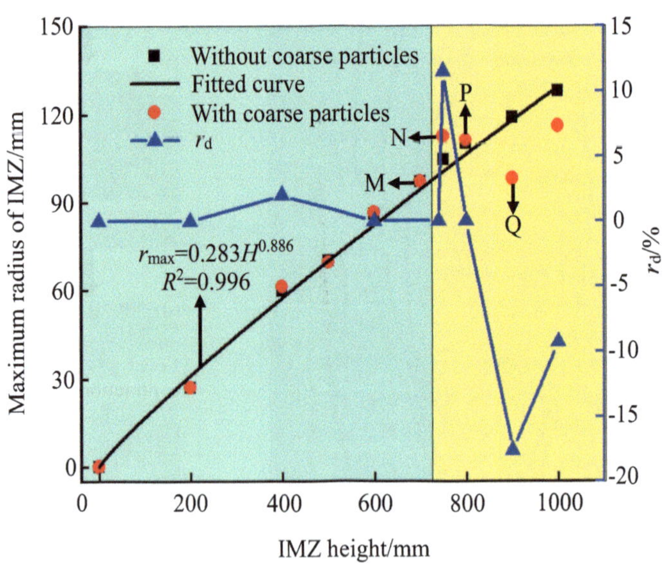

Fig. 5.41 Relationship among IMZ's height, maximum radius and r_d in Test 3

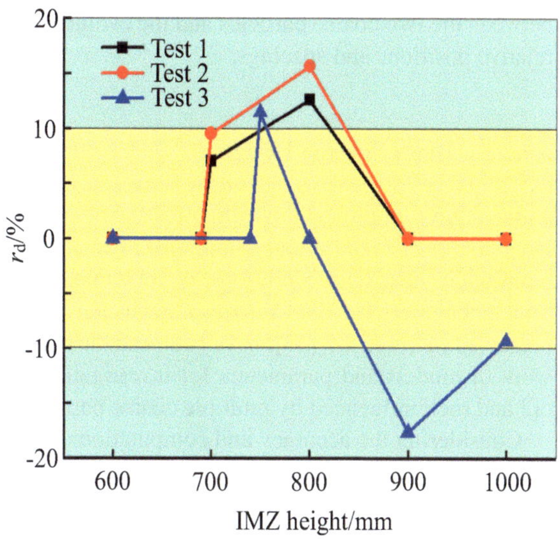

Fig. 5.42 Evolution of increment of maximum IMZ radius on the side with coarse particles r_d in Tests 1–3

the following conclusions can be drawn: (1) Taking the increase in the maximum radius of the IMZ $r_d \pm 10\%$ as the benchmark, different quantities of coarse particles have a significant impact on the shape of the IMZ. (2) The height evolution ranges of the IMZ affected by 1, 2, and 4 coarse particles are approximately 750–850 mm, 700–850 mm, and 750–1000 mm, respectively, indicating that the coarser particles there are, the longer the draw process they affect. (3) The positive maximum increase in the radius of the IMZ across the three sets of tests shows minimal variation, with a range of approximately 11–16%. Given that the size ratios of coarse to fine particles in the three sets of tests are 6.0, 5.4, and 4.0 (refer to Table 5.3), it is suggested that a smaller size ratio of coarse to fine particles can also lead to the early movement of local fine particles. In other words, as the quantity of coarse particles increases, the size ratio of coarse particles that significantly affects the IMZ gradually decreases.

5.4.2 Numerical Simulation Study of the Influence of the Relative Position and Spacing of Coarse Particles on the Flow Characteristics of Caved Ore and Rock

The influence of varying quantities of coarse particles on the flow characteristics of caved ore and rock represents a macroscopic outcome of the interactions between coarse particles and their surrounding fine particles. Due to the limitations of physical draw tests in acquiring the structural and mechanical parameters of the caved ore and rock system, a series of three-dimensional numerical tests were conducted. These tests utilized two coarse particles positioned at the bottom of a single drawpoint, based on the rolling resistance model [101]. The study focused on the separation between the two coarse particles and the evolution of the IMZ shape under different relative positions and spacings.

5.4.2.1 Draw Scheme Design

Authors have employed the rolling resistance model within the PFC [103] to simulate the impact of the actual geometries of various materials flowing toward a drawpoint under near-field conditions, thereby enhancing both reliability and computational efficiency. The relevant meso-mechanical parameters for spheres and walls, as presented in Table 4.5 [33], have been established, providing a foundational framework of models and parameters for investigating the flow characteristics of caved ore and rock influenced by multiple coarse particles.

Considering the accuracy and computational efficiency of this numerical test for draw, Fig. 4.15a illustrates the single-drawpoint numerical draw model utilized in this simulation, with dimensions of length × width × height = 40 × 40 × 80 m^3. The red area at the bottom of the model denotes the drawpoint. In Fig. 5.43, the green particles represent the coarse particles employed in this simulation, while the remaining red and blue particles signify fine particles. The total column height of 80 m comprises a height of 60 m for the draw (the red particles in Fig. 5.43) and an overburden height of 20 m (the blue particles in Fig. 5.43).

Based on the results of the aforementioned physical test for draw, as shown in Table 5.4, the range of particle size d_f for the fine particles used in this numerical test was set to 0.3–0.8 m, and the ratio of coarse particle size d_c/d_f was set to 6.0. In each group of tests, one of the coarse particles A was fixedly placed at a position in the horizontal distance model with an axis length of 7 m and a vertical distance from the drawpoint of 30 m (Fig. 5.43), and the numerical draw test scheme could be divided into the following two categories according to the different placement positions of another coarse particle B-E: (1) Through Tests 1–4, we investigated the influence of the relative positions of two coarse particles on the flow characteristics of caved ore and rock. Test 1 served as a control, featuring only one coarse particle, designated as A, placed within the model. In Tests 2–4, coarse particles B, C, and D were positioned in contact with the outer surface of coarse particle A, such that the distance between the centers of the two coarse particles L was equal to the diameter d_c of coarse particle A. The lines connecting the centers of the two coarse particles were oriented at angles of 0°, 45°, and 90° relative to the positive direction of the y-axis. (2) Through Tests 2 and 5–8, the influence of different distances between the two coarse particles on the flow characteristics of caved ore and rock was compared and analyzed. The range of the distance L between the centers of the two coarse particles varied from $[d_c, d_c + 3.0d_f]$. In each group of numerical draw tests, the FISH programming language was utilized to monitor the shape of the IMZ at varying heights, as well as the horizontal and vertical distances between two coarse particles. Additionally, the evolution of unbalanced forces, void ratios, and force chains within the local areas of the two coarse particles was tracked. The draw was halted when

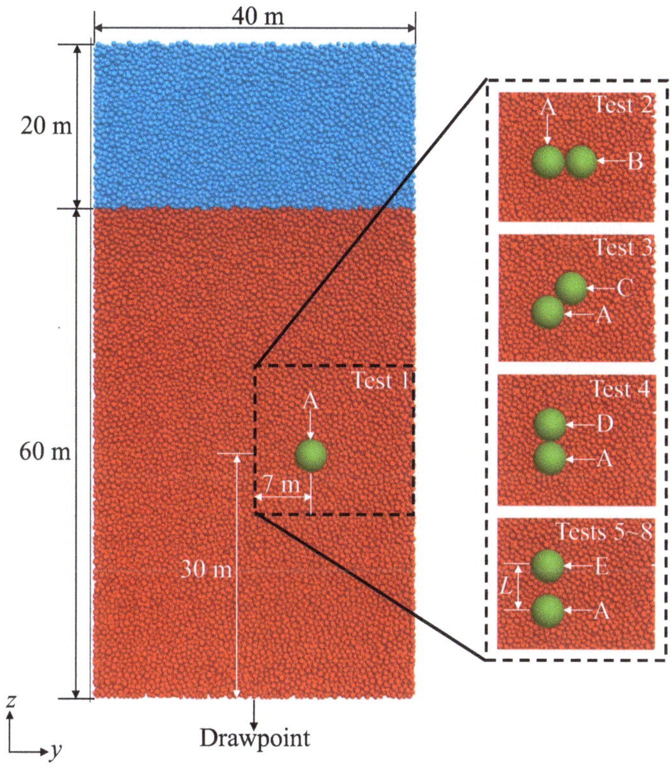

Fig. 5.43 Vertical section of the numerical draw model

the IMZ height of the caved ore and rock reached 60 m, leaving the top 20 m of the overlying rock in the simulation unchanged.

5.4.2.2 Analysis of Numerical Test Results of Draw

Based on the results of Tests 1–8 regarding the release of particles, this study first explores the influence of the relative positions and spacings of various coarse particles on the evolution of the IMZ's shape. It is important to note that in Tests 1–8, the position of coarse particle A remained constant, while the positions of coarse particles B-E varied. Therefore, to ensure data comparability, the analysis of the IMZ's shape evolution in this study focuses on the initial movement of fine particles in the vicinity of coarse particle A. Figure 5.44 illustrates the increase in the maximum IMZ radius influenced by coarse particles positioned at various relative locations and spacings. The increase in the maximum IMZ radius refers to the percentage difference in the maximum radius of the IMZ on both sides due to the presence of coarse particles, denoted as rd. Notably, the red particle assembly represents the typical IMZ shape

Table 5.4 Schemes of numerical draw simulation

Test number	Influencing factors	Size range of fine particle d_f/m	Size ratio of coarse and fine particles d_c/d_f	Position of coarse particles (Fig. 5.43)	Horizontal/vertical spacing between two coarse particles L (Fig. 5.43)
1	Relative position of the two coarse particles	0.3–0.8	6.0	A	d_c
2				A + B	d_c
3				A + C	d_c
4				A + D	d_c
5	Spacing between two coarse particles	0.3–0.8	6.0	A + E	$d_c + 0.5d_f$
6				A + E	$d_c + 1.0d_f$
7				A + E	$d_c + 2.0d_f$
8				A + E	$d_c + 3.0d_f$

observed in Tests 2, 4, and 7. As depicted in Fig. 5.44, the increment of the maximum IMZ radius, r_d, in all eight tests exceeded 10%, indicating that the shape of the IMZ adjacent to the coarse particles experienced substantial variations, specifically, the fine particles surrounding the coarse particles became significantly loosened.. The analysis of Tests 1–4 leads to several conclusions: (1) The morphological influence of two coarse particles in contact is significantly greater than that of a single coarse particle; (2) When the two coarse particles are oriented vertically, their impact on the morphology of the IMZ is maximized, with an increase in the maximum IMZ radius reaching 24.794%. In contrast, when the particles are positioned at a 45° angle, the influence is secondary, and when arranged horizontally, their effect on the IMZ shape is relatively minimal. Furthermore, the analysis of Tests 4–8 indicates that the increase in the maximum IMZ radius diminishes as the spacing between the two coarse particles increases. Specifically, when the center distance L between the two coarse particles reaches $d_c + 3.0d_f$, the increase in the maximum IMZ radius approaches that observed with a single coarse particle, suggesting negligible interaction between the two coarse particles.

Taking Tests 2, 4, and 7 as examples, Fig. 5.45a shows the relationship between the height of the IMZ and the maximum radius on the side with coarse particles. As a comparison, the black scattered points and the curve in Fig. 5.45a correspond to the data of the maximum radius on the side without coarse particles (the side not affected by coarse particles). Similar to the results obtained from the physical test (Fig. 5.41), from Fig. 5.45a, it can be concluded that: (1) When the height of the IMZ is small i.e., the movement range of caved ore and rock is away from the initial position of the coarse particle A (such as the height of the IMZ is less than 30 m), the height of the IMZ in the three tests and the maximum radius on the side with coarse particles all satisfy the power function relationship described by the upside-down drop shape theory [89] (Eq. (5.7)), with a correlation coefficient of R^2 reaching 0.997; when the height of the IMZ is large, i.e., the movement range of caved ore and rock is

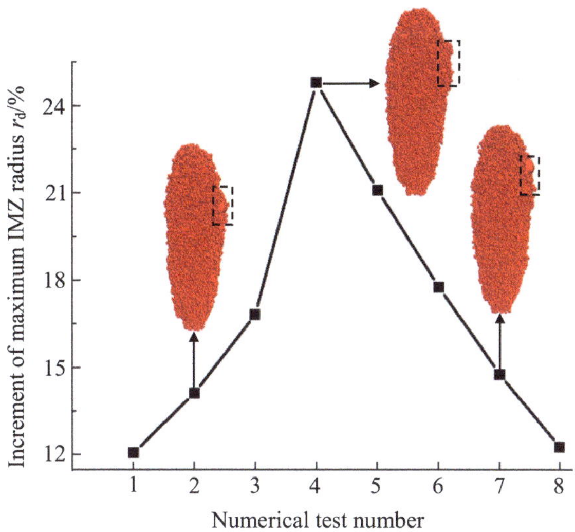

Fig. 5.44 Increment of maximum IMZ radius r_d under the influence of coarse particles with different relative position and spacing

close to or exceeds the point A where the coarse particle is located (such as the height of the IMZ is greater than 40 m), the maximum radius on the side with coarse particles in the three tests significantly increases and no longer satisfies the power function relationship described by the upside-down drop shape theory. On this basis, the increase in the maximum radius r_d of the IMZ at different heights of the IMZ is calculated, as shown in Fig. 5.45b. When the height of the IMZ is within the range of 35–60 m, the radius r_d in the three tests exhibits a trend of initially increasing and then decreasing. The maximum value of r_d occurs when the height of the IMZ is approximately 45 m. Furthermore, using a 10% increase in the maximum radius as a benchmark, it can be inferred from Fig. 5.45b that when the two coarse particles are in contact and arranged vertically, their influence on the shape of the IMZ is the most prolonged, as the draw process significantly affected by the coarse particles lasts the longest. Conversely, when the two coarse particles are not in contact and are vertically distributed, the duration of their influence is moderate. In contrast, when the two coarse particles are in contact and arranged horizontally, their influence on the shape of the IMZ is the shortest.

Figure 5.46 illustrates the evolution of normalized horizontal separation distance (Test 2) and vertical separation distance (Tests 4 and 7) between two coarse particles during the first 20 draw processes, with the height of the IMZ maintained at 60 m. As depicted in Fig. 5.46, using the normalized separation distance of $0.2d_f$ as a benchmark, the analysis indicates that in Test 2, the separation of the two coarse particles occurs earliest, with the maximum separation distance reaching approximately $1.2d_f$ during the same draw stage. Test 7 closely follows this pattern, while in Test 4, the distance between the centers of the two coarse particles remains relatively small (less than $0.2d_f$). It is observed that when the two coarse particles are horizontally or vertically distributed with a large initial distance between their centers,

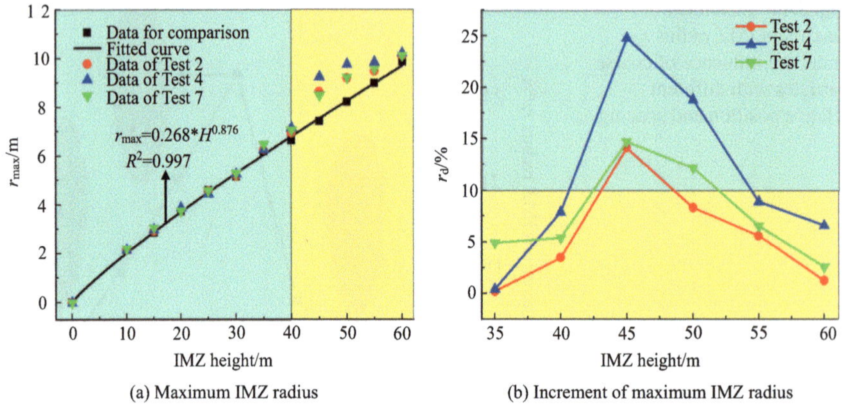

Fig. 5.45 Evolution of IMZ maximum radius and its increment on the side with coarse particles in Tests 2, 4 and 7

Fig. 5.46 Evolutions of normalized horizontal/vertical separation distance between two coarse particles in Tests 2, 4 and 7

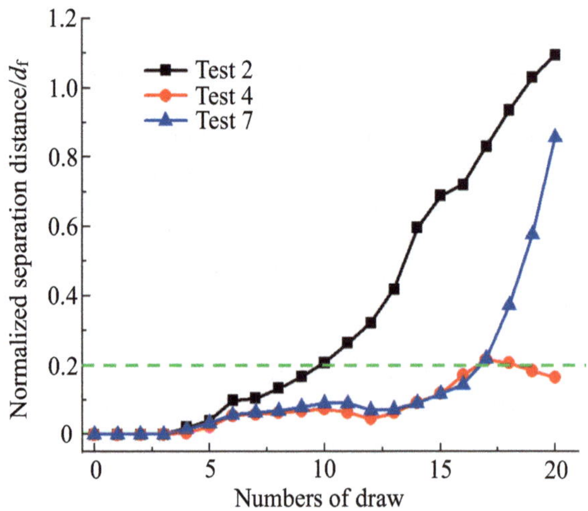

the separation distance increases approximately exponentially with the number of draws. Conversely, when the two coarse particles are vertically distributed and in contact, no significant separation occurs between them.

5.4.3 Influence Mechanism Analysis of Two Coarse Particles Based on Three-Dimensional Force Chain

This study conducts a macroscopic analysis of the separation distance between two coarse particles and examines the evolution of the IMZ shape under varying relative positions and spacings of these particles. It explores the mechanisms by which two coarse particles influence the flow characteristics of caved ore and rock, focusing on the mesoscopic perspective of inter-particle interactions, including unbalanced forces, void ratio, and force chains.

5.4.3.1 Evolutions of Unbalanced Force, Velocity and Void Ratio

This study focuses on the separation of two coarse particles influenced by other coarse particles with varying relative positions and spacing. It analyzes the evolutionary process of the unbalanced forces and velocities experienced by the two coarse particles during a typical single draw period, with the height of the IMZ approximately 45 m. The analysis is conducted both horizontally (Test 2) and vertically (Tests 4 and 7). It is important to note that the positive directions of the horizontal and vertical unbalanced forces and velocities experienced by the two coarse particles are represented by the y and z axes in Fig. 5.43. The differences illustrated in Fig. 5.47 pertain to the unbalanced forces, velocities, and local void ratios of coarse particle A, subtracted from those of coarse particles B through E. Specifically, Fig. 5.47a depicts the difference in unbalanced forces in both horizontal and vertical directions between the two coarse particles during a single draw period. From the analysis of Fig. 5.47, it can be concluded that: (1) As the drawpoint is opened, the caved ore and rock within the IMZ gradually move towards the drawpoint due to the influence of gravity, resulting in a gradual difference in the local void ratios between the two coarse particles. In Test 2, coarse particle A is positioned closer to the central axis of the drawpoint compared to coarse particle B. The local caved ore and rock surrounding coarse particle A are in a more loosely packed state, leading to a significantly greater local void ratio for coarse particle A than for coarse particle B (Fig. 5.47c), with the maximum void ratio difference reaching 0.167. Consequently, during most of the single draw period in Test 2, the absolute values of the differences in unbalanced forces and velocities for coarse particle A are considerably larger than those for coarse particle B (Fig. 5.47a, b), with maximum absolute values of unbalanced forces and velocities reaching 72.107 kN and 0.118 m/s, respectively. This observation also explains why the separation distance between the two coarse particles is greatest in Test 2, as illustrated in Fig. 5.46. In Test 7, the evolution patterns of the two coarse particles reveal that the unbalanced force, velocity, and void ratio of the particles are similar to those in Test 2, albeit with a relatively smaller fluctuation amplitude. Consequently, the vertical separation velocity of the two coarse particles in Test 7 is also slower (Fig. 5.46). In Test 4, the difference in local void ratio between the two coarse particles is minimal, resulting in minor positive and negative fluctuations in the differences in unbalanced force and velocity, which aligns with the findings in Test 4 of Fig. 5.46, where no significant separation between the two coarse particles occurs. (2) As the single draw cycle approaches its conclusion (timestep > 1e4 steps), the system of caved ore and rock gradually tends towards equilibrium. Therefore, the differences in unbalanced force, velocity, and local void ratio of the two coarse particles in the three tests also gradually approached 0.

Fig. 5.47 The difference of unbalanced force, velocity and local void fraction between/around two coarse particles in a single draw period

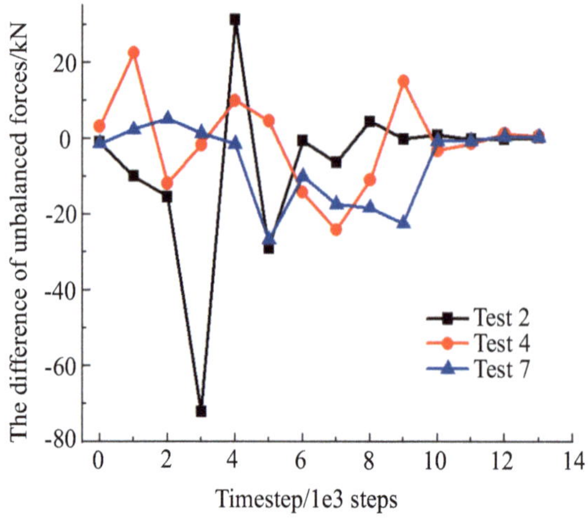

(a) The difference of unbalance force

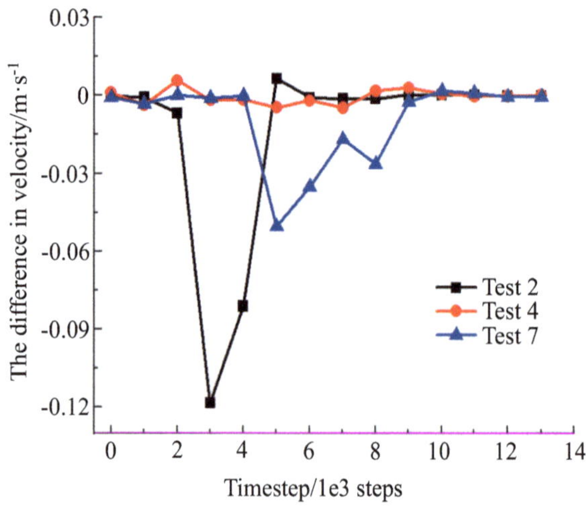

(b) The difference of velocity

5.4.3.2 Macroscopic Distributions Characteristics of Three-Dimensional Force Chains

Based on the test results of the drawing values in Tests 2, 4, and 7, the evolution of the local force chain surrounding coarse particles A was studied. This force chain represents a chain-like structure within the particle system that can bear and transfer forces [104, 105]. As illustrated in Fig. 5.48, three conditions must be satisfied for the particles to form a chain [106]: (1) Contact force criterion: The contact force F between particles is not less than the average contact force $<F>$ within the particle system; (2) Contact angle criterion: The angle θ between the direction vectors of two

Fig. 5.47 (continued)

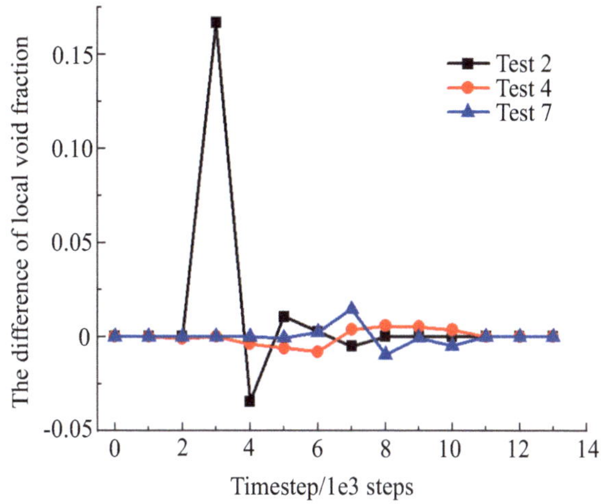

(c) The difference of local void fraction

adjacent contacts is not greater than the angle threshold $\theta_c = 180°/CN$, where CN is the average coordination number within the particle system of caved ore and rock; (3) The number of chain particles N_{fc} is not less than 3. Based on these three chain formation conditions, a three-dimensional force chain identification program, termed FISH, was developed. This program records the ID numbers, three-dimensional coordinates, and intensities of the contacts between particles within a local area of coarse particle A, specifically within a distance not exceeding 6df from the center of coarse particle A, during the first 15 consecutive drawing processes in the draw test. Consequently, it facilitates the automatic identification of three-dimensional force chains, visualization of the chain particles, and extraction of characteristic parameters such as the number, average intensity, and straightness of the force chains.

Fig. 5.48 The identifying criterions and sketch of force chain

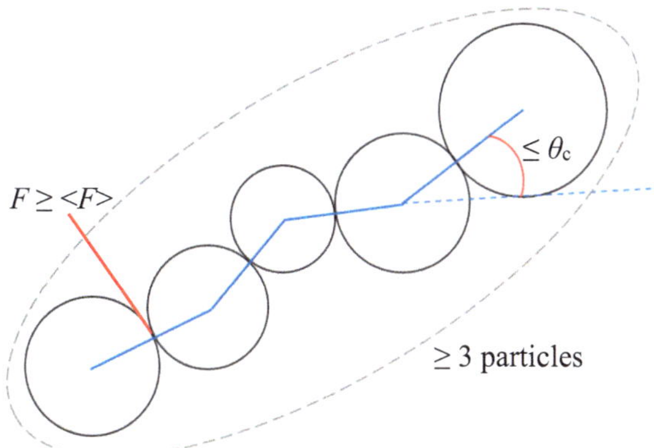

Figure 5.49 illustrates the typical macroscopic distribution of three-dimensional force chains within the local area of coarse particle A, under varying relative positions and spacings of the coarse particles. The numbers in the upper right corner indicate the corresponding number of linked particles N_{fc}. From the analysis presented in Fig. 5.49, several conclusions can be drawn: (1) When the movement range of caved ore and rock has not yet reached the position of coarse particle A (as seen in No. 6 of the draw), all three tests indicate that coarse particle A forms force chains with another coarse particle, with the number of linked particles in the local area remaining relatively constant. (2) As coarse particle A becomes increasingly loose (from No. 10 to No. 14 of the draw), there is a significant decrease in the number of linked particles in the local area of coarse particle A in Tests 2 and 7. In Test 2, the force chain between coarse particle A and another coarse particle has failed, while in Test 4, the number of linked particles in the local area of coarse particle A is 40, suggesting that it can still maintain a relatively high bearing capacity.

5.4.3.3 Evolution Laws of Numbers and the Maximum Horizontal Position of Transformed Particles

Based on the analysis of Fig. 5.49 from the previous section, it is evident that the force chain aggregation effect of coarse particle A gradually diminishes as the draw process progresses, particularly due to its loosening. Consequently, the fine particles that are locally chained around coarse particle A will also gradually transition into non-force chain particles and undergo early movement, resulting in variations in the shape of the local IMZ. In this context, particles that shift from force chain particles to non-force chain particles are defined as transformed particles. Figure 5.50 illustrates the evolution of both the number and maximum horizontal position of transformed particles around coarse particle A, influenced by two coarse particles with different relative positions and spacings. Figure 5.51 depicts the typical distribution of transformed particles around coarse particle A during No. 12 of the draw in Tests 2, 4, and 7, where blue particles represent transformed particles within the local vicinity of coarse particle A. Analysis of Figs. 5.50 and 5.51 reveals that during the early stage of drawing (the first 4–5 draw processes), the caved ore and rock in the local area surrounding coarse particle A across the three tests maintained an approximate equilibrium state, resulting in no generation of transformed particles. With the development of the movement zone during the 6–15 draw processes, the number of transformed particles within the local area of coarse particle A in the three sets of tests exhibited an approximately linear increase. Additionally, the maximum horizontal position of transformed particles demonstrated a step-like increase. Notably, at the same draw stage, both the quantity and the maximum horizontal position of transformed particles were greatest in Test 4, followed by Test 7, and smallest in Test 2. This observation aligns with the evolutions of the IMZ maximum radius and its increment on the side with coarse particles, influenced by varying relative positions and spacings, as depicted in Fig. 5.45.

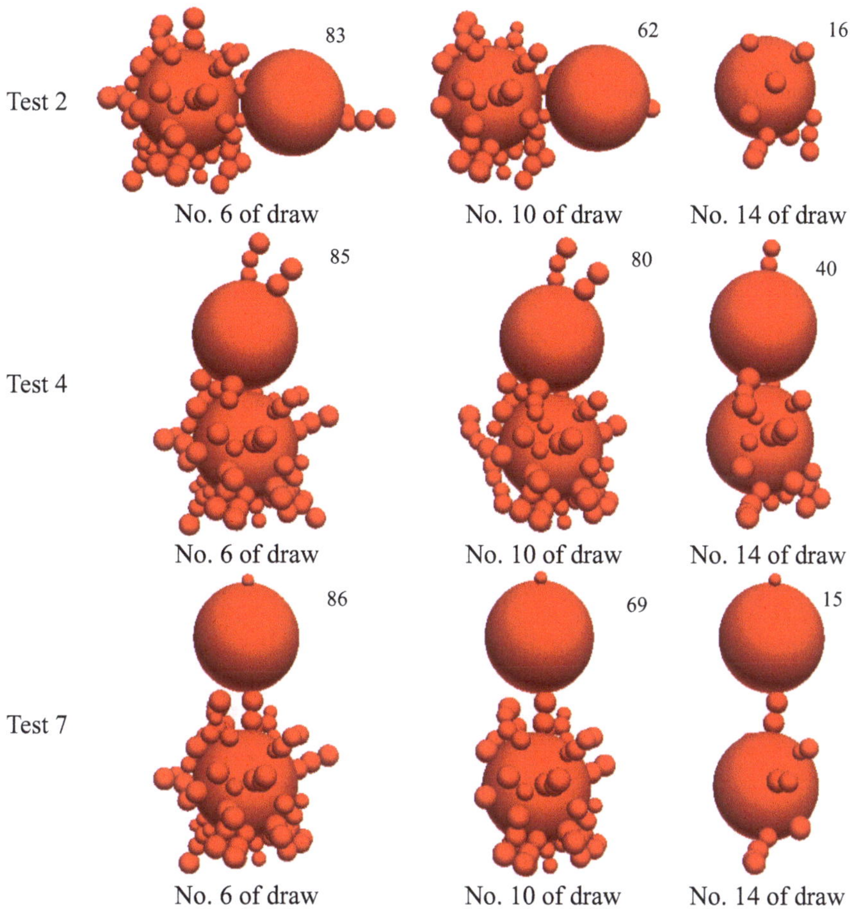

Fig. 5.49 Macroscopic distributions of typical three-dimensional force chains around coarse particle A in Tests 2, 4 and 7

5.4.3.4 Evolutions of Number, Average Strength and Collimation of Three-Dimensional Force Chains

Figure 5.52 shows the evolutions of number, average strength, and collimation of three-dimensional force chains around coarse particle A in Tests 2, 4, and 7. The analysis is as follows:

(1) Number of force chains

The analysis presented in Fig. 5.52a aligns with the findings from Fig. 5.50. In the initial phase of the draw process (specifically during the first six draws), the caved ore and rock around coarse particle A remain intact, resulting in no significant changes in the number of three-dimensional force chains across the three tests, which remains

Fig. 5.50 Evolution of numbers and the maximum horizontal position of transformed particles around coarse particle A in Tests 2, 4 and 7

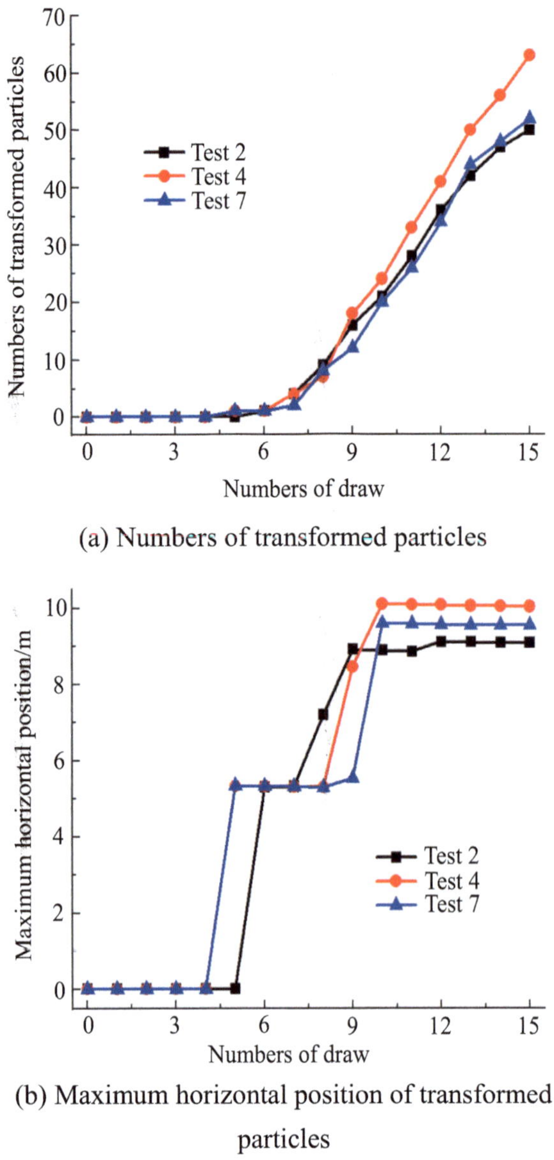

between 50 and 60. As the height of the IMZ reaches 30–42 m (during the 10–15th draw processes), coarse particle A and its associated fine particles gradually enter the movement zone, leading to a decrease in their aggregating effect on the force chains. Consequently, the number of three-dimensional force chains in all three tests experiences a sharp decline. Notably, during this draw process, Test 4 exhibits the highest number of three-dimensional force chains, followed by Test 7, while Test 2

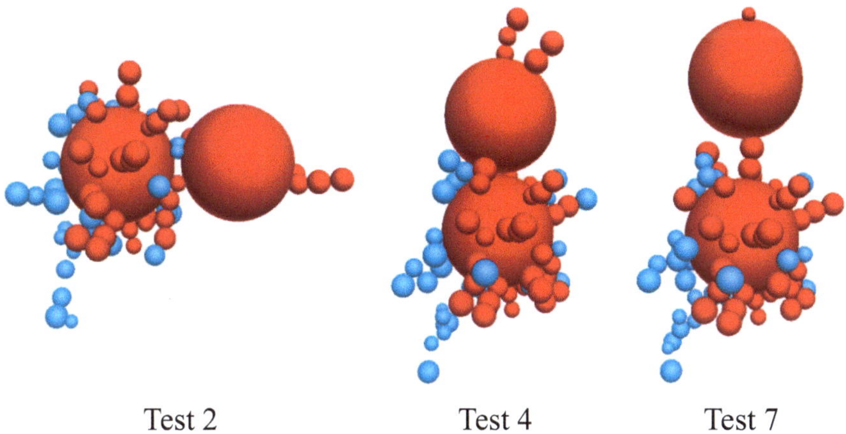

Fig. 5.51 Distributions of transformed particles around coarse particle A in No. 12 of draw in Tests 2, 4 and 7

records the lowest. For instance, after the 15th draw, the two coarse particles in Test 4 do not exhibit significant separation, which enhances their interaction and promotes the 'attraction' of local fine particles over a prolonged draw process, resulting in the formation of a stable 'Clusters' structure with 16 three-dimensional force chains. In contrast, Tests 2 and 7 show an increasing separation distance between the two coarse particles, leading to only 3 and 6 three-dimensional force chains, respectively, indicating that coarse particle A has no significant aggregating effect on the local force chains.

(2) Average strength of the force chain

The average strength of the force chain is defined as the ratio of the cumulative strength of the force chain to the total number of force chains. As illustrated in Fig. 5.52b, during the initial stage of the drawing process, the average strength of the local force chains across all three tests exceeded 1.0 MPa, exhibiting a gradual decline. Notably, Test 2 recorded the highest average strength among the tests. During drawing times 11–15, this trend was consistent with the evolution of the number of force chains depicted in Fig. 5.52a. In all three tests, the average strength of the local force chains displayed a rapid downward trend, with Test 4 exhibiting a higher average strength than both Test 2 and Test 7. This indicates that coarse particles A, along with their adjacent fine particles, can sustain a greater load-bearing capacity throughout an extended drawing process. By analyzing the evolutionary patterns of the number and average strength of the local force chains of coarse particle A, as illustrated in Fig. 5.52a, b, one can elucidate the differences in the duration of the influence exerted by varying relative positions and spacings of coarse particles on the configuration of the IMZ, as shown in Fig. 5.45.

(3) Force chain alignment property

Fig. 5.52 Evolutions of number, average strength and collimation of three-dimensional force chains around coarse particle A in Tests 2, 4 and 7

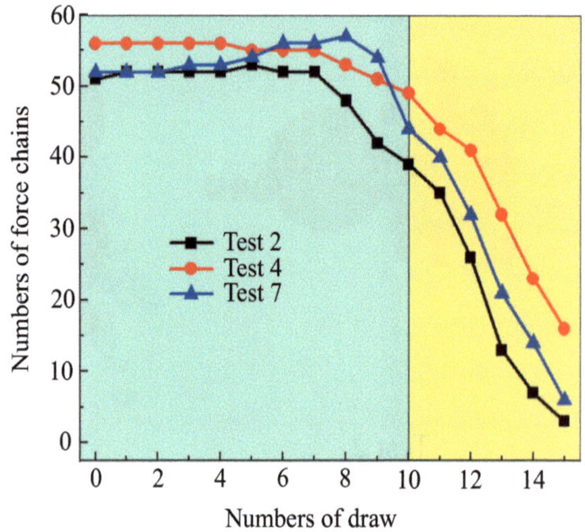

(a) Numbers of force chains

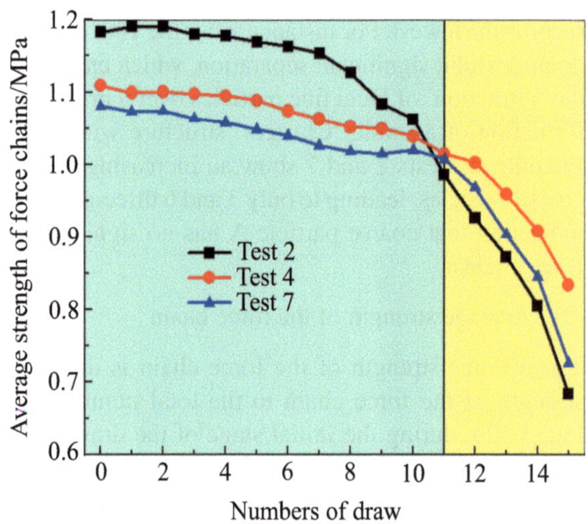

(b) Average strength of force chains

Generally speaking, the closer the force chain structure is to a straight line, the greater its stability. The force chain alignment coefficient δ (Eq. (5.8)) can be utilized to quantitatively analyze this relationship:

$$\delta = 1 - \frac{\sum_{i=1}^{N_c} \alpha_i}{180^\circ \times N_{ac}} \tag{5.8}$$

Fig. 5.52 (continued)

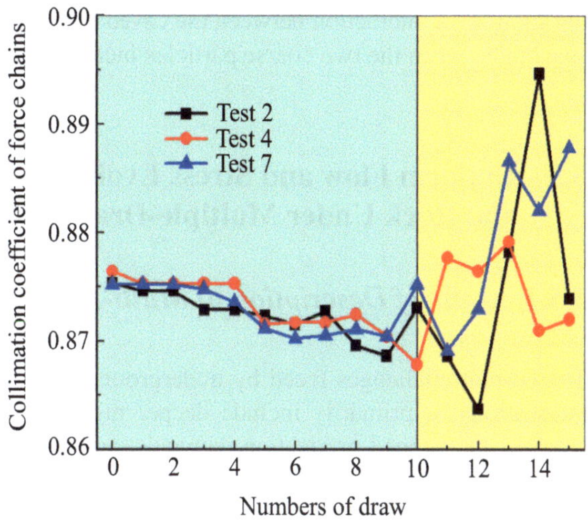

(c) Collimation coefficient of force chains

Here, α_i represents the angle between the *i-th* adjacent normal contact; N_{ac} is the total number of adjacent contacts, and its value is N_c-2 (where N_c is the number of contact links).

The average collimation coefficient of the local force chains, defined as the ratio of the cumulative collimation coefficient of the force chains to the number of force chains, was calculated across the three tests. The evolution of the force chain collimation coefficient is illustrated in Fig. 5.52c. Analysis of Fig. 5.52c reveals that during the initial stage of the draw process, the local force chain collimation coefficients across the three tests do not exhibit significant differences; all demonstrate a gradually fluctuating downward trend within the range of 0.87–0.88. This indicates that variations in the relative positions and spacings of coarse particles have minimal impact on the collimation of local force chains. However, as the coarse particles and local fine particles enter the movement zone, the fluctuations in the local force chain collimation coefficients across the three tests increase significantly, ranging from 0.86 to 0.90. Moreover, this fluctuation is negatively correlated with the evolution of the number of force chains, suggesting that a decrease in the number of force chains corresponds to an increase in the fluctuation of the force chain collimation coefficient.

In conclusion, when comparing the scenario where two coarse particles are horizontally distributed and in contact with each other to the case where they are vertically distributed and in contact, it is evident that the force chain aggregation effect is more pronounced in the latter arrangement. Specifically, the local force chain in the vertical configuration can maintain a higher level of stability, characterized by a greater quantity and average strength, over an extended period of draw. This results in a more substantial impact on the shape of the IMZ and prolongs its duration. Furthermore, the significant interaction between the caved ore and rock diminishes as the vertical spacing between the two coarse particles increases.

5.5 Uniform Flow and Stress Evolution Laws of Caved Ore and Rock Under Multiple-Drawpoint Condition

5.5.1 A Brief Description of Multi-Drawpoint Draw Research

The current challenges faced by underground metal mines, both domestically and internationally, primarily include deeper mining environments, increased ground stresses, heightened production demands, and rising mining costs [107, 108]. For mines employing the block caving method, as mining depth increases, both the height of the rock layer and the load it exerts on the bottom structure of the stope continuously rise. Consequently, establishing reasonable stope structure parameters and effective draw control measures—particularly concerning the spacing of drawpoints and the draw method—becomes increasingly critical. This is essential for understanding the flow characteristics of caved ore and rock, ensuring the stability of the stope's bottom structure, and minimizing ore loss. Therefore, it is imperative to investigate the interaction dynamics between caved ore and rock under varying drawpoint spacing and draw method conditions.

The primary factors [3, 12] currently assessed to determine the pros and cons of structural parameters in stopes and to establish control strategies for block caving operations largely include: the uniform descent of the ore-waste interface, the formation of a mass flow of caving materials, and the interaction between the granular materials above neighboring drawpoints. Achieving a uniform flow in the rock layer is highly advantageous for several reasons: it minimizes the leftover ore between adjacent drawpoints, enhancing recovery rates; it decreases the movement and early mixing of fine waste rock, thereby lowering ore dilution; and it ensures an even distribution of stress across the stope's bottom structure, mitigating the risk of instability due to stress concentration. However, existing research presents two differing perspectives on the optimal drawpoints spacing necessary for achieving a uniform descent of the ore-waste interface or uniform flow. (1) Laubscher [12] proposed based on the three-dimensional physical draw test results of Marano [109] using sand as the medium and his own experience, suggested that when multiple drawpoints operate simultaneously, the spacing D_s between them should be set at 1.5 times the maximum width of the IMZ w_{IMZ} from a single drawpoint, When $D_s <=$ $1.5w_{IMZ}$, two adjacent IMZs can exhibit intersecting phenomena. Hancock [110], utilizing the discrete element software ESyS-Particle, conducted numerical simulations on multi-drawpoint scenarios with spherical particles and similarly concluded that a uniform flow of caved ore and rock occurs when the drawpoint spacing is at or below 1.5 times the maximum width of the IMZ. (2) Conversely, Trueman, Castro [75], performed extensive three-dimensional physical tests with coarse crushed schist aggregates and discovered that under multi-drawpoint conditions, a spacing of 1.2 times the maximum width ($D_s = 1.2w_{IMZ}$) results in no intersection between adjacent IMZ. They noted that the physical draw model's material was steel [111], which has a smoother surface compared to the actual draw boundary that exhibits a higher friction coefficient, leading to a more pronounced stress arching effect between the caved ore and rock. Consequently, Trueman et al. posited that if the drawpoint spacing between adjacent drawpoints exceeds the maximum width of the IMZ ($D_s > 1.0w_{IMZ}$), the two IMZ may not intersect.

The stress distribution at various locations within the stope during the draw process, particularly its impact on the bottom structure and its developmental patterns, serves as a crucial method for examining the stress arching phenomenon and the movement dynamics between caved ore and rock. This analysis is vital for refining the arrangement of drawpoints and the techniques employed. Lorig [111] and Pierce [112] investigated the quantitative relationship between the vertical stress acting on the bottom structure of the stope and factors such as the accumulation density of caved ore and rock, friction coefficient, and hydraulic radius of the draw zone based on the Janssen formula [113] and the limit equilibrium method. Castro et al. [30] focused on the large-scale underground copper mine EI Teniente in Chile, conducting a physical draw test with a 1:200 similarity ratio to analyze the changes in induced vertical stress across various draw methods. Furthermore, Tao et al. [114] and colleagues formulated a mathematical model to depict the movement laws of caved ore and rock under multi-drawpoint condition, enhancing the method for determining the dispersion void diffusion coefficient in the kinematic model of draw, and identifying the movement paths of granular materials, IEZ, and the shapes of residual zones. Sun et al. [115] developed a draw model using discrete element theory and the PFC^{3D} software to investigate the flow behaviors of granular materials under multi-drawpoint conditions, visualizing the transformations in the shapes of IEZ throughout the draw process. They found that the shape of the IEZ varied significantly due to the interactions between drawpoint, deviating from a regular ellipsoid. Chen et al. [40, 116, 117] conducted quantitative analyses using physical tests and numerical simulations to investigate the flow behaviors of caved ore and rock beneath flexible isolation layers. Their study examined the factors influencing the movement dynamics and tensile force characteristics of the isolation layer, as well as the evolution of force chain properties, including length, quantity, strength, and alignment within the particle system under multi-drawpoint conditions.

Ongoing debates persist in contemporary studies regarding the optimal spacing of drawpoints necessary to ensure a uniform flow of caved ore and rock. Furthermore, there is a notable lack of comprehensive research on the stress conditions and their variations at various locations within the stope, which are influenced by different drawpoint spacings and draw modes. Consequently, this study employs a numerical testing approach utilizing a rigid block model to quantitatively analyze the uniform flow and stress evolution laws of caved ore and rock under multiple drawpoint conditions. The findings provide valuable insights into existing draw theories and offer practical guidance for the design of stope structures, support parameters, and the management of draw in block caving operations.

5.5.2 Interaction Dynamics of Caved Ore and Rock with Varying Drawpoint Spacing

The distance between adjacent drawpoints is a crucial factor that influences the uniform flow of caved ore and rock. To establish the minimum spacing required for drawpoints to ensure a uniform flow of these materials—specifically where two adjacent IMZ intersect—it is essential to first determine the maximum width of the IMZ. The morphological changes of the IMZ during physical draw tests are often challenging to observe directly; however, numerical tests can effectively illustrate its shape. This study concentrates on the particle system of caved ore and rock generated during natural caving or blasting in underground mines. Initially, by determining the maximum width of the IMZ, a series of numerical tests with multiple drawpoints at varying spacings were conducted to investigate the threshold spacing and the stress evolution among caved ore and rock in different zones.

5.5.2.1 Design of Numerical Draw Tests

(1) Draw numerical test scheme

The authors previously carried out a physical test involving a isolated draw using a self-designed large-scale three-dimensional test platform (refer to Fig. 4.1a). For the first time, a numerical model of the drawpoint was developed utilizing the rigid block model [32] (see Fig. 5.1c). By comparing and analyzing the outcomes of the physical test with the simulation results, the study demonstrated the reliability and advantages of the rigid block model in examining the flow characteristics of caved ore and rock. Additionally, the maximum width of the IMZ [55] was established, providing a foundational model for the forthcoming multi-drawpoint numerical test.

Figure 5.53 illustrates the relationship between the height and the maximum width of the IMZ, which is derived from previous numerical simulations conducted using the rigid block model [55]. The figure clearly shows that at a height of 50 m, the maximum width reaches approximately 16 m. Consequently, this simulation has established four distinct numerical tests with drawpoint spacings set at 12 m, 16 m, 20 m, and 24 m (Table 5.5), respectively. These tests aim to investigate the critical spacing of drawpoints necessary for achieving a uniform flow of caved ore and rock within the 50-m height range, as well as to analyze the evolution of stress.

(2) **Construction of the numerical draw model**

Considering the various numerical testing methods associated with different drawpoint spacings and the effectiveness of numerical computations, a numerical model featuring dimensions of 80 m × 40 m × 70 m has been developed, as illustrated in Fig. 5.54. The red area at the base of the model in Fig. 5.54a indicates the three drawpoints, with the distance between adjacent drawpoints adjustable based on various simulation strategies. To enhance the visualization of flow conditions, such as the

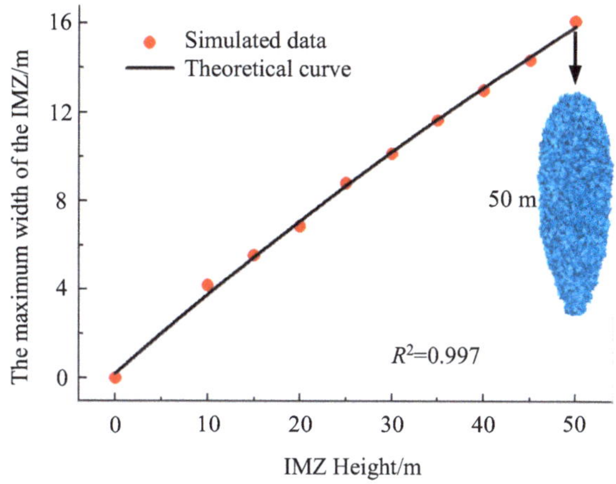

Fig. 5.53 Comparison between the data of numerical draw test and theoretical curve for the relationship between the height and maximal width of IMZ (modified from literature [55])

Table 5.5 Schemes of numerical simulation with different drawpoint spacings

The maximum width of the IIMZ w_{IMZ}/m	Test number	The ratio of the spacing between adjacent drawpoints to the maximum width of the IMZ D_s/w_{IMZ}	Adjacent drawpoint spacing D_s/m	Draw method	Conditions for stopping draw
16	1	0.75	12	Simultaneous draw	The height of IMZ reaches 50 m
	2	1.00	16		
	3	1.25	20		
	4	1.50	24		

uniform flow of caved ore and rock, particles at varying heights within the model are represented in two distinct colors, as depicted in Fig. 5.54b with a 10 m unit scale. Furthermore, prior to the start of the draw, 15 measurement spheres, each with a diameter of 6 m, are incorporated into the model, as shown in Fig. 5.54b, to track the changes in coordination numbers and both horizontal and vertical stresses across different spatial regions. The measurement areas labeled 1–3, 7–9, and 13–15 are positioned at heights of 10 m, 25 m, and 40 m above drawpoints 1–3, respectively. Meanwhile, measurement areas 4–6 and 10–12 are symmetrically arranged along the midline between adjacent drawpoints, with the distance between measurement areas at the same elevation matching half the spacing of the adjacent drawpoints, specifically $0.5D_s$.

(3) Numerical test process

The complete numerical testing procedure for the draw is outlined as follows: (1) A numerical draw model is created using the rigid block model (refer to Fig. 5.1c) in conjunction with the rainfall method [33], as illustrated in Fig. 5.54b. The relevant meso-mechanical parameters for both the wall and the rigid block are assigned (see

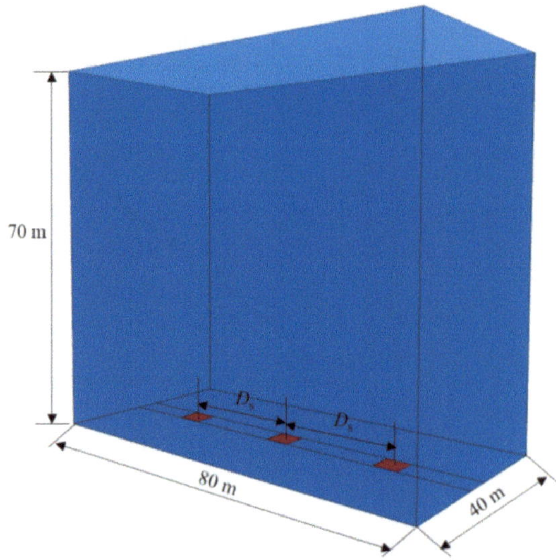

(a) Walls

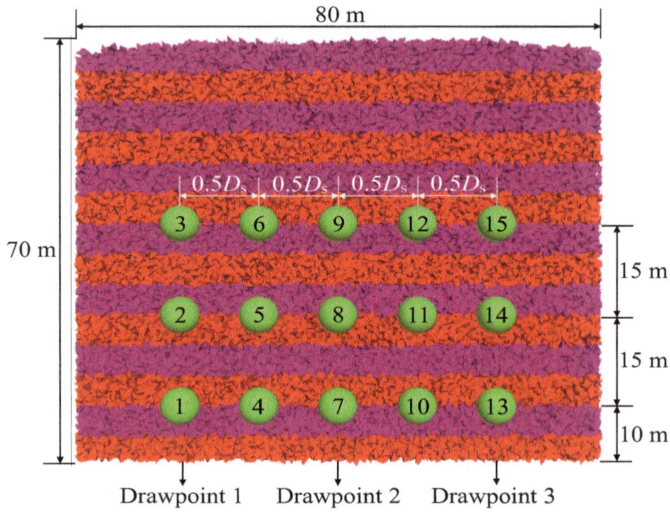

(b) Longitudinal profile and measurement sphere layout

Fig. 5.54 The numerical draw model with multiple drawpoints

Table 4.5). It is noteworthy that the specifics regarding block shape, gradation, and meso-mechanical parameters are detailed in the referenced literature [55] and will not be reiterated here to avoid redundancy. Once the model achieves gravitational equilibrium, the initial horizontal stress $\sigma_{h,0}$, and vertical stress $\sigma_{v,0}$ at various locations within the model are observed and recorded statistically; (2) Simultaneously open Drawpoints 1–3 to facilitate draw, while monitoring and documenting the movement of caved ore and rock, the coordination number *CN* in different spatial areas, and the variations in horizontal stress σ_h and vertical stress σ_v experienced by the rigid block throughout the draw process; (3) Draw is stopped when the IMZ height reaches 50 m, and a constant 20 m overburden height is maintained during the simulation process.

5.5.2.2 Analysis of the Draw Test Results

Upon completing the four tests concerning draw, a comparative study was conducted to examine the interaction between thc caved ore and rock at various drawpoint spacings. This analysis identified the critical spacing threshold that facilitates a uniform flow of caved ore and rock.

(1) Analysis of the gravity flow

Figure 5.55 illustrates the longitudinal profiles depicting the movement of caved ore and rock when the IMZ height reaches 50 m and draw ceases. In Fig. 5.55a, b, with drawpoint spacings D_s of 12 m ($0.75w_{IMZ}$) and 16 m ($1.0w_{IMZ}$), distinct horizontal declines are observed in the layers of caved ore and rock at heights of 35 m and 45 m, suggesting that the adjacent IMZ have interacted. Conversely, Fig. 5.55c, d reveal that at drawpoint spacings D_s of 20 m ($1.25w_{IMZ}$) and 24 m ($1.5w_{IMZ}$), the layers of caved ore and rock maintain a wave-like configuration at the same height, with no consistent decline noted across different heights, indicating a lack of interaction between adjacent IMZ. This analysis suggests that the drawpoint spacing threshold for achieving a uniform flow of caved ore and rock lies between 1.0 and 1.25 times the maximum width of the IMZ.

Using Test 4 as a reference, the shape characteristics of the three IMZs exhibit similar patterns. The maximum widths of the IMZs at various heights, along with their average values, were statistically analyzed. The cross-sectional representation of the IMZ shapes at a drawpoint spacing of 24 m, along with the fitted curve relating the height of the IMZ to their average maximum width, is illustrated in Fig. 5.56. As depicted in Fig. 5.56, at a height of 50 m, all three IMZ take on an upside-down drop shape. The correlation between the height of the IMZ and their maximum width aligns with the power function relationship outlined in the upside-down drop shape theory [8] (refer to Eq. (5.9)), with a correlation coefficient R^2 exceeding 0.995:

$$w_{\max} = a \times H^b \tag{5.9}$$

Here, $w_{\max}$ represents the maximum width of the IMZ, *H* represents the height of the IMZ, and *a* and *b* are fitting coefficients. Additionally, Fig. 5.56 also quantitatively proves that when the spacing between the drawpoints is large (such as $D_s > 1.5w_{\max}$), the adjacent IMZ are independent of each other, and their shape evolution laws are the same as those of the IMZ under the condition of single drawpoint, still conforming to the upside-down drop shape theory.

(2) Analysis of stress evolution laws

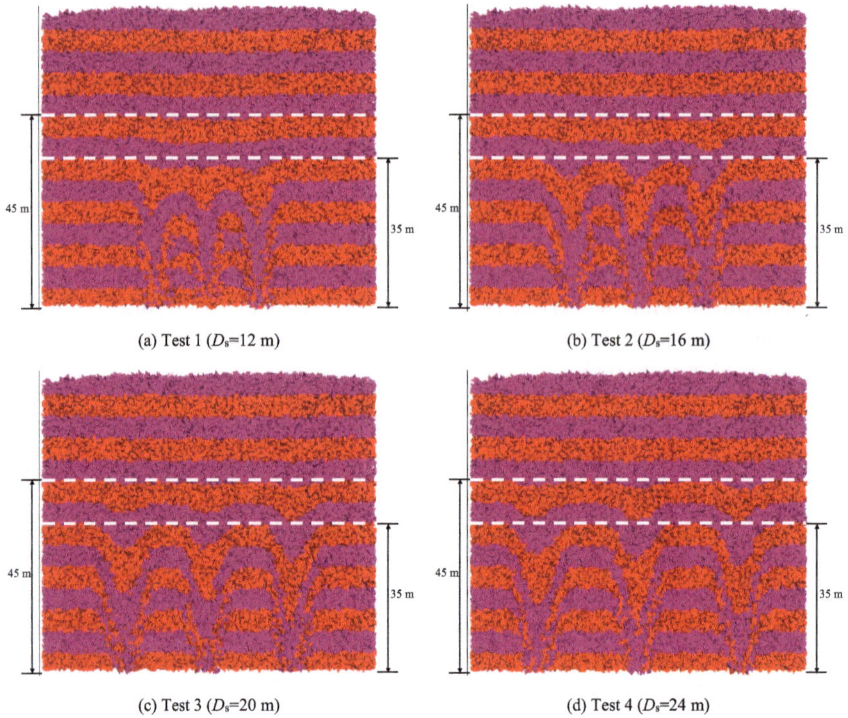

Fig. 5.55 Longitudinal profiles of movement of caved ore and rock with different drawpoint spacings

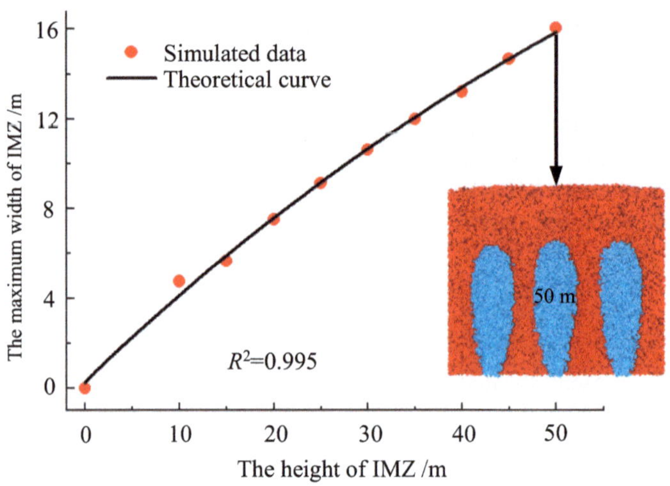

Fig. 5.56 Comparison between the data of numerical draw test and theoretical curve for the relationship between the height and average maximal width of MZs obtained in Test 4

This study conducts a statistical analysis of the evolution of coordination numbers between caved ore and rock, as well as the horizontal and vertical stresses experienced within different measurement spherical domains (see Fig. 5.54b) across four numerical draw tests. Notably, Fig. 5.57 illustrates the variations in vertical stress within the fourth measurement domain at different distances between drawpoints. The analysis reveals that: (1) Initially, during the four draw tests, the fourth measurement domain is located in a region that remains intact between two adjacent drawpoints. Here, the vertical stress σ_{V} experiences a rapid increase compared to the initial vertical stress $\sigma_{\mathrm{V},0}$ prior to drawing. As the distance between drawpoints decreases, as observed in Test 1 and Test 2, the movement zone extends into the fourth measurement domain, resulting in a gradual reduction of vertical stress to approximately 0.6 times the initial value, as illustrated by the black and blue curves in Fig. 5.57. Conversely, in scenarios where the drawpoint distance is larger, such as in Test 4, the fourth measurement domain remains intact, sustaining a vertical stress of around 1.2 times the initial stress, as depicted by the green curve in Fig. 5.57. (2) Additionally, a smaller distance between drawpoints is associated with higher vertical stress in the intact region and an earlier occurrence of maximum vertical stress. For example, with a drawpoint distance D_{s} of 12 m in Test 1, the maximum vertical stress in the eighth measurement domain approaches 1.8 times the initial vertical stress, as shown by the black curve in Fig. 5.57. This analysis indicates that excessively close drawpoint distances may compromise the stability of the underlying structure in the stope located between the two drawpoints.

To delve deeper into the differences and interactions between stress evolutions in both the movement and intact regions, we use Test 4, which features a greater drawpoint distance, as a case study. Figure 5.58 illustrates the variations in horizontal

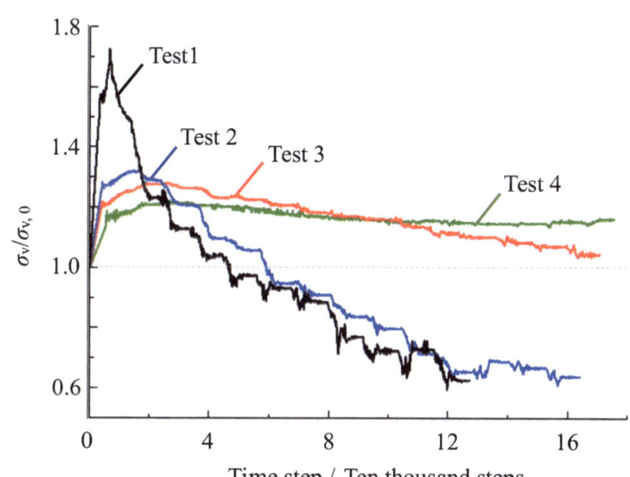

Fig. 5.57 Variations of vertical stress within measurement region No. 4 under different drawpoint spacings

and vertical stresses within the 5th and 8th measurement zones derived from Test 4. Analysis of Figs. 5.54b and 5.58 reveals the following: (1) The 8th measurement zone, located directly above drawpoint 2, remains unaffected by the adjacent drawpoints. Consequently, the stress variation in this zone aligns with the stress change pattern observed under isolated draw conditions [55]. As the movement zone approaches the height of the 8th measurement zone from below, the horizontal stress (represented by the green dotted line in Fig. 5.58) initially increases before decreasing, while the vertical stress (depicted by the blue solid line in Fig. 5.58) consistently diminishes. Once the movement zone surpasses the height of the 8th measurement zone, the caved ore and rock in this region become relatively loose, resulting in fluctuations of both horizontal and vertical stresses within a lower range. (2) In the 5th measurement zone, the horizontal stress (indicated by the red dotted line in Fig. 5.58) consistently declines, while the vertical stress (depicted by the black solid line in Fig. 5.58) remains relatively stable during the early draw phase. At time t_1, when the IMZ area reaches the height of the 8th measurement zone, the vertical stress begins to rise gradually, suggesting that the vertical stress above the IMZ progressively transfers to the non-loosened regions on either side throughout the draw process.

Figure 5.59 illustrates the variations in coordination numbers across the 4–9th measurement zones derived from Test 4. From this analysis, the following conclusions can be drawn: (1) During the draw operation, the coordination numbers *CN*, represented by the magenta, yellow, and green lines in Fig. 5.59, in the 4–6th measurement zones—areas that remain unaffected—consistently range from 8 to 9 without significant variations. Additionally, it is observed that a greater height of the overlying rock layer corresponds to a higher coordination number. (2) As the movement zone extends into the 7–9th measurement zones, the coordination numbers *CN*, depicted

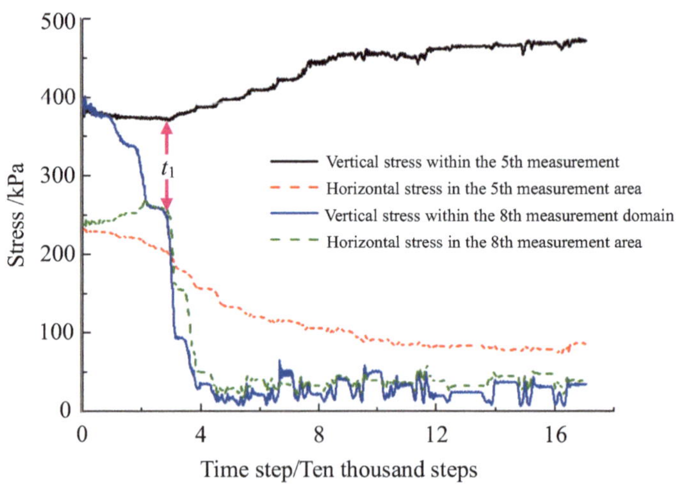

Fig. 5.58 Variations of horizontal and vertical stresses within measurement regions Nos. 5 and 8 obtained in Test 4

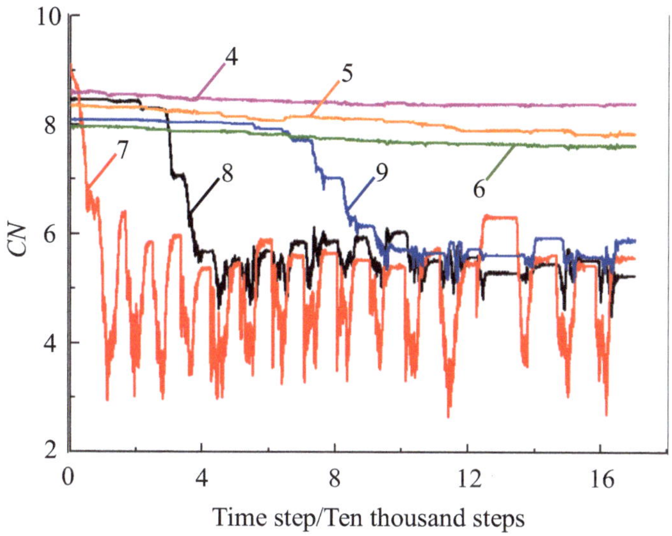

Fig. 5.59 Variations of coordination number within measurement regions Nos. 4–9 obtained in Test 4

by the red, black, and blue lines in Fig. 5.59, progressively decline. Furthermore, in contrast to the stable coordination numbers in the intact regions, a lower spatial height in the IMZ area correlates with a reduced coordination number, albeit with increased variability. This indicates that proximity to the drawpoint results in greater looseness of caved ore and rock, leading to a higher degree of disorder in their movement and interactions.

5.5.3 *Interactions of Caved Ore and Rock Under Different Draw Modes*

The presence of irregular horizontal surfaces at the ore-waste interface, coupled with the hang-up of certain drawpoints in real block caving mines, results in inconsistent simultaneous draw. Therefore, numerical tests were conducted to investigate the interactions of caved ore and rock across different draw modes by examining the dynamics and stress evolutions at varying drawpoint distances.

5.5.3.1 Design of Numerical Draw Test

To prevent excessive leftover ore and minimize losses between adjacent drawpoints, the spacing of drawpoints in a block caving mine must facilitate a uniform flow of caved ore and rock at the interaction of two adjacent IMZ [35]. Drawing from prior research, the current numerical tests for various draw modes maintained a fixed drawpoint spacing of 16 m (1.0w_{IMZ}), as detailed in Table 5.6. Specifically, Test 2 in Table 5.6, which features a 16-m drawpoint spacing and simultaneous draw, will serve as the control for this numerical draw test. Additionally, the numerical model and its mesoscopic parameters align with the design used in previous sections for different drawpoint spacings, as illustrated in Fig. 5.54 and Table 4.5.

Table 5.6 Schemes of numerical simulation using different draw modes

Test number	Draw sequence	Drawpoint 1	Drawpoint 2	Drawpoint 3	Draw termination criterion
Control group	Simultaneous draw from Drawpoints 1, 2, and 3	Draw commences right from the initial stage	Draw commences right from the initial stage	Draw commences right from the initial stage	The IMZ height reaches 50 m
1	Sequential draw from Drawpoints 1 → 2 → 3	Draw commences right from the initial stage	When the IMZ height above the Drawpoint 1 reaches 30 m, the draw will commence	When the IMZ height above the Drawpoint 2 reaches 30 m, the draw will commence	The IMZ height above Drawpoint 3 reaches 50 m
2	Sequential draw from Drawpoints 1 → 3 → 2	Draw commences right from the initial stage	When the IMZ height above the Drawpoint 3 reaches 30 m, the draw will commence	When the IMZ height above the Drawpoint 1 reaches 30 m, the draw will commence	The IMZ height above Drawpoint 2 reaches 50 m
3	Simultaneous draw from Drawpoints 1, and 3, while intermittent draw from Drawpoint 2	Draw commences right from the initial stage	Drawing commences when the IMZ height above Drawpoint 1 and Drawpoint 3 increases by 10 m. This process persists until the IMZ height above Drawpoint 2 also increases by 10 m, at which point draw ceases. This cycle is repeated	Draw commences right from the initial stage	The IMZ height above Drawpoint 2 reaches 50 m

5.5.3.2 Analysis of the Draw Test Results

Upon concluding the 3 draw tests, a comparative analysis was conducted on the behaviors of caved ore and rock. This analysis focused on particle movement and the evolution of stress, considering various draw modes.

(1) Analysis of flow Laws

Figure 5.60 illustrates the longitudinal profiles of caved ore and rock movement at the cessation of draw, utilizing various draw modes. In comparison to the particle movement depicted in Fig. 5.55b for the simultaneous draw mode, both sequential and intermittent draw modes exhibit interactions among the rock layers at different elevations, such as 35 m and 45 m, indicating the intersection of neighboring IMZs. However, a key distinction is that not all layers at the same height descend in a nearly horizontal manner; rather, the different draw modes result in varying degrees of tilt or irregular descent patterns. This analysis reveals that during the actual draw process, issues such as hang-ups at certain drawpoints and inadequate management can lead to uneven contact surfaces between caved ore and rock. By adjusting the draw order of each drawpoint, it is possible to influence the migration paths of caved ore and rock, thereby enhancing the contact surface and potentially increasing ore recovery rates or minimizing the mixing of waste rock.

(2) Analysis of stress evolution laws

A statistical analysis was conducted on the evolutions of horizontal and vertical stress across various measurement domains (refer to Fig. 5.54b) during four numerical tests, including a control scenario. Figure 5.61 illustrates the variations in vertical stress within the 4th measurement domain for different draw modes. The trends indicated by the curves within the magenta circles in Fig. 5.61 reveal two key points: (1) Initially, across all four draw tests, the vertical stress σ_v in the 4th measurement domain showed a notable increase compared to the baseline vertical stress $\sigma_{v,0}$ prior to draw. (2) In contrast, during the initial phases of draw tests 1–3, only Drawpoint 1 was activated while Drawpoint 2 remained closed. As a result, the vertical stress above Drawpoint 1 was the sole contributor to the stress transferred to the intact zone of the 4th measurement domain, leading to a maximum vertical stress approximately 1.3 times greater than the initial value. Conversely, in the control test, both Drawpoints 1 and 2 were opened simultaneously, which facilitated the transfer of vertical stress from both Drawpoints to the intact zone. This condition resulted in a slightly elevated maximum vertical stress in this area compared to tests 1–3. This analysis indicates that, with consistent spacing between Drawpoints, the variation in draw modes does not significantly affect the maximum vertical stress at the stope base.

Figure 5.62 illustrates the evolution of horizontal stress in the 3rd and 8th measurement zones under various draw conditions. The curves highlighted in the red and yellow circles of Fig. 5.62a indicate that as the movement zone approaches the 3rd measurement area, the horizontal stress in tests 1–3 (represented by the red, blue, and green curves) experiences a notable increase, while the control test (depicted by the black curve) shows no significant change. This discrepancy arises from the differing

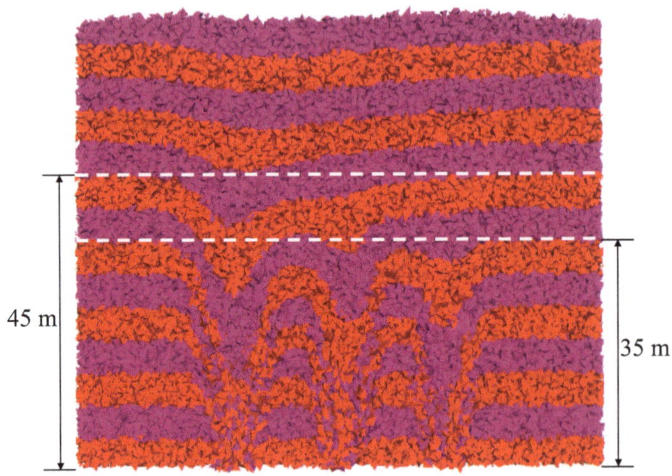

(a) Test 1

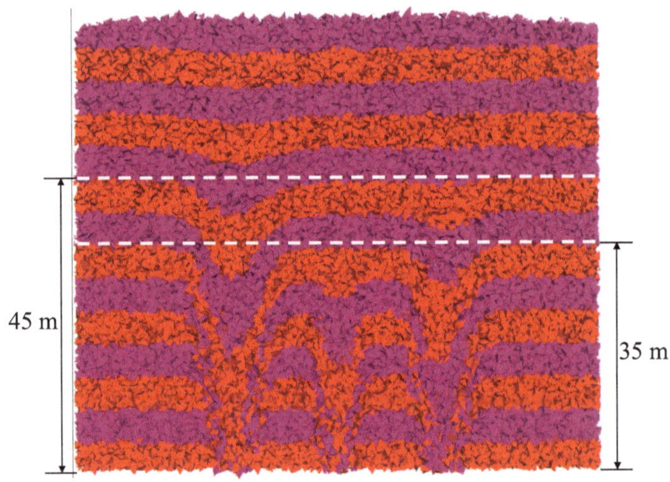

(b) Test 2

Fig. 5.60 Longitudinal profiles of movement of caved ore and rock with different draw modes

draw modes employed. In tests 1–3, the draw commenced at Drawpoint 1 and subsequently progressed through Drawpoints 2 and 3. Consequently, as the movement zone nears the 3rd measurement area, the surrounding regions remain intact, resulting in a dense contact state among the caved ore and rock, and a pronounced stress arch effect [36]. This leads to a clear transfer of vertical stress and a marked increase in horizontal stress in that area. In contrast, the control test utilized a simultaneous draw mode, which created a uniform flow around the 3rd measurement zone, resulting in a

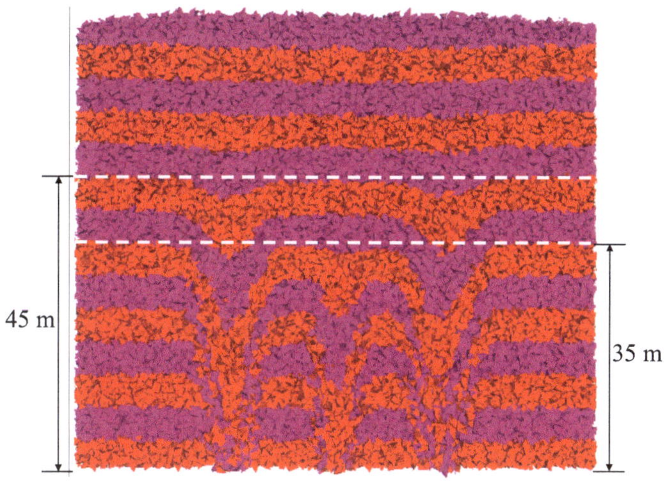

(c) Test 3

Fig. 5.60 (continued)

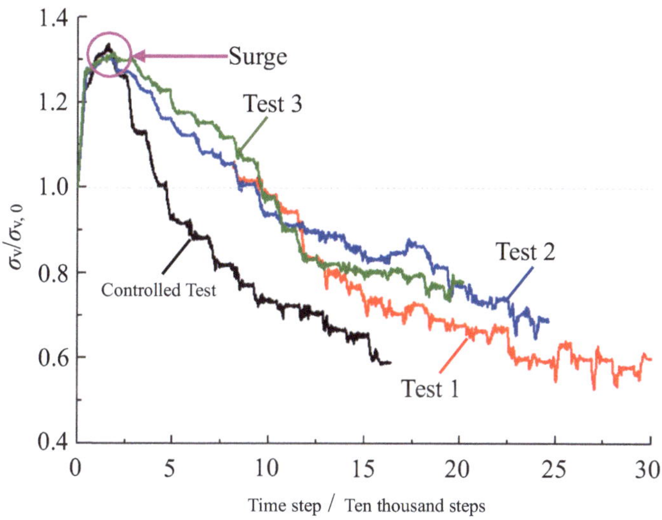

Fig. 5.61 Variations of vertical stress within measurement region No. 4 under different draw modes

loose contact state among the caved ore and rock and a lack of significant stress arch effect, thereby preventing a notable increase in horizontal stress. Similarly, the trend observed in the red circle of Fig. 5.62b illustrates that different draw modes significantly influence the stress distribution around the 8th measurement zone. Consequently, across the four tests, the horizontal stress in the 8th measurement domain exhibits varying degrees of increase at different draw stages. This analysis indicates that the selection of draw mode has a substantial impact on the stress distribution at various times and locations within the stope, ultimately affecting the movement range and flow dynamics of the entire particle system of caved ore and rock.

It is well established that various forms and intensities of arch structures, or stress arches, exist within the particle flow system of caved ore and rock [30, 55]. Furthermore, there exists a significant correlation mechanism between the interaction

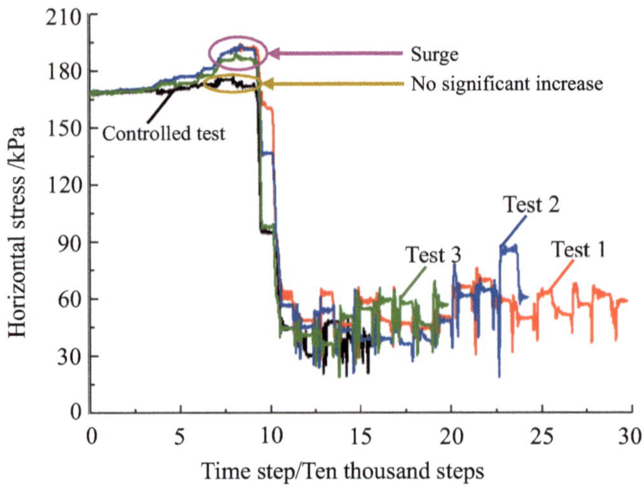

(a) Measurement region No. 3

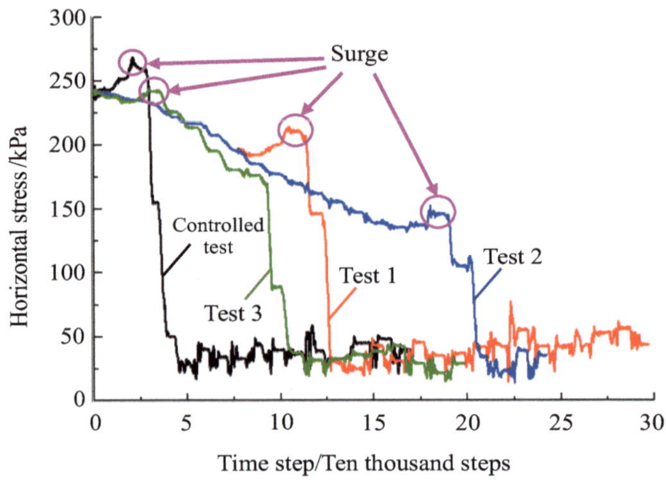

(b) Measurement region No. 8

Fig. 5.62 Variations of horizontal stress within measurement regions Nos. 3 and 8 under different draw modes

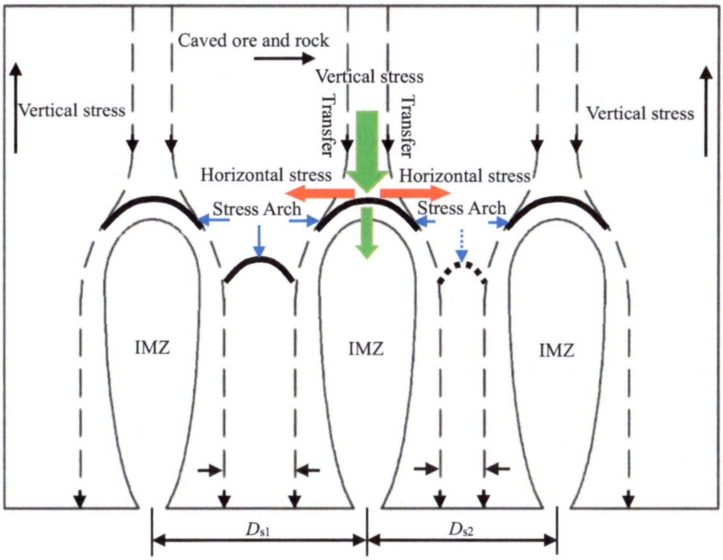

Fig. 5.63 Schematic of stress arch and stress transfer within the particle flow system of caved ore and rock with different drawpoint spacings

rules of caved ore and rock across multiple drawpoints and the arch structures, as well as stress transfer. Therefore, this study will explore topics such as the threshold for drawpoint spacing and the patterns of stress evolution, with a focus on arch formations and stress transfer, as illustrated in Fig. 5.63.

5.5.4 Discussions

(1) The spacing threshold D_s for a drawpoint that facilitates a uniform flow of caved ore and rock—where two neighboring IMZs meet—ranges from 1.0 to 1.25 times the maximum width w_{IMZ} of the IMZ. This differs from Laubscher's [12] finding of $1.5w_{IMZ}$, derived from laboratory tests using sand. The reason lies in the nature of the caved ore and rock generated during underground mining, which results from blasting or natural caving. This material system is intricate and exhibits a diverse particle size distribution [58], varied shapes, and greater strength and internal friction angles compared to agricultural products [118] like corn, wheat, and soybeans, as well as materials like sand. Consequently, a significant stress arch effect exists among the caved ore and rock. When the drawpoint spacing is set at 1.5 times the maximum width of the IMZ, the vertical stress from above the adjacent IMZ is redirected to the intact zone in between, leading to a substantial increase in vertical stress there, as illustrated by the green curve in Test 4 of Fig. 5.57. In scenarios where sand is the medium, this

induced vertical stress can cause the intact zone between the drawpoints to move and facilitate a uniform flow. Conversely, when caved ore and rock are used, the arch structure formed in the intact zone can support a certain level of vertical stress, thereby maintaining stability.

(2) The occurrence of a uniform flow of caved ore and rock is influenced by the spacing of the drawpoints, specifically when it is equal to 1.0 times the maximum width of the IMZ, which is tied to the conditions that create a stress arch. As illustrated in Fig. 5.63, a larger drawpoint spacing (like D_{s1}) allows the caved ore and rock between the drawpoints to establish a relatively stable arch structure. However, as spacing narrows, the length, span, and stability of the stress arch diminish. When the spacing reaches a critical threshold or less (for instance, D_{s2}), the vertical stress from the overburden and the stress transferred from the adjacent IMZ can exceed the capacity of the stress arch, resulting in its collapse [55]. This poses a challenge for the caved ore and rock to maintain a stable arch structure, as indicated by the black curve in the control test shown in Fig. 5.62a, where the horizontal stress exhibits minimal increase. Consequently, the behavior of caved ore and rock under varying drawpoint spacings and draw modes can fundamentally be understood as a manifestation of the mechanical interactions within the particle flow system of caved ore and rock.

Subsequent authors will systematically analyze the types of arch structures present in the particle flow system of caved ore and rock, categorizing them into simple arches and complex arches. They will also examine the proportions of these structures, along with geometric characteristics such as arch length, horizontal extension, spatial distribution, and evolution characteristics. Additionally, they will investigate inter-particle arching and instability capabilities, which include factors such as particle size, quantity, and the spatial relationships among particles forming arch structures, as well as their corresponding thresholds [119–122]. This analysis will employ various methodologies, including indoor tests, numerical simulations, and statistical analyses. Furthermore, the authors will explore the internal mechanisms linking inter-particle mechanical behavior to the stress arch effect, ultimately revealing the mechanical mechanisms underlying the migration and evolution of caved ore and rock.

5.6 Conclusions

In relation to the research on the flow characteristics of near/far field caved ore and rock based on the rigid block model [55], the main conclusions are as follows:

(1) A comparison of the outcomes from physical tests and numerical simulations demonstrates that the rigid block model is both reliable and effective for investigating gravity flow. This model is applicable for examining the structural features and mechanisms behind the formation of the particle flow system of caved ore and rock in more complex scenarios.

(2) The alterations in the shape of the IMZ, observed under both near and far field scenarios, align with the principles of the upside-down drop shape theory. During the initial phase of the draw process, the IMZ's maximum width expands rapidly as height increases, adhering to a power function. After this stage, the increase in maximum width becomes nearly linear with respect to height. Consequently, under far-field conditions, the approximate linear relationship between height and maximum width of the IMZ is particularly beneficial for guiding the optimization of stope structural parameters in large-scale block caving mines.
(3) During gravity flow, a significant stress arch phenomenon is observed. As the movement zone of ore and rock expands, the vertical stress near the IMZ shows a pronounced decrease, while the horizontal stress gradually increases, resulting in a sudden spike just prior to the movement zone. In contrast, both horizontal and vertical stresses within the IMZ experience a sharp decline to markedly lower values.

In relation to the study of gravitational flow resulting from the non-uniform distribution of particle sizes [80], the key findings are summarized as follows:

(1) Among the various examined particle size distributions, the mean particle size at a specific draw height is greater, and the maximum dimension of the IEZ is also larger.
(2) As the height of the particle packing system of caved ore and rock increases, the void ratio also rises, while the average coordination number of the particles decreases. The distribution of the average coordination number adheres to a Gaussian pattern, with typical values peaking between 7 and 11.
(3) The structural anisotropy of the particle packing system in caved ore and rock is significantly influenced by variations in the spatial configuration of the IEZ. As the packing state transitions from a natural to a denser configuration, the degree of anisotropy in the packing system structure intensifies. This transition alters the shape of the IEZ from an upside-down drop shape to a standard drop shape, while simultaneously reducing its maximum width. To minimize ore dilution and loss rates, as well as the frequency of hang-ups, mines employing the block caving method should strive to avoid the extremely dense packing state of caved ore and rock, which can be exacerbated by factors such as blasting effects or prolonged packing.

In relation to the investigation of how a single coarse particle influences the shape changes of the IMZ [101], the key findings are summarized as follows:

(1) An analysis of the IMZ shapes obtained from both manual sketching and the IPP software reveals a strong correlation between the IMZ's height and its maximum radius. This confirms the reliability of the IPP software in the draw test from both qualitative and quantitative perspectives. Furthermore, the IPP software exhibits improved efficiency and convenience in the automatic sketching of the IMZ, highlighting its superior capabilities.
(2) The presence of coarse particles significantly affects the movement of neighboring fine particles, with this influence being more pronounced on the fine

particles located beneath the coarse ones compared to those situated above. As the IMZ traverses a single coarse particle, the displacement of particles in the region above transitions from a power-law distribution to a logistic distribution, while the displacement of the coarse particle in the region below shifts from a power-law distribution to a more erratic and variable distribution.

(3) A single coarse particle can significantly influence the local structure of the IMZ. When positioned at the edge of the IMZ, this coarse particle progressively alters the shape of the IMZ, exhibiting features such as concavities or irregular bends, until the material fully envelops the coarse particle. In tests involving fine particles sized between 3 and 8 mm, a size ratio of coarse to fine particles d_c/d_f exceeding 6.0 leads to noticeable alterations in the shape of the IMZ. In scenarios where multiple coarse particles are present within the stope, these localized changes can collectively result in an irregular IMZ shape, which subsequently affects the design parameters of the stope. Therefore, it is crucial to optimize the blasting parameters, manage the outcomes of blasting, and minimize the presence of coarse particles.

(4) The presence of coarse particles accelerates the movement of fine particles towards the drawpoint, with 17.02% and 31.42% of particles in the upper and lower regions, respectively, arriving sooner. When coarse particles are situated near the waste rock zone, they not only facilitate the passage of smaller particles, leading to ore dilution, but also prompt the earlier release of surrounding waste rock, which negatively impacts economic returns. Therefore, it is crucial to regulate the blasting effects to prevent the emergence of coarse particles.

Physical draw tests based on the IPP method and numerical draw tests utilizing the rolling resistance model were conducted to investigate the evolution of coarse particle separation and the shape of the IMZ under varying conditions, including the number, relative position, and spacing of coarse particles. Additionally, the influence mechanism of multiple coarse particles on the flow characteristics of caved ore and rock was analyzed from the perspective of interparticle interactions, based on the evolution of unbalanced forces, void fractions, and force chains [123]. The main conclusions of this study are as follows:

(1) In this study, within the range of values examined, as the number of coarse particles increases from 1 to 4, the size ratio of coarse to fine particles, which significantly influences the movement of local fine particles surrounding the coarse particles, decreases from 6.0 to 4.0. Furthermore, the relationship between the maximum radius of the IMZ on the side with coarse particles and its height no longer adheres to the power function described by the upside-down draw shape theory.

(2) When the initial positions of two coarse particles are aligned either horizontally or vertically, and the distance between their centers is substantial, the gap between them tends to increase as the draw progresses. Specifically, in the case of horizontal alignment, separation occurs more rapidly, resulting in a greater distance between the particles at that stage of the draw. Conversely, when the

particles are arranged vertically and are in contact, they do not experience significant separation. Additionally, the local void ratio varies considerably for the two coarse particles depending on their relative positions and distances, leading to differences in the unbalanced forces and velocities they encounter, which ultimately affects the degree of their separation.

(3) The arrangement of two coarse particles in contact significantly influences the shape of the IMZ. The most pronounced effect is observed when these particles are vertically aligned, followed by an arrangement at a 45° angle, while horizontal alignment has a comparatively lesser impact. As the vertical distance between the two coarse particles increases, their influence on shapc diminishes. Specifically, if the spacing exceeds three times the average size of the fine particles, the interaction between the coarse particles can be considered negligible. Furthermore, when comparing vertical and horizontal arrangements, the vertical configuration demonstrates a more pronounced force chain aggregation effect, resulting in greater stability in the local force chain during prolonged draw operations. Consequently, this leads to a more significant and lasting impact on the shape of the IMZ. The interaction between the caved ore and rock weakens as the vertical separation between the coarse particles increases.

Other interesting and important findings from this study that should be considered are as follows:

(1) This study reveals that the influence of the number of coarse particles on the flow characteristics of caved ore and rock is not merely a simple linear superposition of the individual effects of these particles. Instead, due to variations in the relative positions and spacings of the coarse particles, different local interparticle interactions arise, which subsequently affect the shape of the IMZ. This phenomenon is referred to as 'emergence' in the field of complex systems [124, 125], indicating that the particle system of caved ore and rock represents a non-equilibrium complex system. Consequently, the author plans to employ theories and methodologies such as cellular automata models, network models, and statistical analysis from the domain of complex systems to further investigate the emergent outcomes of the flow characteristics of caved ore and rock influenced by multiple coarse particles, thereby enhancing the existing draw theory.

(2) This study also found that the initial relative positions and spacings between different coarse particles do not lead to significant changes or only gradual increases in the spacing between the two coarse particles during the draw process. Specifically, it is posited that, similar to other particle flow systems, the interaction between two coarse particles and their surrounding fine particles during the drawing process results in long-range attraction or repulsion between coarse particles positioned at certain distances apart in the particle flow [126, 127]. Consequently, the authors will conduct a comprehensive analysis of the local structure, motion, and mechanical properties of coarse particles with varying diameters, relative positions, and quantities. This analysis will include determining the correlation length between the coarse particles and elucidating

the origins of the long-range attraction or repulsion observed. Furthermore, additional research will focus on the flow characteristics of caved ore and rock, the percolation behavior of local fine particles [24], and the issue of hang-up [83].

In relation to the investigation of the uniform movement of caved ore and rock across several drawpoints, along with the evolution of stress [128], the key findings are summarized below:

(1) The spacing of drawpoints and the draw modes significantly affect the gravity flow and stress evolution within the stope. Under multi-drawpoint conditions, the threshold spacing for drawpoints that can facilitate uniform flow of caved ore and rock lies between 1.0 and 1.25 times the maximum width of the IMZ. This finding holds considerable significance for optimizing the structural parameters of stopes in actual block caving mines.
(2) A notable stress arch phenomenon occurs within the particle flow system of caved ore and rock. As draw progresses, the vertical stress from the IMZ progressively shifts to the intact regions on either side. In the analyzed area, the maximum vertical stress at the stope base has reached nearly 1.8 times the original vertical stress. Consequently, if the distance between neighboring drawpoints is insufficient, it could jeopardize the stability of the underlying structure in the stope situated between these drawpoints.
(3) In a simultaneous draw mode, the flow of caved ore and rock layers at an equivalent elevation tends to create a nearly level downward flow. In contrast, sequential or intermittent draw modes may result in varying degrees of tilt or irregular downward movement among the caved ore and rock layers at the same height. Therefore, during actual operations, it is crucial to consider factors such as the effects of caved ore and rock extraction and the organization of production processes. Adjusting the draw order for each drawpoint can effectively manage the flow patterns of caved ore and rock, thereby enhancing the interaction surface between these materials, which can lead to improved ore recovery rates and minimize the early mixing of waste rock.
(4) In the stope, the coordination number for regions that remain intact during draw is approximately between 8 and 9, increasing with greater spatial height. Conversely, in the movement zones, a decrease in spatial height corresponds to a reduced coordination number, although the variability in this number tends to be greater. Essentially, closer to the drawpoint inside the IMZ, the caved ore and rock become increasingly loose, leading to more chaotic particle flow and contact rearrangement.

Note. Figures 5.25 in Sect. 5.3, Figs. 5.38, 5.39, 5.40, 5.41, 5.42, 5.43, 5.45, 5.46, 5.47, 5.48, 5.49, 5.50, 5.51, 5.52 and Tables 5.3 and 5.4 in Sect. 5.4 are reprinted from the authors' previously published work [123]:

Sun H, Zhou S, Chen S, et al. Flow characteristics of caved ore and rock under the influence of multiple coarse particles: insight from interparticle interaction. Granular Matter, 2024, 26: 82.

References

1. Shen NS, Gu XC, Yin SH (2009) The technical status of block caving mining method at home and abroad. Min Technol 9(04):1–4+15
2. Chitombo GP (2010) Caving mining-16 years after Laubscher's 1994 paper 'Cave mining-state of the art'. In: Proceedings of the 2nd international symposium on block and sublevel caving, Perth, pp 45–61
3. Pierce M (2010) A model for gravity flow of fragmented rock in block caving mines. The University of Queensland, Brisbane
4. Wang CH (1982) Draw engineering. Metallurgical Industry Press, Beijing
5. Li RF, Guo JP (2016) Quasi-ellipsoid ore-drawing theory and verification of ore-drawing theory. Metallurgical Industry Press, Beijing
6. Ren FY (1994) Stochastic medium draw theory and its applications. Metallurgical Industry Press, Beijing
7. Fröström J (1970) Examination of equivalent model materials for development and design of sublevel caving. Royal Institute of Technology, Stockholm (in Swedish)
8. Jin AB, Sun H, Wu SC et al (2017) Confirmation of the upside-down drop shape theory in gravity flow and development of a new empirical equation to calculate the shape. Int J Rock Mech Min Sci 92:91–98
9. Kvapil R (1965) Gravity flow of granular material in hoppers and bins. Int J Rock Mech Min Sci 2(1):35–41
10. Kvapil R (1965) Gravity flow of granular material in hoppers and bins in mines—II. Coarse material. Int J Rock Mech Min Sci 2(3):277–292
11. Janelid I, Kvapil R (1966) Sublevel caving. Int J Rock Mech Min Sci 3(2):129–132
12. Laubscher DH (2000) Block cave manual, design topic: drawpoint spacing and draw control. The Julius Kruttschnitt Mineral Research Centre (JKMRC), The University of Queensland, Brisbane, Australia
13. Power G (2004) Modelling granular flow in caving mines: large scale physical models and full scale experiments. University of Queensland, Brisbane, pp 57–168
14. Castro R, Trueman R, Halim A (2007) A study of isolated draw zones in block caving mines by means of a large 3D physical model. Int J Rock Mech Min Sci 44(6):860–870
15. Tao GQ, Yang SJ, Ren FY (2009) Experimental research on granular flow characters of caved ore and rock. Rock Soil Mechanics 30(10):2950–2954
16. Wang HJ, Ying SH, Wu AX et al (2010) Experimental study of the factors affecting the ore flow mechanism during block caving. J China Univ Min Technol 39(5):693–698
17. Wang YP, Yu J (2014) Optimization of breaking interval in non-pillar sublevel caving mining. J Central South Univ (Sci Technol) 45(2):603–608
18. Shao AL (2012) Experimental research on mullock movement in the side drawing. Min Metall Eng 32(03):1–3
19. Xu S, An L, Li YH et al (2012) Optimization of caving space for different angles of end-wall during pillarless sublevel caving. J Northeast Univ (Nat Sci) 33(1):120–123
20. Castro R, Pineda M (2015) The role of gravity flow in the design and planning of large sublevel stopes. J South Afr Inst Min Metall 115(2):113–118
21. Sun H, Jin AB, Gao YT et al (2016) Testal research on the expectation body theory and optimization of the rate of advance during ore breaking in side drawing. Chin J Eng 38(9):1197–1203
22. Cundall PA, Strack ODL (1980) A discrete numerical model for granular assemblies. Geotechnique 30(3):331–336
23. Zhu HC (2006) PFC and application case of caving study. Chin J Rock Mech Eng 25(9):1927–1931
24. Hazizan MHM (2011) Particle percolation in block caving mines. The University of New South Wales, Sydney
25. Song ZY, Wei WJ, Zhang JW (2018) Numerical investigation of effect of particle shape on isolated extracted zone (IEZ) in block caving. Arab J Geosci 11(12):310

26. Hu JH, Guo FZ, Luo XW et al (2015) Simulation of ore flow behavior and optimization of discharge parameters for caving method in gently inclined medium thickness ore-body. J Central South Univ (Sci Technol) 46(5):1772–1777
27. Sun H, Jin AB, Gao YT et al (2015) Influencing factors on the flow characteristics of an isolated extraction zone in caving mining. Chin J Eng 37(9):1111–1117
28. Sun H, Jin AB, Gao YT et al (2015) Flow characteristics of caved ore and rock under complex boundary conditions. Chin J Eng 46(10):3782–3788
29. Sun H, Jin AB, Gao YT et al (2016) Research of the isolated extraction zone form and determination of optimal independent advance under different end wall angles. Chin J Eng 38(2):159–166
30. Castro R, Gómez R, Pineda M et al (2020) Experimental quantification of vertical stresses during gravity flow in block caving. Int J Rock Mech Min Sci 127:104237
31. Rafiee R, Ataei M, Khalookakaie R et al (2018) Numerical modeling of influence parameters in cavabililty of rock mass in block caving mines. Int J Rock Mech Min Sci 105:22–27
32. Itasca Consulting Group (2019) Particle flow code (PFC) version 6.0. Itasca Consulting Group Inc., Minneapolis
33. Sun H, Jin AB, Elmo D et al (2020) A Numerical based approach to calculate ore dilution rates using rolling resistance model and upside-down drop shape theory. Rock Mech Rock Eng 53:4639–4652
34. Sun H (2019) Study on migration and evolution mechanism of caved ore and rock based on the particle flow theory. University of Science and Technology Beijing, Beijing
35. Castro R (2006) Study of the mechanisms of gravity flow for block caving. University of Queensland, Brisbane
36. Arévalo R, Maza D, Pugnaloni LA (2006) Identification of arches in two-dimensional granular packings. Phys Rev E 74(2):021303
37. Eadie B (2003) A framework for modelling fragmentation in block caving. The University of Queensland. Julius Kruttschnitt Mineral Research Centre, Brisbane, pp 1–5
38. Peters DC (1984) Physical modeling of the draw behavior of broken rock in caving. Colorado School of Mines Quarterly 79(1)
39. Wu AX, Wu LC, Liu XH et al (2012) Study on structural parameters of sublevel caving. J Central South Univ (Sci Technol) 43(5):6
40. Chen QF, Wang SP, Qin SK et al (2020) Discrete element simulation for evolution characteristics of multi-funnel mineral-rock force chain under flexible isolation layer. Chin J Eng 42(9):1119–1129
41. Zhu S, Zhong CX, Zheng XL et al (2018) Filling standards and gradation optimization of rockfill materials. Chin J Geotechn Eng 40(1):108–115
42. Yao XK, Bian J, Wang XD, et al (2018) Testal study on influences of scale effect on mechanical properties of railway ballast. Chin J Geotechn Eng 40(S2):230–234+245
43. Xing YY, Wu HB, Liu XJ et al (2018) Grain composition on solid-state-sintered sic ceramics. J Inorg Mater 33(11):1167–1172
44. Zheng GB, Kang TH, Chai ZY et al (2006) Applied the Rosin-Rammler distribution function to study on the law of coal dust particle-size distribution. J Taiyuan Univ Technol 37(3):317–319
45. Zhu J, Liu ZH, Yang YF et al (2015) Applying the Rosin-Rammler function to study on drop size distribution in spray tower. J Chem Eng Chinese Universities 29(5):1059–1064
46. Li CN, Gong XN (2002) Study on the non-uniformity of loose body of ore rock. Min Metall Eng 2002(2):37–39+44
47. Li X, Jia MT, Wang LG et al (2009) Study of ore fragment size prediction in block caving based on Monte Carlo stochastic simulation. Rock Soil Mech 30(4):1186–1190
48. Zhang JW, Wang JC, Wei WJ et al (2019) Testal investigation on the effect of size distribution on the flow characteristics of loose top coal. J China Coal Soc 44(4):985–994
49. Ren FY, Liu Y, He RX, et al (2020) Flow-through features of tailings in subsidence area under draw conditions. J Northeast University (Nat Sci) 41(6):858–862+887
50. Li T (2018) Study on the migration law of granular ore rock under draw by caving method. University of Science and Technology Beijing, Beijing

51. Liu XB, Wang LC (2017) Simulation model of granular flow of the non-uniform caved ore and rock. J Northeast Univ (Nat Sci) 38(8):1167–1172
52. Sun H, Gao YT, Elmo D et al (2019) A study of gravity flow based on the upside-down drop shape theory and considering rock shape and breakage. Rock Mech Rock Eng 52(3):881–893
53. Jin AB, Chen SJ, Sun H et al (2020) Characteristics of particle percolation based on inhomogeneous particle size distribution. J Central South Univ (Sci Technol) 51(6):1673–1681
54. Elmo D, Rogers S, Dorador L, et al (2015) An FEM-DEM numerical approach to simulate secondary fragmentation. In: Computer methods and recent advances in geomechanics: proceedings of the 14th international conference of international association for computer methods and recent advances in geomechanics (IACMAG 2014). Taylor & Francis Books Ltd, pp 1623–1628
55. Sun H, Chen SJ, Gao YH et al (2021) Research on near/far-field flow characteristics of caved ore and rock based on rigid block model. Chin J Eng 43(02):205–214
56. Zhao SW (2018) Discrete element investigation of physical and mechanical properties of granular materials. South China University of Technology, Guangzhou
57. Kou BQ, Cao YX, Li JD et al (2017) Granular materials flow like complex fluids. Nature 551(7680):360–363
58. Sun QC, Liu XX, Zhang GH et al (2017) The mesoscopic structures of dense granular materials. Adv Mech 47:263–308
59. Lu KQ, Liu JX (2004) Static and dynamic properties of granular matter (I). Physics 33(9):629–635
60. Lu KQ, Liu JX (2004) Static and dynamic properties of granular matter (II). Physics 33(10):713–721
61. Bi D, Zhang J, Chakraborty B et al (2011) Jamming by shear. Nature 480(7377):355–358
62. Tao GQ, Ren FY, Liu ZD et al (2010) Improvement of random medium drawing theory. J Min Saf Eng 27(2):239–243
63. Brunton ID, Sharrock G, Lett J (2012) Full scale near field flow behaviour at the Ridgeway Deeps Block Cave Mine. In: Proceedings of the 6th international conference and exhibition on mass mining (MassMin 2012), Canadian Institute of Mining, Metallurgy and Petroleum, Sudbury, Ontario, Canada
64. Yang H, Zhang GH, Wang YJ et al (2018) Measurement techniques of grain motion and inter-grain structures in dense granular materials. Adv Mech 48:541–590
65. Slominski C, Niedostatkiewicz M, Tejchman J (2007) Application of particle image velocimetry (PIV) for deformation measurement during granular silo flow. Powder Technol 173(1):1–18
66. Katsuragi H, Abate AR, Durian DJ (2010) Jamming and growth of dynamical heterogeneities versus depth for granular heap flow. Soft Matter 6(13):3023–3029
67. Liu JG, Sun QC, Jin F et al (2009) Review of contact force measurements in granular materials. Rock Soil Mech 30(S1):121–128
68. Chen FX, Zhuang Q, Wang RL (2016) A preliminary study of dynamic failure mechanism of force chain in a particle system based on digital image correlation. Rock Soil Mech 37(2):563–573
69. Zhang P, Zhao XD, Zhang GH et al (2016) Acoustic detection and nonlinear response of granular materials under vertical load. Acta Physica Sinica 65(2):214–220
70. Xia CJ (2018) Structural study of granular system as a model glass system. Shanghai Jiao Tong University, Shanghai
71. Scheel M, Seemann R, Brinkmann M et al (2008) Morphological clues to wet granular pile stability. Nat Mater 7(3):189–193
72. Xia CJ, Zhu K, Cao YX et al (2014) X-ray tomography study of the random packing structure of ellipsoids. Soft Matter 10(7):990–996
73. Zhang XD, Xia CJ, Xiao XH et al (2014) Fast synchrotron X-ray tomography study of the packing structures of rods with different aspect ratios. Chin Phys B 23(4):377–380
74. Neudecker M, Ulrich S, Herminighaus S et al (2013) Jammed frictional tetrahedra are hyperstatic. Phys Rev Lett 111(2):028001

75. Trueman R, Castro R, Halim A (2008) Study of multiple draw-zone interaction in block caving mines by means of a large 3D physical model. Int J Rock Mech Min Sci 45(7):1044–1051
76. Melo F, Vivanco F, Fuentes C et al (2007) On drawbody shapes: From Bergmark-Roos to kinematic models. Int J Rock Mech Min Sci 44(1):77–86
77. Melo F, Vivanco F, Fuentes C et al (2008) Kinematic model for quasi static granular displacements in block caving: Dilatancy effects on drawbody shapes. Int J Rock Mech Min Sci 45(2):248–259
78. Melo F, Vivanco F, Fuentes C et al (2009) Calculated isolated extracted and movement zones compared to scaled models for block caving. Int J Rock Mech Min Sci 46(4):731–737
79. Chen QF, Chen QL, Zhong JY et al (2016) Flow pattern of granular ore rock in a single funnel under a flexible isolation layer. Chin J Eng 38(7):893–898
80. Sun H, Zhu DF, Jin AB et al (2022) Flow characteristics of caved ore and rock based on inhomogeneous particle size distribution. Chin J Nonferrous Metals 32(8):2433–2445
81. Chen XR, Lu DY, Sun YF et al (2015) Characterization of macro pore structure of foam geopolymers by image analysis technology. Mater Rep 29(S2):113–117
82. Yan JG, Shen FY, Thayer J et al (2017) Repair of the musculocutaneous nerve using the vagus nerve as donor by helicoid end-to-side technique: an experimental study in rats. J Neurosci Res 95(12):2493–2499
83. Castro R, López S, Gómez R et al (2021) Experimental study of the Influence of drawbell geometry on hang-ups in cave mine applications. Rock Mech Rock Eng 54:1–10
84. Kvapil R, Baeza L, Rosenthal J et al (1989) Block caving at El Teniente mine, Chile. Trans Inst Min Metall Section A: Min Technol 98:A43–A56
85. Chen G (1997) Stochastic modelling of rock fragment flow under gravity. Int J Rock Mech Min Sci 34(2):323–331
86. Verdugo R, Ubilla J (2004) Geotechnical analysis of gravity flow during block caving. In: Proceedings of the MassMin 2004: 4th international conference and exhibition on mass mining, Santiago, Chile, 22–25 August 2004, pp 195–200
87. Yashar P (2013) Mathematical programming for sequence optimization in block cave mining. The University of Alberta, Edmonton
88. Jin AB, Sun H, Ma GW et al (2016) A study on the draw laws of caved ore and rock using the discrete element method. Comput Geotech 80:59–70
89. Zadeh AA, Bares JT, Brzinski TA et al (2019) Enlightening force chain: a review of photoelasticimetry in granular matter. Granular Matter 21(4):83
90. Wang D, Ren J, Dijksman JA et al (2018) Microscopic origins of shear jamming for 2D frictional grains. Phys Rev Lett 120:208004
91. Wang MM, Wang D, Socolar JE (2019) Jamming by shear in a dilating granular system. Granular Matter 21:102
92. Wang YQ, Wang YJ, Zhang J (2020) Connecting shear localization with the long-range correlated polarized stress fields in granular materials. Nat Commun 11:4349
93. Wang YQ, Shang J, Wang YJ et al (2021) Contact force measurements and local anisotropy in ellipses and disks. Phys Rev Res 3:043053
94. Wang JA, Yang L, Li F et al (2020) Force chains in top coal caving mining. Int J Rock Mech Min Sci 127:104218
95. Imseeh WH, Alshibli KA (2018) 3D finite element modelling of force transmission and particle fracture of sand. Comput Geotech 94:184–195
96. Hou SQ, Wang W, Wang ZY et al (2018) Force chain characteristics and effects of a dense granular flow system in a third body interface during the shear dilatancy process. J Appl Mech Tech Phys 59:153–162
97. Díaz-Melián VL, Serrano-Muñoz A, Espinosa M et al (2020) Rolling away from the Wall into granular matter. Phys Rev Lett 125(7):078002
98. Espinosa M, Díaz-Melián V, Serrano-Muñoz A, et al (2021) Intruders cooperatively interact with a wall into granular matter 24:39
99. Dhiman M, Kumar S, Reddy KA et al (2020) Origin of the long-ranged attraction or repulsion between intruders in a confined granular medium. J Fluid Mech 886:A23

100. Merceron A, Sauret A, Jop P (2018) Cooperative effects induced by intruders evolving through a granular medium. EPL 121(3):34005
101. Jin AB, Yin ZS, Sun H et al (2023) Experimental study on shape evolution characteristics of IMZ under influence of single coarse particle. J Central South Univ (Sci Technol) 54(04):1531–1542
102. Rawlinson A, Elcock C, Cheung A et al (2005) An in-vivo methodology study of alveolar bone measurement using extra-oral radiographic alignment apparatus, Image Pro-Plus® software and a subtraction programme. J Dent 33(9):781–788
103. ITASCA CONSULTING GROUP (2016) PFC3D (particle flow code in 3 dimensions) user's guide in PFC2D version 5.0. Minneapolis: Itasca Consulting Group Inc.
104. Zhang LR, Nguyen Ngh, Lambert S, et al (2017) The role of force chains in granular materials: from statics to dynamics. Eur J Environ Civil Eng 21(7–8):874–895
105. Sun QC, Jin F, Liu JG (2010) Understanding force chains in dense granular materials. Int J Mod Phys B 24(29):5743–5759
106. Wang T, Wautier A, Liu SH et al (2022) How fines content affects granular plasticity of under-filled binary mixtures. Acta Geotech 17:2449–2463
107. Cai MF, Xue DL, Ren FH (2019) Current status and development strategy of deep mining in metal mines. Chin J Eng 41(4):417–426
108. Fairhurst C (2017) Some challenges of deep mining. Engineering 3(4):527–537
109. Marano G (1980) The interaction between adjacent draw points in free flowing materials and it application to mining. Chamber Mines J 22:25–32
110. Hancock W (2013) Gravity flow of rock in caving mines: Numerical modelling of isolated, interactive and non-ideal draw. The University of Queensland, Brisbane
111. Lorig L (2000) Relation between caved column height and vertical stress at the cave base. JKMRC and Itasca Consulting Group Inc.
112. Pierce ME (2019) Forecasting vulnerability of deep extraction level excavations to draw-induced cave loads. J Rock Mech Geotech Eng 11(3):527–534
113. Janssen HA (1895) Tests regarding grain pressure in soils. Zeitschrift Des Vereines Deutscher Ingenieure 39(35):1045–1049
114. Tao GQ, Ren QY, Ma JY (2012) Draw law of multiple leaks based on kinematic model. J China Coal Soc 37(3):407–410
115. Sun H, Jin AB, Gao YT et al (2015) Flow characteristics of caved ore and rock in the multiple draw-point condition. Chin J Eng 37(10):1251–1259
116. Chen QF, Chen QL, Zhong Y, et al (2017) Orthogonal simulation test on flow characteristics and influencing factors of multi funnel ore and rock under flexible isolation layer. Scientia Sinica Technologica 47(9):923–932
117. Chen QF, Chen QL, Wang YD et al (2018) Interface movement law and tension characteristics of flexible isolation layer in multi funnel draw. J Basic Sci Eng 26(5):1101–1109
118. Wang LJ, Zhou WX, Ding ZJ et al (2017) Experimental determination of parameter effects on the coefficient of restitution of differently shaped maize in three-dimensions. Powder Technol 284:187–194
119. Mehta A, Barker GC, Luck JM (2004) Cooperativity in sandpiles: statistics of bridge geometries. J Stat Mech: Theory Test 2004(10):P10014
120. Mehta A, Barker GC, Luck JM (2009) Heterogeneities in granular materials. Physical Today 62(5):40–45
121. Jenkins MC, Haw MD, Barker GC et al (2011) Does gravity cause load-bearing bridges in colloidal and granular systems? Phys Rev Lett 107(3):038302
122. Cao YX (2017) Synchrotron X-ray imaging study of properties of packing structure and rheology of granular materials. Shanghai Jiaotong University, Shanghai
123. Sun H, Zhou S, Chen S et al (2024) Flow characteristics of caved ore and rock under the influence of multiple coarse particles: insight from interparticle interaction. Granular Matter 26:82
124. Anderson PW (1972) More is different: broken symmetry and the nature of the hierarchical structure of science. Science 177(4047):393–396

125. Holland JH (1995) Can there be a unified theory of complex adaptive systems. Addison Wesley, New Jersey
126. Zuriguel I, Boudet JF, Amarouchene Y et al (2005) Role of fluctuation-induced interactions in the axial segregation of granular materials. Phys Rev Lett 95(25):258002
127. Pacheco VF, Ruiz-Suárez JC (2010) Cooperative dynamics in the penetration of a group of intruders in a granular medium. Nat Commun 1(8):123
128. Sun H, Su N, Jin AB et al (2022) Uniform flow and stress evolution laws of caved ore and rock particles between multiple drawpoints. J Northeast Univ (Nat Sci) 43(5):724–732

Chapter 6
The Hang-Up Characteristics of Caved Ore and Rock

The study of the hang-up characteristics of caved ore and rock represents a critical aspect of researching the draw issue in caving mining methods. This chapter will present an investigation into the influence of multiple parameters on hang-ups, utilizing numerical simulation and complex network analysis. Additionally, it will include an identification and characterization analysis of vital particles within the hang-up system of caved ore and rock, approached from a complex network perspective.

6.1 Numerical Study of the Influence of Multiple Parameters on Hang-Ups

The caving mining technique presents several advantages, including reduced explosive usage, enhanced production efficiency, minimal labor intensity, and lower rates of ore loss and dilution. It is recognized as the most cost-effective underground method and is the simplest to implement for automated mining [1, 2]. As related theories, technologies, and intelligent mining processes have advanced, the range of applications and the scale of mining utilizing the caving method have significantly expanded. This technique has been successfully applied in mines across various countries, including Australia, South Africa, Indonesia, the United States, Chile, Cuba, and China.

Current investigations into gravity flow primarily focus on the evolution of the shape of IMZ and IEZ [3–5], the secondary fragmentation of caved ore and rock [6, 7], calculations of ore loss and dilution index [8, 9], the design and planning of stopes [10, 11], as well as issues related to hang-ups [12–15] in cave mining contexts. Hang-ups are common in the operations of block caving mines; however, their detection and removal present significant challenges. The presence of hang-ups directly affects the accessibility of drawpoints and the flow dynamics of caved ore

H. Sun et al., *Study on Flow and Hang-up Characteristics of Caved Ore and Rock*,
https://doi.org/10.1007/978-981-92-0335-2_6

and rock, significantly impacting mining productivity and the structural parameters of stope design. A hang-up is defined as an interlocking structure of fragments that obstructs the gravity flow of caved ore and rock, either by spanning the drawpoint or blocking ore passes. This phenomenon is largely influenced by the geometrical and mechanical properties of caved ore and rock, including factors such as size distribution, shape, friction, strength, column height, moisture content, drawpoint dimensions, and the draw schedule, among others [14, 15].

Numerous laboratory experiments and numerical simulations have been conducted by researchers to study hang-up phenomena across various fields. Independent investigations by Lozano et al. [16, 17] and Gella et al. [18] focused on how factors such as obstructions above the drawpoint, vertical vibrations, and outlet dimensions influence the stability of clogging arches within silos, utilizing a two-dimensional test apparatus for quantitative assessment. Fang et al. [19] explored the arch mechanism among particles through a two-dimensional loading model test employing photoelastic particles. They identified two types of arches: stress arches and internal arches, establishing that both geometric factors and stress conditions contribute to the development of stress arches. Park et al. [20] conducted a digital RGB photoelasticity test and found that the formation of arches results from the transmission of force chains, rather than variations in the force magnitude exerted on the particles. Furthermore, Meng et al. [21] investigated the influence of the friction coefficient on arch networks by integrating physical modeling with discrete element method simulations of granular layers. They found that contact sliding serves as the primary driving force behind arch collapse. As the friction coefficient increases, particle rearrangement shifts from sliding to rotational movements, resulting in the formation of longer and more resilient arches. Bao et al. [22] conducted a series of discrete numerical simulations based on the plane-strain laboratory trapdoor test to explore the interactive effects of the double-arch design on stress distribution, particle motion, contact force anisotropy, and energy state. In the context of underground cave mining, Hadjigeorgiou et al. [23] conducted a survey that investigated the design and performance of ore passes in various mines located in Quebec and Ontario. Their review highlighted common issues encountered and the methods employed to restore material flow, revealing that a prevalent occurrence in these mines is the interlocking hang-ups of coarse materials. To further explore material flow within an ore pass, they performed numerical studies that considered factors such as ore pass shape, rock fragment geometry, and size distribution. Vo et al. [24] analyzed hang-ups in ore passes caused by cohesion and suction using methods that assess discontinuous stress and velocity, deriving straightforward algebraic formulas to describe the stability of these hang-ups. Castro et al. [14] conducted gravity flow tests under vertical load, examining how fragmentation, average vertical pressure, drawing strategies, and humidity influence hang-up characteristics. Their findings on flowability mechanisms were illustrated through a state graph and a graph depicting hang-up frequency. Hekmatnejad et al. [25] developed a hybrid approach to predict hang-up frequency in large-scale block cave mining at the El Teniente mine in Chile.

This approach provides valuable insights for the systematic characterization of fractured rock masses, design assessment, production rate evaluation, and risk analysis in cave mining.

Currently, there is a significant gap in comprehensive research addressing the challenges encountered during the caving mining process, as existing studies are frequently limited to a phenomenological perspective. Additionally, a coherent understanding of the mesoscopic mechanisms that govern various macroscopic factors is lacking. Therefore, this study conducts isolated numerical draw tests, considering variables such as particle size, friction properties, and overburden stress, to quantitatively evaluate the conditions under which caved ore and rock experience hang-ups, guided by the rolling resistance model. It provides a thorough analysis of the hang-up conditions of caved ore and rock influenced by multiple factors. Building on this foundation, the study employs statistical physics techniques from other granular domains to investigate the structural and mechanical properties of the local accumulation system of ore particles under varying flow and blockage conditions. This includes examining parameters such as porosity, Minkowski tensors, bond orientation order, coordination number, count of force chains, force chain strength, and alignment coefficients of the force chains. The research explores the relationships and sensitivities connecting various macroscopic factors with microscopic structural and mechanical parameters.

6.1.1 *Description of Numerical Scheme*

In numerous engineering applications utilizing the discrete element method, it is vital to consider both the accuracy of calculations and computational efficiency. The rolling resistance contact model effectively integrates the effects of particle shape while maintaining computational efficiency [26, 27]. Consequently, this research employs spherical particles in conjunction with the rolling resistance contact model [28] to simulate realistic caved ore and rock. A series of numerical tests based on the isolated draw model are conducted to investigate how various parameters influence hang-ups and the underlying mesoscopic influence mechanisms.

6.1.1.1 Numerical Draw Model

The researchers conducted physical draw experiments using a self-constructed three-dimensional draw model (see Fig. 4.1a) and performed numerical draw assessments based on the rolling resistance model (refer to Fig. 4.14). By evaluating and comparing the results from both experimental types, they validated the reliability and advantages of the rolling resistance model in numerical draw analyses [29]. Additionally, the relevant micromechanical parameters for the balls and walls, as outlined in Table 4.5, have been established, providing a foundational basis for the model and parameters relevant to the investigation of hang-ups influenced by various factors.

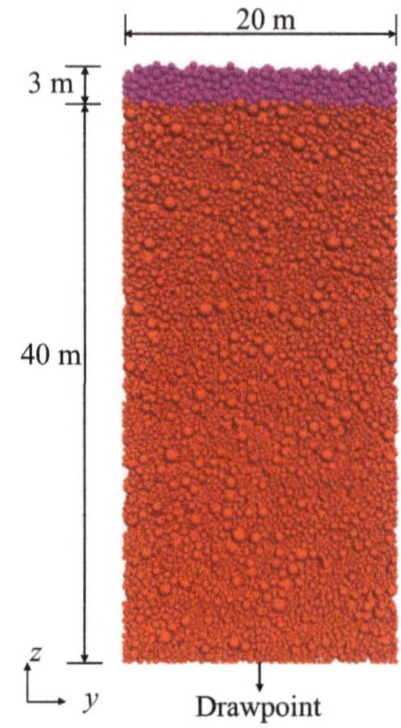

Fig. 6.1 Vertical section of the numerical draw model

Depicted in Fig. 4.15a is the standalone numerical draw model, which measures $20 \times 20 \times 50$ m^3 (length × width × height). The red square orifice situated at the base of the model is referred to as the drawpoint, with a side length of $D = 5$ m. The initial packing height is set at 40 m (illustrated by the red balls in Fig. 6.1b), with approximately 10,000 balls of extremely high density (represented by the purple balls in Fig. 6.1) densely arranged at the top of the model. In accordance with the established research methodology [30], this study modifies the average stress within the particle packing system of caved ore and rock by adjusting the density of the overburden particles.

6.1.1.2 Schemes and Processes of Numerical Draw Tests

This numerical study aims to investigate the effects of three macro-factors on the occurrence of hang-ups: specifically, the ratio of drawpoint size to average particle size D/d, the particle friction coefficient μ, and the overburden stress p. Understanding the intricate mechanisms by which these factors contribute to hang-ups is critical. To ensure a robust and thorough analysis, the ranges of values for these three factors in the numerical simulations are deliberately set larger than those typically found in actual conditions. Firstly, the distribution of particle sizes is modeled using the Rosin–Rammler function, which features a non-uniformity coefficient of 2.0 [15]. The ratio of drawpoint size to particle size D/d is manipulated by varying the average particle size while maintaining a constant drawpoint size. In actual caving mines, D/d typically ranges from 4.5 to 8.5 [15]. However, for the purposes of this study, the values of D/d are extended to a range of 4–10, with increments of 1.0, as indicated in Table 6.1. The typical grading curves relevant to caved ore and rock specimens with *D*/*d* values ranging from 4 to 10 are illustrated in Fig. 3. Previous research has established that the friction coefficient and the rolling friction coefficient for caved ore and rock generally vary between 0.3 and 0.7 [8, 15]. In this numerical analysis, these coefficients are varied from 0.2 to 0.8 in increments of 0.1, as shown in Table 6.1. For the purposes of this study and to simplify calculations, it is assumed that the particle friction coefficient and the rolling friction coefficient remain constant throughout the tests. Finally, the overburden stress σ_v is calculated using Janssen's formula [31], as presented in Eq. (6.1):

$$\sigma_v = \frac{R_h * \rho_o * g}{\tan(\phi) * k}\left(1 - \exp^{\frac{-k*\tan(\phi)*z}{R_h}}\right) \tag{6.1}$$

Here, R_h represents the hydraulic radius of the area of draw under analysis, which can be estimated by the mass drawn and draw rate; ρ_o represents the solid density of ore, while g represents the gravitational acceleration. The ratio of the horizontal and vertical stresses k, is given by k = 1-sinϕ, where ϕ represents the angle of internal friction. Furthermore, z is the depth of the particle packing system of caved ore and rock, which ranges from the production level to the surface. Here, R_h is set to 50 m, ϕ is 38° [32],k = 1-sinϕ = 0.38. The mean vertical stress and solid density of overburden particles at various depths are calculated using Eq. (6.1), as presented in Table 6.2. It can be observed from Table 6.2 that when the overburden stress σ_v reaches 5.0 MPa, the vertical depth of the drawpoint approaches 800 m. Consequently, the range of overburden stress p in these numerical tests is established as 0.1–5.0 MPa. In total, 343 numerical draw tests were conducted (Fig. 6.2).

The specific process of the numerical draw test is described as follows:

(1) Initial packing stage

In accordance with the pre-established particle size distribution curve, the initial generation of particles within the model was conducted randomly. This initial

Table 6.1 The range of values for *D*/*d*, μ and p

Influencing factors	Ratio of drawpoint size to average particle size *D*/*d*	Friction coefficient of particle μ	Overburden stress p/MPa
Value range	4	0.2	0.1
	5	0.3	0.5
	6	0.4	1
	7	0.5	2
	8	0.6	3
	9	0.7	4
	10	0.8	5

Table 6.2 The overburden stress and the corresponding solid density of overburden particle calculated based on Janssen's formula

Depth (m)	Mean vertical stress (MPa)	Solid density of overburden particle (kg/m^3)
50	1.01	1.06×10^4
100	1.82	6.38×10^4
200	3.01	1.70×10^5
500	4.63	4.89×10^5
800	5.03	8.08×10^5
1000	5.19	1.02×10^6
1500	5.26	1.55×10^6

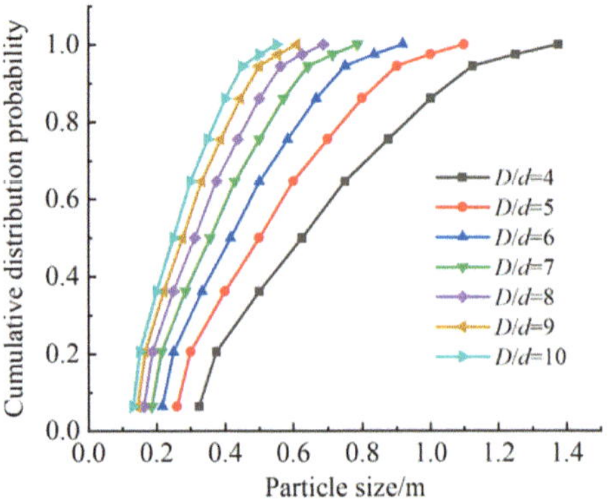

Fig. 6.2 The typical grading curves of caved ore and rock with $D/d = 4$–10 used in the numerical draw tests

randomness in particle creation establishes a foundation for a more systematic analysis of the packing system. Subsequently, a comprehensive model representing the packing system of caved ore and rock was developed, utilizing the principles of the rolling resistance model. This model also incorporates the effects of gravity settling, as referenced in previous literature [8]. To ensure realistic behavior and interactions within the system, both the walls and the particles were assigned specific meso-mechanical parameters. These parameters are detailed in Table 4.5, providing essential information for understanding the interactions occurring within the model.

(2) Stage of applying overburden stress

Approximately 10,000 high-density overburden particles were generated above the particle packing system of the caved ore and rock. These particles ultimately settled into a state of equilibrium under the influence of gravity, resulting in a final packing height of approximately 3 m. Additionally, the size of the drawpoint was found to be roughly seven times larger than the average size of the overburden particles. To accurately evaluate the average stresses that developed within the packing system of the caved ore and rock, various solid densities were assigned to the overburden particles, as detailed in Table 6.2.

(3) Draw stage

Once the particle packing system of caved ore and rock reaches a state of equilibrium, the drawpoint is activated to initiate the draw process and monitor for potential hang-ups. Hang-ups are identified based on the following criterion: if no particles are drawn within a specified time frame, indicating a mass drawn of zero, then a hang-up is deemed to have occurred. Conversely, if the draw height exceeds 10 m, it can be concluded that a hang-up has not occurred. In this case, the drawpoint is subsequently closed, and the model is operated until the packing system of caved ore and rock attains equilibrium once more. The 10-m threshold mentioned has been established through trial and error. Observations from all conducted numerical draw tests consistently demonstrate that when the draw height surpasses this 10-m threshold, hang-ups do not manifest. This finding reinforces the reliability of using the 10-m draw height as a benchmark for assessing the occurrence of hang-ups during the draw process, thereby ensuring efficient operations in managing caved ore and rock dynamics.

6.1.2 *Effect Analysis of Multiple Parameters on Hang-Ups*

The qualitative and quantitative analysis of the effects of the drawpoint size to average particle size ratio *D*/*d*, particle friction coefficient μ, and overburden stress *p* on hang-ups was performed. Furthermore, a sensitivity analysis was conducted to explore the extent of influence that various macro factors have on hang-ups.

6.1.2.1 The Influence of Multiple Parameters on Hang-Ups

Based on the numerical test outcomes observed under various combinations of three macro factors specifically, the ratio of drawpoint size to average particle size *D*/*d*, the particle friction coefficient μ, and the overburden stress *p*, a dataset concerning hang-up instances was compiled. Hang-ups were recorded in 153 numerical tests, representing 44.6% of the total test count, which allowed for the identification of parameter ranges contributing to these hang-ups. It is important to clarify that the term ‘hang-up frequency’ is not utilized in this context, as it typically refers to the tons of caved ore and rock accumulated between hang-ups during draw activities. To enhance clarity in visualizing the findings, the critical hang-up ratio of drawpoint size to average particle size—referred to in this paper as $(D/d)_{hmax}$ was defined as the highest observed value of *D*/*d* during hang-ups, with the particle friction coefficient and overburden stress held constant. Figure 6.3 illustrates the three-dimensional fitting surface diagram of hang-ups characterized by $(D/d)_{hmax}$, μ, and *p*. The inner region of the 3D surface delineates the value ranges for each parameter associated with hang-ups, while the outer region indicates the value ranges for parameters related to non-hang-ups. The black scatter point marks the instances of hang-ups.

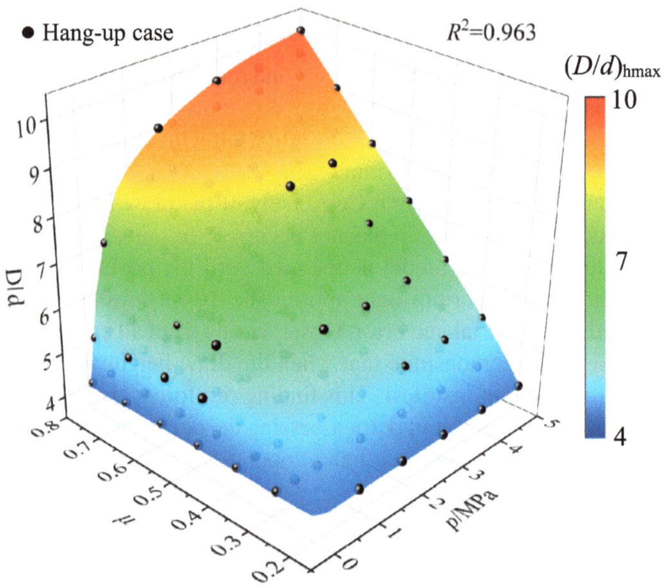

Fig. 6.3 3D fitting surface diagram of hang-ups determined by $(D/d)_{\text{hmax}}$, μ and p

The DoseResp2D function was employed to fit $(D/d)_{\text{hmax}}$, μ, and p, achieving a correlation coefficient R^2 of 0.963.

$$(D/d)_{\text{h max}} = k_1 + \frac{k_2}{\left[1 + (p/k_3)^{k_4}\right]\left[1 + (\mu/k_5)^{k_6}\right]} \tag{6.2}$$

Here, the fitting parameters k_1 tok_6 are 3.806, 12.083, 0.523, – 0.926, 0.712, and – 2.264.

The evaluation illustrated in Fig. 6.3 indicates that, within the specified parameter ranges of these numerical draw tests, an increase in the ratio of drawpoint size to average particle size D/d leads to a progressive reduction in the ranges of the other two parameters associated with hang-ups. Specifically, hang-ups are observed when $D/d = 4$, regardless of the particle friction coefficient and overburden stress. Conversely, at $D/d = 10$, hang-ups are nearly absent, suggesting that a higher D/d ratio enhances the flowability of caved ore and rock. Moreover, the maximum critical hang-up ratio $(D/d)_{\text{hmax}}$ tends to rise as the particle friction coefficient μ and overburden stress p increase, following distinct functional relationships. The applications of Fig. 6.3 can encompass various aspects of cave mining, including the development of strategies for drawpoint layout and draw techniques.

In order to analyze the impact of the particle friction coefficient μ and the overburden stress p on hang-ups, Fig. 6.4 presents the relationship curves between these two variables and the critical hang-up ratio $(D/d)_{\text{hmax}}$. In Fig. 6.4a: (1) Under conditions of near-field draw ($p \leq 1$ MPa), the critical hang-up ratio $(D/d)_{\text{hmax}}$ exhibits a power function increase with rising particle friction coefficient μ, with more rapid increases observed as overburden stress rises. (2) In the context of far-field draw ($p \geq 2$ MPa), there is a linear increase in the critical hang-up ratio $(D/d)_{hmax}$ with an increase in the particle friction coefficient μ; notably, a larger overburden stress correlates with a more rapid rise, suggesting a heightened likelihood of hang-ups. (3) For overburden stresses ranging from 3 to 5 MPa, the fitting curves exhibit a near overlap. This suggests that when the overburden stress exceeds 3 MPa (with column heights surpassing 200 m), merely increasing the column height does not significantly improve the frequency of hang-up occurrences. Additionally, the R^2 values for each curve in Fig. 6.4a are all above 0.991. Based on the analysis presented in Fig. 6.4b, several observations can be made: (1) The critical hang-up ratio $(D/d)_{hmax}$ exhibits a power function growth pattern in relation to the increase in overburden stress p, which is influenced by the varying friction coefficients of caved ore and rock. As the friction coefficient of the particles increases, the rate of ascent of the corresponding fitting curve also accelerates, indicating a heightened likelihood of hang-ups within the particle packing system of caved ore and rock. (2) In cases where μ falls between 0.2 and 0.3, the maximum ratio of $(D/d)_{hmax}$ consistently ranges from 4 to 5 across various overburden stress scenarios. This observation suggests that a lower particle friction coefficient is associated with a reduced likelihood of hang-ups. (3) For overburden stress levels ranging from 3 to 5 MPa, the increase in the maximum height ratio $(D/d)_{hmax}$ diminishes across various particle friction coefficients. This observation aligns with the findings presented in Fig. 6.4a, where an increase in load or overburden stress leads to more challenging flow conditions. Additionally, when the overburden stress exceeds 3 MPa, particularly in large-scale gravity flows (for instance, when the column height surpasses 200 m), the height of the column no longer significantly affects the occurrence of hang-ups. The correlation coefficients R^2 for each curve illustrated in Fig. 6.4b are all above 0.987. Therefore, it is vital to conduct a more thorough analysis of the influence of three macro factors: the ratio of drawpoint size to average particle size D/d, the particle friction coefficient μ, and the overburden stress p on the incidence of hang-ups.

6.1.2.2 Sensitivity Analysis

The approach used for analyzing system reliability is applicable for determining the probability of hang-ups. Specifically, sensitivity analysis, particularly sensitivity S_i, investigates how uncertainties in vital variables affect the hang-up reliability index or the likelihood of failure. In this research, the Sobol method [33] is employed to calculate the sensitivity S_i associated with three major factors influencing hang-ups. The Pareto diagram [34] effectively illustrates the overall impact S of these three factors on the hang-up properties of caved ore and rock:

$$S = \frac{|S_i|}{\sum_{i=l}^{l} |S_i|} \tag{6.3}$$

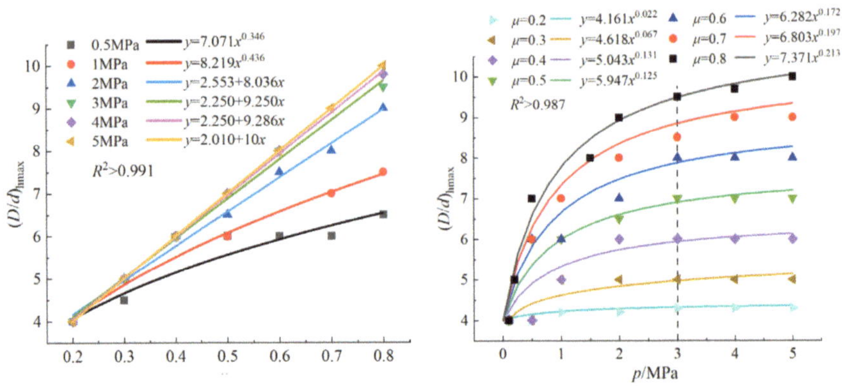

(a) Different overburden stresses (b) Different particle friction coefficients

Fig. 6.4 The relationship curves between the two factors of overburden stress p, particle friction coefficient μ, and the critical hang-up ratio $(D/d)_{\text{hmax}}$

Here, l is the total number of variables, 3.

Figure 6.5 illustrates the likelihood of hang-up occurrences influenced by the normalized parameters $[D/d]$, $[\mu]$, and $[p]$. It is evident from Fig. 6.5 that as $[D/d]$ increases, the frequency of hang-up events decreases linearly, while an increase in $[\mu]$ is associated with a linear rise in hang-up events. Notably, the slope reflecting the relationship between hang-up events and $[\mu]$ is smaller in absolute magnitude compared to the slope derived from the fitting line for $[D/d]$. Additionally, the relationship between hang-up events and $[p]$ exhibits a trend consistent with a power function as $[p]$ ascends. The R^2 values for each curve depicted in Fig. 6.5 exceed 0.981, and the variations are in alignment with the patterns illustrated in Figs. 6.3 and 6.4. Figure 6.6 presents a Pareto analysis regarding the effects of D/d, μ, and p on hang-ups. The insights derived from this analysis indicate that, within the evaluated parameter range, the ratio D/d significantly impacts hang-ups, accounting for 54.562% of the variance. This is followed by the particle friction coefficient μ, which contributes 33.569%, and the overburden stress p, which has a lesser impact at 11.869%. These findings underscore that the rate of hang-up incidents is most sensitive to variations in the D/d ratio, with the particle friction coefficient being the next most influential factor, and the overburden stress showing the least sensitivity. Therefore, it is imperative for block caving operations to prioritize the design of drawpoint arrangements and assess rock mass capabilities to effectively mitigate the likelihood of hang-up incidents.

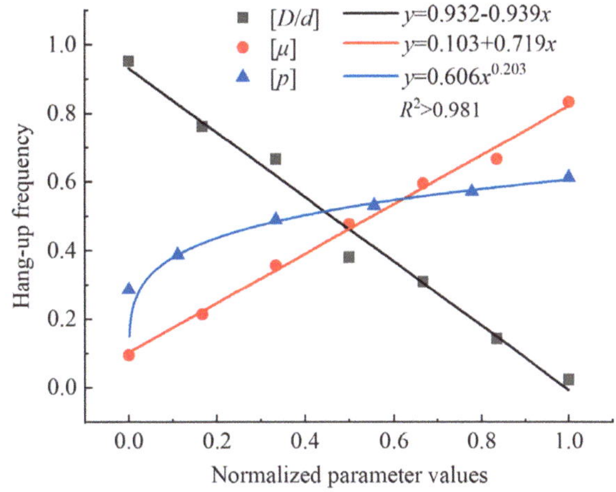

Fig. 6.5 Hang-up frequency under the influence of normalized parameters [D/d], [μ] and [p]

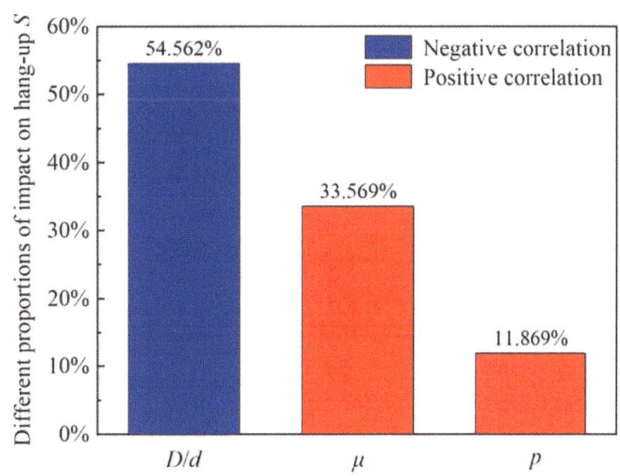

Fig. 6.6 Pareto diagram of the impact of D/d, μ and p on hang-ups

6.1.3 Structural and Mechanical Characteristics Analysis

To examine the mesoscopic influences of varying ratios of drawpoint size to the average particle size D/d, as well as the particle friction coefficient μ and overburden stress p on hang-up occurrences, we carried out an analysis focusing on the structural and mechanical properties of the particle packing system of caved ore and rock both prior to draw and following a hang-up event. Through a process of trial and error, our attention centered on the caved ore and rock within a 2.5 average particle size range surrounding the drawpoint for the analysis of hang-up characteristics (denoted as red particles in Fig. 6.7a). For the numerical assessment where no hang-up occurred, we designated the caved ore and rock from the same area prior to drawing as the subject of analysis (also represented as red particles in Fig. 6.7b).

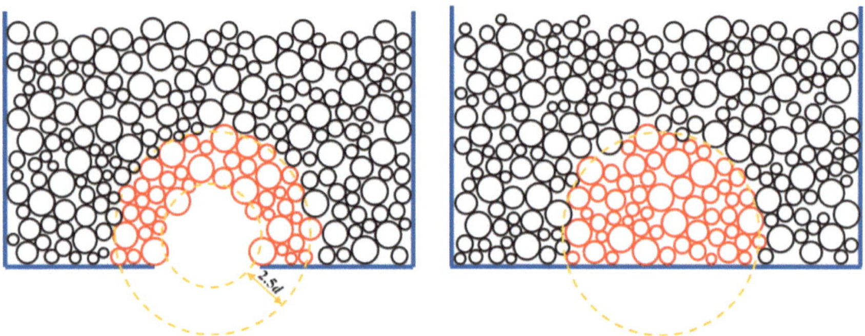

(a) Hang-up (b) Non hang-up (before draw)

Fig. 6.7 Schematic diagrams of local regions for hang-up characteristic analysis

6.1.3.1 Structural Characteristics Analysis

Using parameters such as void ratio, Minkowski tensor, and bond orientational order, this study analyzes the structural characteristics of the local packing system of caved ore and rock in terms of compactness, anisotropy, and degree of order. The Minkowski tensor specifically describes the morphology of Voronoi cells constructed using the free radical Voronoi partition method within the particle packing system of caved ore and rock [35–38]. Specific definitions are provided in Eqs. (4.28) and (4.29).

Figure 6.8 illustrates the variations in local meso-structural parameters under different conditions of *D*/*d*, μ, and *p*, with the orange area highlighting the range of local meso-structural parameters associated with the occurrence of hang-ups. (1) By using the numerical test results for $\mu = 0.5$ and $p = 3$ MPa as a reference, the analysis presented in Fig. 6.8 reveals that the ratio of drawpoint size to average particle size *D*/*d* typically ranges from 4.0 to 6.5 during hang-up events. Importantly, regardless of the presence of hang-ups, there is no significant variation in the structural parameters of the particle packing system across different *D*/*d* values. The void ratio *v*, the average anisotropy coefficient $<\beta>$, and the bond orientational order Q_6 are found to be around 0.425, 0.625, and 0.297, respectively. This implies that the local structural features of both the caved ore and the overlying rock at the drawpoint are largely consistent, indicating that these three meso-structural indicators do not directly influence the occurrence of hang-ups. (2) Considering the numerical test results for a *D*/*d* value of 6 and $p = 3$ MPa, the analysis shown in Fig. 6.8 reveals that the friction coefficient of caved ore and rock μ tends to vary approximately between 0.45 and 0.80 when hang-ups occur. As the particle friction coefficient μ rises, the void ratio *v* demonstrates an increase, with a notable rise of 20.142% from $\mu = 0.3$ to $\mu = 0.8$. Simultaneously, there is a decrease of 5.779% in the average anisotropy coefficient $<\beta>$, while the bond orientational order Q_6 remains fairly stable at around 0.298. This suggests that an increase in the friction coefficient enhances the resistance of both caved ore and rock to deformation. Consequently, this leads to higher void ratios in the local particle packing system and an increase in overall structural anisotropy

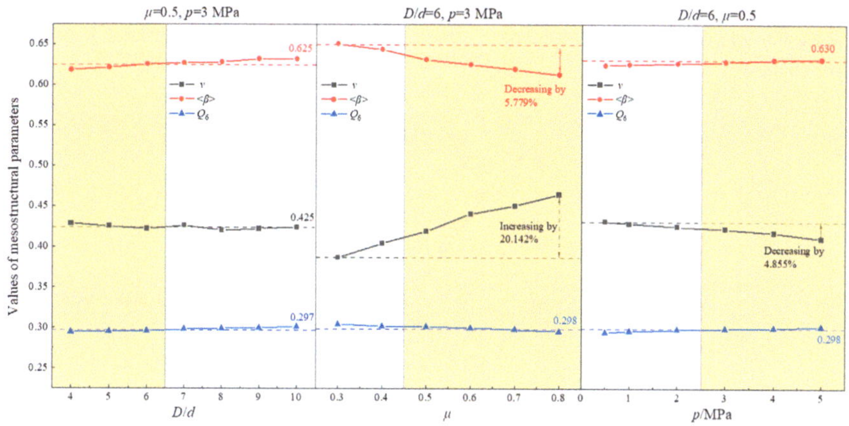

Fig. 6.8 Evolution of local meso-structural parameters under different *D*/*d*, μ and *p*

to a certain extent. (3) Taking the numerical test results for the scenario where *D*/*d* = 6 and $\mu = 0.5$ as reference points, the analysis depicted in Fig. 6.8 suggests that overburden stress *p* generally fluctuates between 2.5 and 5.0 MPa during hang-up events. With an increase in overburden stress, there is a corresponding decrease of 4.855% in the void ratio *v* of the local particle packing system. However, no notable variations are observed in the average anisotropy coefficient $<\beta>$ or the bond orientational order Q_6, implying that overburden stress minimally affects the local structural characteristics of the caved ore and rock above the drawpoint. In conclusion, neither the ratio of drawpoint size to average particle size nor the overburden stress significantly influences the local structural traits of caved ore and rock situated above the drawpoint. The maximum fluctuations in void ratio, average anisotropy coefficient, and bond orientational order remain below 5%. Meanwhile, the friction coefficient of caved ore and rock plays a vital role in affecting hang-ups by altering the local void ratio and structural anisotropy. The high frictional properties, considerable variability in particle size, and the presence of complex external stress conditions within caved ore and rock result in a local particle packing structure that exhibits a low degree of order. Consequently, its effect on hang-ups is not distinctly pronounced. Thus, a more in-depth analysis of the meso-mechanical properties of the local particle packing system in caved ore and rock is warranted.

6.1.3.2 Mechanical Characteristics Analysis

The force chain describes a structural arrangement formed by particles in contact with one another, enabling the bearing and transmission of forces [27]. It serves as a mesoscopic link that connects microscale individual particles to macroscale particle assemblies. As illustrated in Fig. 6.9, the chaining between particles must satisfy three specific conditions [39]:

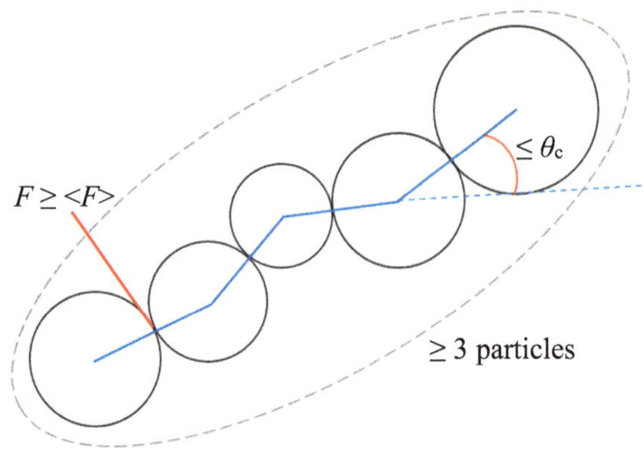

Fig. 6.9 Identification criteria and sketch of the force chain

(1) Contact force criterion: The force F acting among the particles must not fall below the average contact force $< F >$ present within the particle system.
(2) Criterion for contact angle: The angle θ between the directional vectors of two neighboring contacts should not exceed the threshold angle $\theta_c = 180°/CN$, where CN represents the average coordination number within the particle system.
(3) The number of particles N_{fc} constituting the force chain should be no fewer than 3.

Given these three conditions for chain formation, this research identifies the three-dimensional force chain located within the typical local hang-up region above the drawpoint and collects its contact data, which includes contact ID, spatial coordinates, direction, and strength. Additionally, the analysis emphasizes the meso-mechanical properties of the caved ore and rock present in the local hang-up area. Vital characteristics analyzed include the coordination number, count of force chains, strength, and collimation. Generally, the force chain structure exhibits enhanced stability as it approaches a linear form. The collimation coefficient δ of the force chain can be mathematically expressed as follows (Eq. 6.4):

$$\delta = 1 - \frac{\sum_{i=1}^{N_c} \alpha_i}{180° \times N_{ac}} \tag{6.4}$$

Here, α_i represents the angle between the i-th adjacent normal contacts. N_{ac} is the total number of adjacent contacts, and its value is equal to the number of chain contacts minus 2 ($N_{ac} = N_c - 2$).

Figure 6.10 illustrates the variations in meso-mechanical parameters within the localized region above the drawpoint, which are affected by different ratios of drawpoint size to average particle size D/d, particle friction coefficient μ, and overburden stress p. The orange area emphasizes the range of local meso-mechanical parameters linked to the occurrence of hang-ups. (1) In Fig. 6.10a, it is observed that the average coordination number CN fluctuates between approximately 3.6 and 4.8. This value demonstrates a slight increase with higher *D*/*d* ratios and overburden stress *p*, exhibiting growths of 4.144% and 5.389%, respectively. Conversely, an increase in the particle friction coefficient μ results in a decrease of 10.691%. Overall, variations in *D*/*d*, μ, and *p* have minimal impact on the average coordination number in the local region, suggesting that this parameter, which defines short-range interactions, is not a primary factor in the occurrence of hang-ups. (2) From Fig. 6.10b, c, it is evident that the force chain number *n* ranges from approximately 90 to 200. This number increases as the *D*/*d* ratio rises, yet it shows a marked decline with increasing μ and *p*. In contrast, the average strength of force chains σ_p ranges from roughly 0.5 to 5.5 MPa. Here, the average strength decreases with an increase in *D*/*d* but rises with higher values of μ and *p*. Furthermore, when analyzing single factors in isolation, the force chain count associated with hang-up events in the local region is lower than that observed during non-hang-up situations. Conversely, the average strength of force chains during hang-up events in the local region exceeds that of non-hang-up regions. This indicates a notable negative correlation between the prevalence of force chains and their average strength. Force chains serve as the fundamental medium for force transmission and support within the particle packing system of caved ore and rock. A higher bearing capacity of these force chains correlates with a reduced total number of chains in the local area. This phenomenon enhances mechanical stability, which consequently increases the likelihood of hang-up occurrences and diminishes the flowability of the caved ore and rock. Hence, to grasp the mechanisms behind hang-ups, it is essential to examine the medium-range mechanical structure of force chains rather than focusing solely on short-range particle contact. As illustrated in Fig. 6.10d, the collimation coefficient δ of the force chain fluctuates between approximately 0.82 and 0.88. It experiences a modest reduction of 3.750% with the rising ratio of drawpoint size to average particle size *D*/*d*, while showing increases of merely 6.252% and 1.608% with the escalation in friction coefficient μ and overburden stress *p*, respectively. Consequently, the collimation coefficient is not deemed a primary meso-mechanical index influencing hang-ups. Rather, the count of force chains and their average strength are considered the pivotal meso-mechanical parameters impacting these occurrences. Additional research is necessary to explore their relationships with macroscopic variables and meso-structural characteristics.

6.1.3.3 Correlation and Sensitivity Analysis

The Pearson correlation coefficient method [27] was employed to assess and compute the relationships among various macroscopic factors, including the ratio of drawpoint size to average particle size *D*/*d*, the friction coefficient μ, and overburden stress *p*. Additionally, meso-structural parameters such as void ratio *v*, the average anisotropy coefficient $< \beta >$, and bond orientational order Q_6 were analyzed alongside meso-mechanical parameters like the average coordination number *CN*, the number of force chains *n*, the average strength of force chains σ_p, and the collimation coefficient of force chains δ. Illustrated in Fig. 6.11 is the correlation hotspot map, which depicts

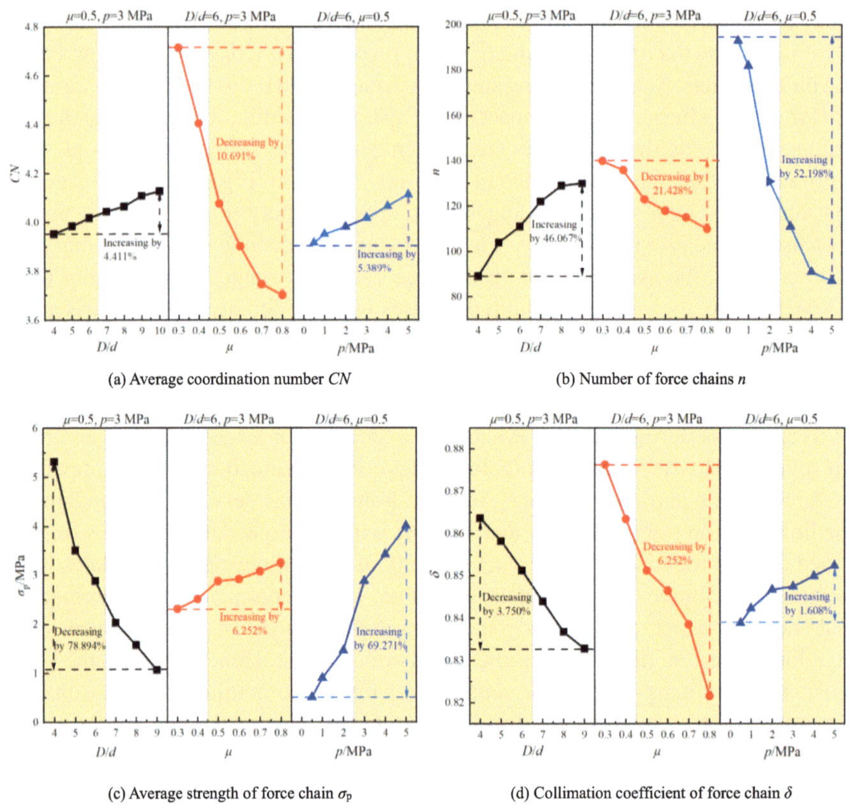

(a) Average coordination number *CN*

(b) Number of force chains *n*

(c) Average strength of force chain σ_p

(d) Collimation coefficient of force chain δ

Fig. 6.10 Evolution of local meso-mechanical parameters under different *D*/*d*, μ and *p*

the various macroscopic influencing factors related to hang-up, as well as the meso-structural and mechanical parameters within the localized hang-up region.

Typically, a Pearson correlation coefficient with an absolute value below 0.2 indicates no correlation between two variables. Conversely, a substantial correlation is identified when the absolute value exceeds 0.6. As illustrated in Fig. 6.11: (1) In the context of various macro influencing factors, the ratio of drawpoint size to average particle size *D*/*d* demonstrates a strong negative correlation (− 0.638) with the average force chain strength σ_p, while a significant correlation is not observed with other structural and mechanical parameters. The particle friction coefficient μ shows robust correlations with the collimation coefficient of the force chain δ, void ratio *v*, average coordination number *CN*, and average anisotropy coefficient $< \beta >$, exhibiting correlation coefficients of − 0.766, 0.945, − 0.868, and − 0.816, respectively. Overburden stress *p* reveals strong positive and negative correlations (0.681 and − 0.876) with average force chain strength σ_p and the number of force chains *n*, respectively. These results imply that the ratio of drawpoint size to average particle size *D*/*d* serves as a macroscopic factor that chiefly influences the average force

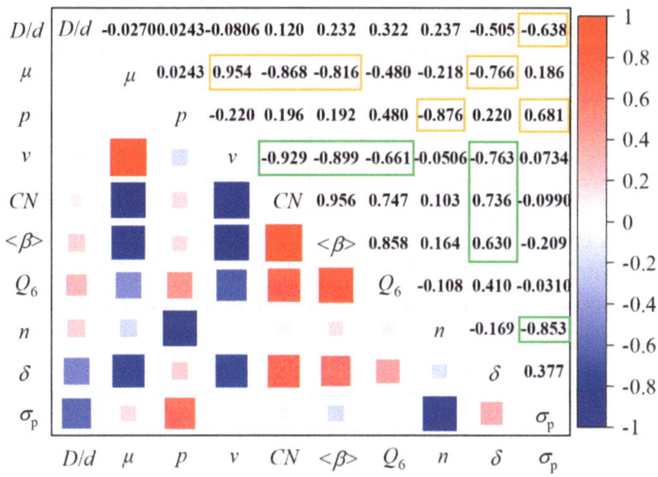

Fig. 6.11 Correlation hotspot map of different macroscopic influencing factors of hang-up, meso-structural and mechanical parameters in the local region of hang-up

chain strength σ_p, whereas the particle friction coefficient μ significantly impacts the collimation coefficient δ, void ratio v, average anisotropy coefficient $<\beta>$, and average coordination number CN. Additionally, overburden stress p predominantly affects average force chain strength σ_p and the number of force chains n. (2) The Pearson correlation coefficients for various meso-structural and mechanical parameters revealed that between void ratio v and the average coordination number CN, average anisotropy coefficient $<\beta>$, and bond orientational order Q_6, the values were -0.929, -0.899, and -0.661, respectively. These values suggest a strong negative correlation, aligning with the data depicted in Fig. 6.8, which shows that void ratio is a vital meso-structural factor. Moreover, for the collimation coefficient δ of the force chain regarding void ratio v, average anisotropy coefficient $<\beta>$, and average coordination number CN, the corresponding Pearson correlation coefficients were -0.763, 0.736, and 0.630. This indicates a significant correlation between the stability of the force chain and the parameters of void ratio v, average anisotropy coefficient $<\beta>$, and average coordination number CN. Additionally, the Pearson correlation coefficient between the number of force chains n and the average strength of the force chain σ_p was -0.853, demonstrating a strong negative correlation, which is consistent with the findings presented in Fig. 6.10b, c. In conclusion, various macroscopic influencing factors demonstrate varying levels of correlation with meso-structural and mechanical parameters, collectively impacting the complexity and extent of the associated hang-ups. As with many engineering challenges, identifying the most significant factors among the myriad influences is essential, as this aids in uncovering the fundamental mechanisms and developing appropriate treatment strategies.

To further explore how various meso-structural and mechanical parameters affect hang-ups, sensitivity coefficients were determined for each parameter. The Pareto diagram presented in Fig. 6.12 demonstrates the effects of these parameters on hang-ups. It is noted that among the parameter values assessed in these numerical draw tests, the average strength of the force chain σ_p exerts the most substantial influence on hang-ups, followed by the number of force chains n, the collimation coefficient δ of the force chain, the void ratio v, the average anisotropy coefficient $< \beta >$, and the bond orientation sequence Q_6. The influence proportions for these parameters are 30.809%, 23.868%, 16.085%, 15.544%, 10.544%, 1.563%, and 1.521%, respectively. Hence, it can be deduced that both the average force chain strength σ_p and the quantity of force chains n are the vital meso-mechanical parameters influencing hang-ups. Conversely, with regard to meso-structural parameters, the void ratio v has the strongest effect, whereas the average anisotropy coefficient $< \beta >$ and the bond orientational order Q_6 show minimal impact on hang-ups, which is consistent with the analysis results in Fig. 6.8. The packing system's structure for caved ore and rock exhibits short-range disorder, primarily due to variations in particle size and the random nature of the packing. This disorder complicates the establishment of a typical local meso-structure, significantly affecting the occurrence of hang-ups. Furthermore, the evaluations presented in Figs. 6.6, 6.11, and 6.12 reveal that the size ratio of the drawpoint to the average particle size D/d significantly influences hang-ups by altering the average force chain strength σ_p. Additionally, the particle friction coefficient μ plays a vital role in affecting hang-ups through its impact on the collimation coefficient δ of the force chain and the void ratio v. The overburden stress p also influences hang-ups by affecting both the average force chain strength σ_p and the total number of force chains n. As a result, the characterization and effect of hang-ups heavily rely on the local mechanical properties of caved ore and rock and their evolution, instead of the structural features of the local particle packing system. Following this, we will delve deeper into the correlation between the physical and mechanical attributes of caved ore and rock, the external stress conditions, and the flow and hang-up characteristics of the caved ore and rock, focusing on the mechanical structure and the topological properties of interconnected particles, such as contact networks [40] and contact force loops [41].

6.2 Vital Particles Identification and Characteristics Analysis Within Hang-Up System of Caved Ore and Rock from Complex Network Perspective

6.2.1 Hang-Up Characteristics

Hang-up, a prevalent and complex issue encountered in caving mines, poses significant challenges during the drawing process. It directly affects the efficiency of ore extraction and the flow capacity of caved ore and rock, ultimately influencing the mine's productivity. Additionally, it is a critical consideration in the design of drawpoint structures and scheduling. The formation of a stress arch, created by the

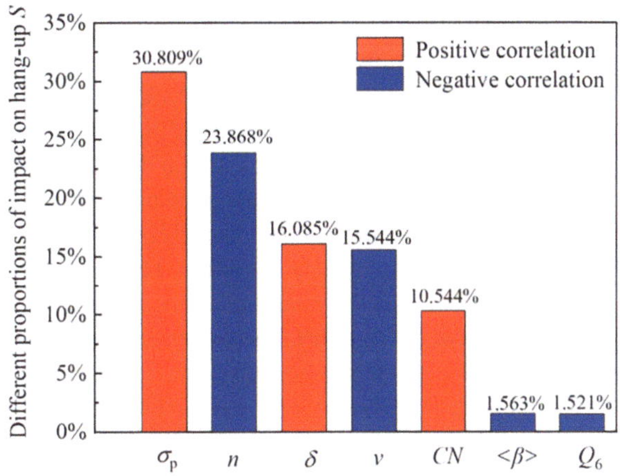

Fig. 6.12 Pareto diagram of the impact of different meso-structural and mechanical parameters on hang-ups

interlocking of caved ore and rock positioned above the drawpoint, is primarily responsible for hang-up, which obstructs the typical gravitational flow of materials [14, 15]. This phenomenon exemplifies a jamming phase transition within a localized particle system [42], wherein the local caved ore and rock transition from a mechanically unstable flowing state to a mechanically stable jamming state. Such a transition involves a complex alteration in the structural and mechanical properties of the particle system.

Investigations involving both physical and numerical experiments have been conducted to examine the characteristics of hang-up in caved ore and rock. A self-designed three-dimensional model was employed by Castro et al. [14] to study gravity flow under specific drawing and pressure conditions, resulting in the creation of a diagram that reflects hang-up frequency. Meanwhile, Hadjigeorgiou et al. [23], Sato et al. [43], and Sun et al. [44] performed discrete element numerical simulations to evaluate how factors such as particle shape, size distribution, density, friction coefficients, and various geometric, physical, and mechanical properties interact with external elements, including drawpoint dimensions, overburden pressure, and vibration frequency, to influence the hang-up behavior of caved ore and rock. Additionally, Gómez et al. [45, 46] investigated the hang-up phenomena associated with different caving methodologies in mines, identifying vertical stress and the ratio of drawpoint size to particle size as vital factors that lead to hang-ups. Their predictive model demonstrated that, to initiate the sudden release of fine materials, the removal of blockages is consistently essential. Hekmatnejad et al. [25] presented a comprehensive analytical methodology and applied it to estimate the frequency of hang-ups during mining activities at the El Teniente mine in Chile. In areas where caved ore and rock blockages occur, the geometric properties and spatial arrangement of the particles are critical elements that influence both the frequency and severity of hang-ups. A particularly compelling inquiry that warrants deeper exploration is whether specific critical particles affect hang-up events and what particular characteristics these critical particles may possess. Currently, studies examining the hang-up characteristics of caved ore and rock primarily focus on the statistical and descriptive aspects of broad trends; however, they lack effective analytical techniques for understanding the meso-mechanisms behind hang-up formation, especially concerning vital particles and their properties within the hang-up context of caved ore and rock.

The particle packing system of caved ore and rock, which emerges during underground mining—whether through natural processes or external forces—can be understood as a sophisticated particle system. The application of complex network theory, which conceptualizes the components of a complex system as nodes and their relationships as edges, has illuminated significant statistical features such as 'small world' phenomena, 'scale-free' characteristics, and the property of 'clustering'. This theory has found extensive use across multiple disciplines, including communication networks [47], transportation systems [48], risk evaluation [49], and predictions of epidemic spread [50]. Furthermore, complex network theory provides an interdisciplinary theoretical framework for exploring the mechanical properties of various particle systems. By utilizing methods from graph theory, statistics, and game theory, researchers can represent particles and their interactions as nodes and edges within contact networks. These networks, whether weighted or unweighted, enable the characterization of the multi-scale heterogeneity present in particle contact network topology [40], thereby facilitating deeper insights into system dynamics.

In recent years, researchers have employed advanced network analysis methods to extract and quantitatively analyze the meso-scale contact structures of various particulate materials, including rock, soil, and landslide constituents. Unlike traditional approaches that emphasize the intrinsic properties of individual particles, complex network theory shifts its focus to the interrelationships among particles. This methodology highlights the identification of significant particles within the system through the application of a network framework. The process entails identifying [51], force networks [52], and force loops [41]. Yu et al. [53], Shi et al. [54], and Liu et al. [55] employed the discrete element method in conjunction with complex network theory to construct force chain networks in material mixing environments. They regarded the force chain network as an undirected and unweighted system, investigating its mesoscopic characteristics through network metrics such as degree and clustering coefficient. Furthermore, Liu et al. [56] applied complex network metrics to elucidate the topological features of force chain networks in clay particle slopes, underscoring the predictive capability of the average shortest path in understanding slope failure mechanisms. Arevalo et al. [57] examined the characteristics of packing evolution in two-dimensional disk assemblies subjected to isotropic compression, highlighting the significance of third-order contact loops in elucidating the transition from fluid-like to solid-like behavior within particle systems. Tordesillas et al. [41] examined the topological evolution and properties of contact networks in densely packed two-dimensional particles, identifying significant metrics for evaluating structural changes during deformation. Concurrently, complex network theory, which originates from a different discipline and viewpoint, serves as a valuable tool for investigating critical node issues across diverse fields and networks. Csermely et al. [58] introduced a 'central strike strategy' aimed at targeting essential nodes in flexible networks composed of infectious substances or cancer cells, significantly enhancing the efficacy of drug design. Kitsak et al. [59] and Pei et al. [60] utilized complex network theory to identify vital nodes across various domains, including drug design, infection networks (such as SIS and SIR), as well as social media platforms like Twitter, Facebook, and LiveJournal. These studies provide innovative methodologies for infection control and software enhancement. Furthermore, Motter et al. [61] applied a 'premeditated attack technique' to undermine critical nodes within Internet and power grid networks, demonstrating that redistributing load among nodes could trigger a cascade of overloads, potentially resulting in total network failure.

Although conventional physical experiments and numerical simulations have provided significant insights into hang-up dynamics, they still fall short in bridging essential gaps, particularly in elucidating the mechanisms that contribute to hang-up formation and instability. Consequently, this paper conducts numerical experiments on the draw, examining the effects of various factors such as particle size, friction coefficient, and overburden stress. By focusing on localized particle packing systems under diverse hang-up conditions, we employ complex network analysis techniques to identify optimal indicators for recognizing vital particles. Moreover, we investigate the characteristics of these vital particles, including their geometric shapes, spatial distributions, aggregation patterns, and interconnections under different influencing factors, utilizing the 'premeditated attack method' [62]. The methodologies and findings presented provide a novel analytical framework and theoretical basis for understanding the mechanisms associated with hang-up formation and instability, while also refining strategies for removing hang-ups in block caving mines.

6.2.2 Description of Numerical Scheme

Utilizing the rolling resistance model [51, 52], numerical draw tests were conducted to simulate real caved ore and rock. A series of numerical experiments of isolated bottom draw were carried out. The study aims to identify vital particles and analyze their characteristics within the hang-up system of caved ore and rock from a complex network perspective. The numerical draw model, processes of numerical draw tests, and the definitions of vital node identification indicators are outlined below.

6.2.2.1 Numerical Draw Model

In this section, a series of physical draw experiments were conducted using a self-developed three-dimensional physical test model (see Fig. 4.15a). Additionally, numerical draw tests were performed utilizing the rolling resistance model (refer to Fig. 6.13) [63], which establishes the foundational models and parameters necessary for the identification and characterization analysis of vital particles within the

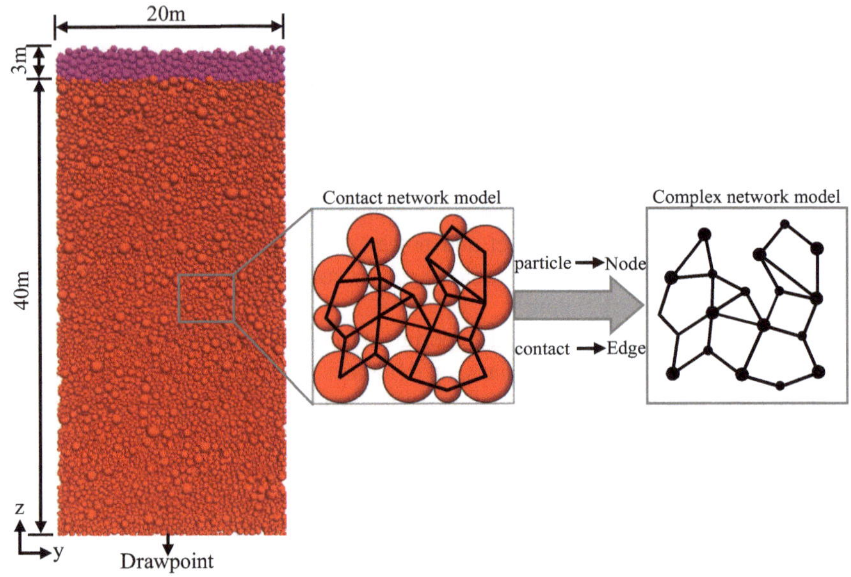

Fig. 6.13 Model vertical section and schematic diagram of the complex network model

hang-up system of caved ore and rock. Following established research methodologies, this study modifies the average stress in the particle packing system of caved ore and rock by varying the density of the overburden particles [64].

As presented in Table 6.1, a comprehensive set of 343 numerical draw tests, calculated as $7 \times 7 \times 7$, has been conducted. The key influencing factors include the ratio of drawpoint size to average particle size *D/d*, the particle friction coefficient μ, and the overburden stress *p*. The *D/d* ratio varies between 4 and 10. The friction coefficient is maintained within a range of 0.2 to 0.8; to streamline the number of variables, the values for both the friction and rolling friction coefficients of the balls are standardized [65]. The overburden stress is evaluated between 0.1 and 5.0 MPa, utilizing the Jensen formula for calculations, where 5.0 MPa corresponds to a maximum column height nearing 800 m. For additional information on the numerical draw model and the mesoscopic parameters associated with the wall and the balls, readers can refer to the existing literature [8, 44, 50].

6.2.2.2 Definitions of Vital Node Identification Indicators

This study analyzes the hang-up system of caved ore and rock by modeling particles as nodes and the interactions between them as edges, as illustrated in Fig. 6.14b. Four key indicators for node identification are examined: Degree [66, 67], H-index [66, 68], Coreness [59, 69], and Cycle ratio [64]. First, it is essential to determine the

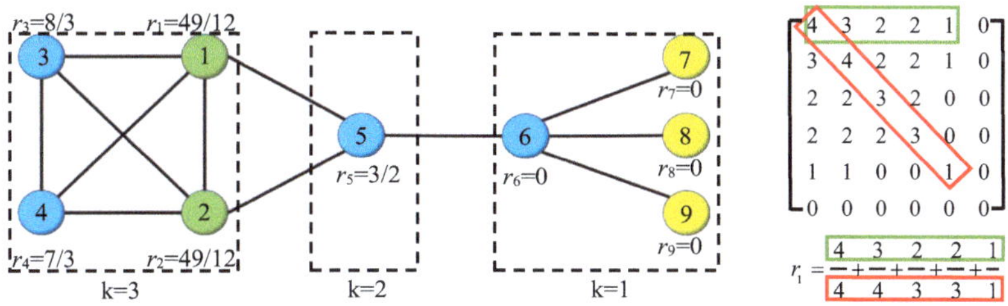

(a) Example network

(b) Cycle number matrix of the example network

Fig. 6.14 Indicators of nodes in an example network

optimal indicators for identifying vital particles. The definitions of the four indicators are as follows:

(1) Degree: This indicator represents the number of adjacent edges connected to a node, analogous to the concept of a particle's coordination number. This metric is crucial for characterizing nodes within a network.
(2) Coreness: By sequentially eliminating nodes that have a Degree less than k, we identify the resulting subgraph as a 'k-core' when every node in this subgraph possesses a degree exceeding k. The Coreness of a node reflects the highest k value linked to the k-core it is a part of. As demonstrated in Fig. 6.14a, this method is similar to the concept of 'peeling an onion'. The process begins with the removal of nodes that possess a Degree of 1 (Nodes 7, 8, and 9). As the procedure advances, new nodes may appear with a Degree of 1 (such as Node 6) within the network and are then removed. This iterative approach persists until there are no remaining nodes with a Degree of 1. The Coreness assigned to a detached node is labeled as 1. The k-core decomposition advances by progressively increasing the deletion threshold, meaning that nodes with Degree of 2, 3, and so on are systematically removed until the process becomes unfeasible. Ultimately, each node is assigned a Coreness value, reflecting its core level within the network. This Coreness value corresponds to the node's position within the structural framework of the network. Figure 6.14a illustrates the Coreness of various nodes.
(3) H-index: This metric is frequently utilized in academic circles to assess the influence of researchers by examining their publications and citation records. The calculation is as follows: if a researcher has published X papers and possesses an H-index of h, it indicates that their$h$ most frequently cited papers have received at least h citations, while the remaining $X–h$ papers have received fewer than h citations. For instance, in Fig. 6.14a, Node 3 connects to Nodes 1, 2, and 4, with the edges assigned values of 3 or greater, leading to an H-index of 3 for Node 3.
(4) Cycle ratio: A cycle within a network is recognized as the fundamental meso-structure and is depicted in Fig. 6.14a with the nodes labeled {1, 2, 3}. The

Table 6.3 Four index values for each node in the example network shown in Fig. 6.14a

Node label	Degree	Coreness	H-index	Cycle ratio
1	4	3	3	49/12
2	4	3	3	49/12
3	3	3	3	8/3
4	3	3	3	7/3
5	3	2	3	3/2
6	4	1	1	0
7	1	1	1	0
8	1	1	1	0
9	1	1	1	0

cycle that has the least number of nodes, involving node *i*, is referred to as the relevant shortest cycle for node *i*, which is simpler termed the shortest cycle of *i*. The cycle ratio is determined by focusing solely on the shortest cycle associated with each node, since larger cycles are typically less significant in relation to network performance [59]. In Fig. 6.14b, the cycle number matrix for node 1 and the process of calculating its cycle ratio are presented. The initial term 4/4 in the equation shown in Fig. 6.14b indicates that all four shortest cycles of node 1 ($\{1, 2, 4\}$, $\{1, 2, 3\}$, $\{1, 3, 4\}$, $\{1, 2, 5\}$) include node 1. The subsequent term, 3/4, signifies that three out of the four shortest cycles of node 2 ($\{1, 2, 4\}$, $\{1, 2, 3\}$, $\{2, 3, 4\}$, $\{1, 2, 5\}$) also include node 1. The cycle ratio for node 1 is determined by summing these values. In essence, r_1 denotes the degree to which node 1 participates in the shortest cycles of other nodes. The respective values for Degree, Coreness, H-index, and Cycle ratio for each node in the sample network shown in Fig. 6.14a are available in Table 6.3.

6.2.2.3 Processes of Numerical Draw Tests

The simulation and data analysis processes for each group of numerical draw tests are described as follows:

(1) Packing and draw stages: In accordance with the rolling resistance model and the principles of gravity settling [30], a system for packing caved ore and rock was developed. This system incorporated various particle sizes, each possessing unique friction coefficients, while also adjusting overburden stress through variations in the densities of the overlying particles. Once this packing arrangement reaches a state of equilibrium, the drawpoint is activated to initiate the draw process and identify any potential hang-ups. If the draw height falls below 10 m, this indicates the presence of a hang-up, prompting the closure of the drawpoint. Subsequently, a model is constructed to evaluate hang-ups to restore balance [44]. Our experimental method focused on caved ore and rock with an

average particle size of approximately 2.5 around the drawpoint, aiming to identify essential particles and assess their attributes (refer to red and blue particles in Fig. 6.15).

(2) Data acquisition and analysis stages: The research elucidated the identification of ID numbers, three-dimensional coordinates, particle sizes, and the interactions among particles within localized particle systems under various hang-up conditions. Subsequently, the modeling of particles and their interactions was conducted using nodes and edges, resulting in the development of a complex network model that represents the local contact network of both caved ore and rock. This modeling was executed using Python programming in conjunction with the Networkx library. Four indices related to complex networks were computed, namely Degree, H-index, Coreness, and Cycle ratio. The methodology known as the 'premeditated attack method' involves prioritizing nodes based on the significance of indices, such as ordering Degree from highest to lowest. This method systematically removes these nodes while monitoring the stability of the system. For instance, as illustrated in Fig. 6.15, this approach entails organizing particles within a yellow dotted circle according to a specified index value in decreasing order. It is crucial to note that N_d represents the top N_d most significant particles within the hang-up structure of caved ore and rock. A higher position of a node in this descending sequence indicates the particle's greater relevance to the hang-up system. The blue particles, deemed vital, are arranged in a descending hierarchy based on this index and are identified as essential particles. Subsequently, particles are removed sequentially from highest to lowest (blue particles 1 → 2 → 3), while continuously monitoring for the occurrence of hang-ups. If hang-up conditions persist after the removal of essential particle 1, the process continues with essential particle 2, and so on. Once the normal gravitational flow of caved ore and rock is restored, the total number of essential particles removed according to this index is recorded as 3.

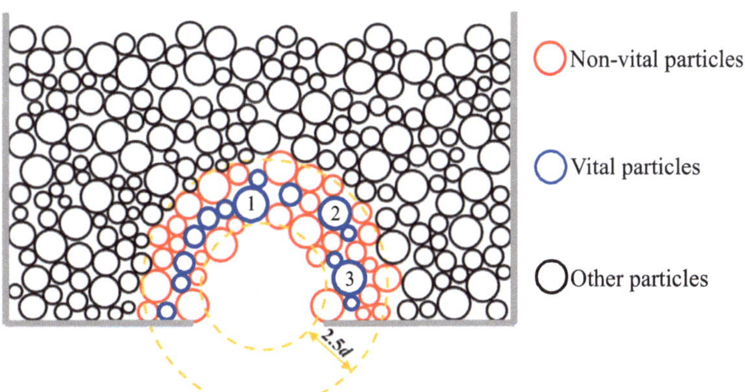

Fig. 6.15 Schematic of local region for vital particles identification and characteristics analysis within hang-up system of caved ore and rock

6.2.3 *Identification and Characteristic Analysis of Vital Particles*

Vital nodes are essential for maintaining network stability, as their intentional removal can significantly compromise network connectivity. This research focuses on investigating the local particle system under various hang-up conditions to identify the most effective index for vital particle identification. The study employs correlation analysis in conjunction with the 'premeditated attack method' to achieve this objective. Furthermore, it incorporates both qualitative and quantitative assessments of geometric properties, spatial arrangements, clustering, and the connectivity of vital particles. The influence of macro factors, including the ratio of drawpoint size to average particle size D/d, particle friction coefficient μ, and overburden stress p, on hang-ups is analyzed from the perspective of complex networks.

6.2.3.1 Determination of Optimal Vital Particle Identification Index

Before establishing the optimal index for vital particle identification, it is essential to first define the type of contact network involving the caved ore and rock. Considering the actual parameters of the base structure, including column height and the properties of the rock in the caving mine—as detailed in the second column of Table 6.4—three primary values are taken into account: the ratio of drawpoint size to average particle size D/d, the particle friction coefficient μ, and the overburden stress p. For instance, in Test 1, characterized by the condition '5–0.6–3', it is indicated that the ratio D/d is 5, the particle friction coefficient μ is 0.6, and the overburden stress p is 3 MPa. Throughout Tests 1 to 7, hang-up models were developed, influenced by these three factors and their respective typical values.

As illustrated in Fig. 6.16, the average degree distribution of the contact network involving caved ore and rock across the seven tests closely aligns with the Poisson distribution function, yielding a strong correlation coefficient of $R^2 = 0.954$. This

Table 6.4 The performance of the four indicators of particle importance on removed vital particles in 7 tests

Test number	Test condition	Degree	H-index	Coreness	Cycle Ratio
Test 1	5–0.6–3	3	10	19	6
Test 2	6–0.6–3	3	4	7	6
Test 3	7–0.6–3	2	4	2	2
Test 4	6–0.7–3	3	10	20	5
Test 5	6–0.8–3	3	6	4	4
Test 6	6–0.6–4	2	2	6	2
Test 7	6–0.6–5	3	9	7	3

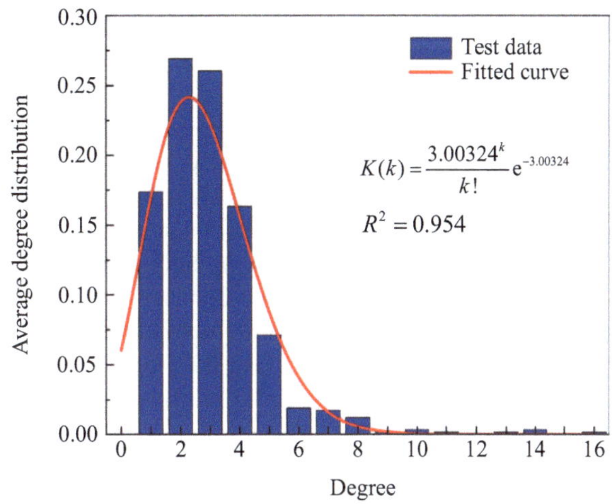

Fig. 6.16 Average degree distribution and Poisson fitting of Tests 1–7

finding suggests that the contact network analyzed in this research exhibits characteristics typical of a random network. Approximately 75% of particle degrees fall between 1 and 4, while the remaining 25% exceed a degree of 4, indicating significant connectivity and a short average path length within the random network structure. According to complex network theory, random networks are susceptible to targeted attacks, whereas small-world networks are characterized by high clustering and shorter path lengths. Furthermore, scale-free networks demonstrate a Degree distribution consistent with a power law. Therefore, this research effectively capitalizes on the vulnerabilities inherent in random networks, which can rapidly compromise the stability and connections among the nodes. Consequently, employing a 'premeditated attack strategy' serves as an efficient means to identify critical nodes. It is important to emphasize that this study focuses on investigating the properties of essential nodes within an undirected network.

The Pearson correlation coefficient technique [70] was used to analyze the relationships among four metrics: Degree, H-index, Coreness, and Cycle ratio. This technique evaluates the linear association between two variables, producing outcome values ranging from − 1 to 1. A positive value reflects a positive correlation, a negative value represents a negative correlation, and a value of 0 denotes the absence of a linear correlation. Figure 6.18 displays the correlation matrix for the four metrics related to particle significance. It is apparent that the correlation coefficients between Coreness and the other three metrics are nearly 0, while the correlation coefficients for H-index with Degree and Cycle ratio exceed 0.72, indicating a relatively strong positive association among these three metrics. This suggests that, compared to Coreness, the H-index, Degree, and Cycle ratio are more likely to exhibit significant similarity in their capacity to identify vital particles and assess particle significance.

Table 6.4 presents the performance metrics of four indicators pertinent to the significance of vital particles removed across seven tests. In Table 6.4, columns 3–6 detail the quantity of vital particles eliminated in a specific order based on their importance, as determined by the four indicators utilizing the 'premeditated attack method' until an instability event occurs within the local system of caved ore and rock hang-ups. The data analysis from Table 6.4 and Fig. 6.17 indicates that Coreness is responsible for the highest total removal of vital particles, followed by the H-index, while the Degree and Cyclic ratio indicators yield comparable results, although Degree demonstrates slightly superior performance. Generally, the removal of approximately three vital particles can lead to instability, with particles identified by Degree playing a crucial role in facilitating hang-up removals. For instance, in Test 1, the network visualization depicted in Fig. 6.18 illustrates graphical representations of the contact network, highlighting the vital particles removed in blue. Notably, particles identified using the Degree metric exhibit stronger connections and clustering compared to those identified by other metrics, suggesting a greater impact on the removal of hang-ups. Furthermore, Degree emerges as a less complex and more economical identification method. In summary, Degree is recognized as the most effective index for pinpointing vital particles in the hang-up system of caved ore and rock within the parameters examined in this study. In comparison to the other three indicators, the Degree metric displays reduced complexity and lower computational costs. Therefore, within the analyzed parameter range of this study, Degree is considered the optimal indicator for determining the hang-up of vital particles. As a result, the subsequent analysis of vital particle features will be based on the rankings established by the Degree metric.

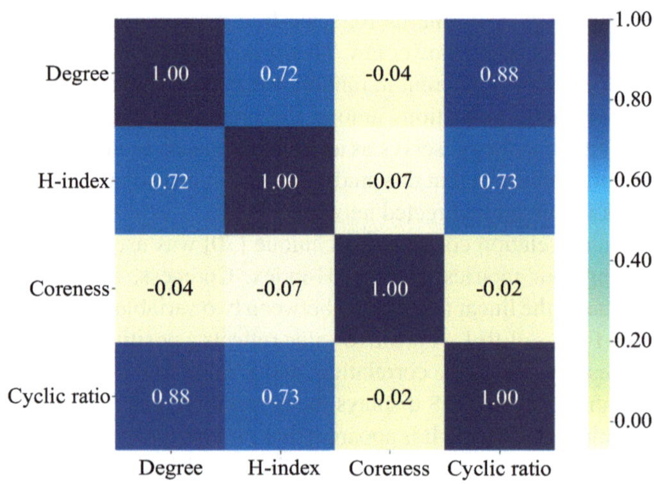

Fig. 6.17 Correlation matrix for the four indicators of particle importance

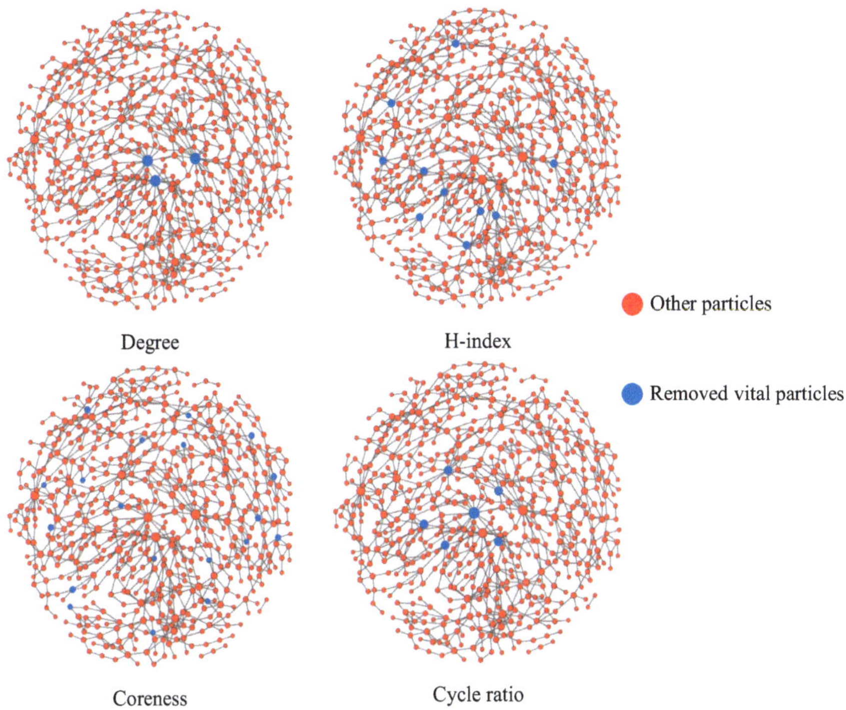

Fig. 6.18 The graph representations of the contact network and removed vital particles in Test 1

6.2.3.2 Size Distribution Characteristics of Vital Particles

The particles were organized based on Degree across seven numerical models, and their sizes were standardized. Figure 6.19 illustrates the size distributions of the top 100 vital particles ranked by Degree in Tests 1–7. In Fig. 6.19, the *Y*-axis represents the ratio of the average size D_d of the top N_d vital particles compared to the average size d of all particles within the system. The light gray region indicates the count of vital particles eliminated during hang-up instability. The analysis presented in Fig. 6.19 shows that in hang-up systems influenced by various factors, the average size of vital particles decreases following a power function pattern as N_d increases. It was observed that the correlation coefficient R^2 is at least 0.981. Moreover, during instances of hang-up instability, the size of removed vital particles exceeds 2.6 times the average particle size. Notably, as R^2 varies from approximately 10 to 20, there is a marked reduction in the size of vital particles by roughly 2.2 to 2.6 times the average particle size. However, as the N_d parameter increases from 20 to 100, the rate at which vital particle size decreases becomes increasingly gradual. At an N_d value of 100, the average size of vital particles remains approximately 1.8 times greater than the average particle size observed in the system, underscoring their importance for stability. In conclusion, it is evident that there is a notable correlation between the

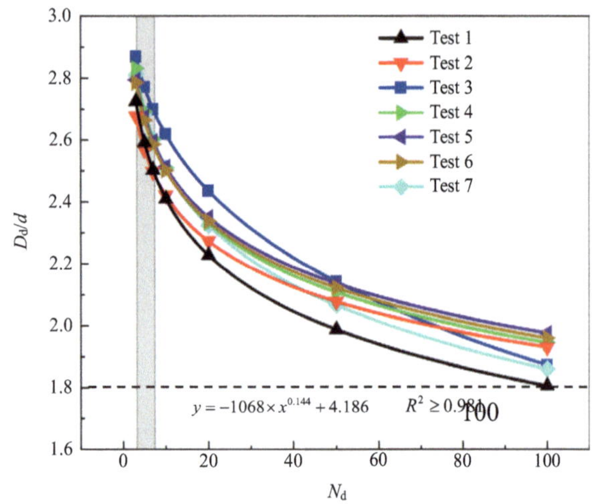

Fig. 6.19 Size distributions of the top 100 vital particles ranked by degree in Tests 1–7

importance ranking of caved ore and rock derived from Degree and the particle size ranking of ore and rock within the packing system, particularly when considering various factors that influence the hang-up systems of caved ore and rock. Generally, larger caved ore and rock play a more vital role in maintaining the stability of local hang-ups. Thus, it is reasonable to assert that effective management of caved ore and rock particle size is vital in practical caving operations, particularly for controlling the incidence of larger caved ore and rock, which is essential for minimizing both the frequency and intensity of hang-ups. Viewed from another angle, emphasizing the 'treatment' of larger caved ore and rock can lead to more efficient resolutions of hang-up incidents.

6.2.3.3 Spatial Distribution Characteristics of Vital Particles

An examination of the essential particle size distribution characteristics led to a further exploration of their spatial distribution patterns. Figure 6.20 depicts the spatial arrangements of essential particles categorized by Degree in Tests 1–7 for N_d values of 3, 5, 7, 10, 20, 50, and 100. The trends in spatial distribution and aggregation of these particles consistently exhibit an arch shape across varying conditions. Notably, when concentrating on a limited number of higher-ranking essential particles (such as $N_d = 3, 5, 7$, and 10), larger particles are generally observed within the 7*d* to 13*d* range from the vertical height of the drawpoint, contributing to the formation of the arch structure. As the quantity of essential particles increases (for example, $N_d = 20$ and 50), a greater number of particles gradually aggregate at the bottom, waist, and top of the arch. Ultimately, at $N_d = 100$, essential particles form a distinct arch shape within a vertical distribution range of approximately 0 to 20*d* from the drawpoint and − 15 to 15*d* horizontally from the center of the drawpoint. The formation of this arch structure, characterized by multiple large particles, serves as a critical indicator of hang-up development.

As illustrated in Fig. 6.21a, the width of the arch, denoted as A_w, is defined as the distance between the centers of the outermost vital particles located at the base of the arch. The coordinate system is established with the origin placed at the center of the drawpoint. In a similar manner, the height of the arch, represented as A_h, is defined as the distance from the centers of the outermost vital particles at the apex of the arch to the same coordinate origin. To better understand the geometrical characteristics of the arch structure, average measurements of both height and width were calculated from observations made in the front and side views. A thorough statistical analysis of these measurements facilitated the creation of an arch geometry histogram, depicted in Fig. 6.21b, which summarizes the findings from Tests 1 through 7. The analysis revealed that the arch height and width, attributed to the accumulation of vital particles across the seven tests, ranged between 17 and 25d for height and between 27 and 33d for width. The data indicated a notable trend: as the ratio of drawpoint size to average particle size, denoted as D/d, was increased across Tests 1, 2, and 3, there was a corresponding increase in the arch width of 22.7% and an 18.0% increase in the arch height. This correlation suggests that a larger ratio results in an increase in the geometric dimensions of the arch, which may ultimately compromise its stability and reduce the extent to which elements become hung up within the packing system of caved ore and rock. Conversely, when the particle friction coefficient, labeled as μ, was increased during Tests 2, 4, and 5, the resulting arch height exhibited negligible

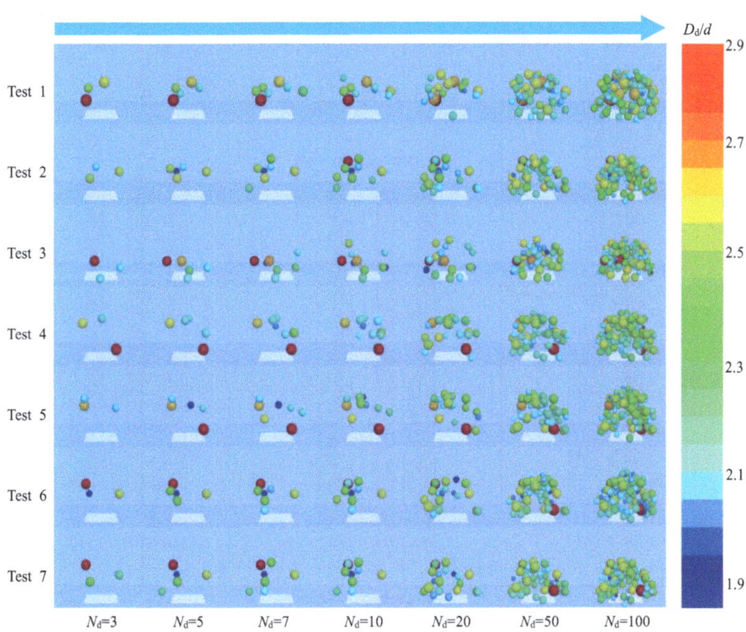

Fig. 6.20 Spatial distributions of the top 100 vital particles ranked by Degree in Tests 1–7

variations, while the arch width experienced a slight decline of 4.6%. Furthermore, an increase in overburden stress, represented as *p*, during Tests 2, 6, and 7 led to minimal changes in the arch structure's geometry. In conclusion, analysis within the studied parameters demonstrates that the primary factor influencing the geometry of the arch is the ratio of drawpoint size to average particle size, *D/d*. This is followed by the particle friction coefficient μ, while the overburden stress *p*, appears to have the least impact. These findings are consistent with previous research on the sensitivity analysis of hang-up frequency in relation to these variables. They indicate that the geometric dimensions of the arch formed by vital particles serve as reliable indicators of both hang-up frequency and extent.

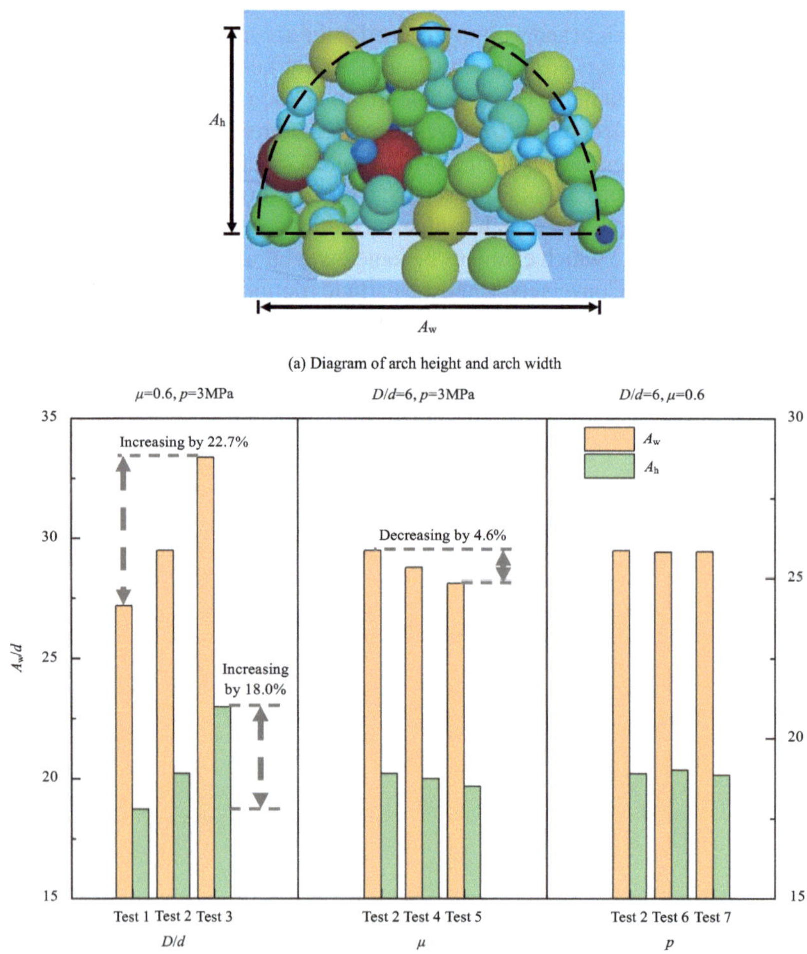

Fig. 6.21 Position distribution and dimensional characteristics of the arch structure in Tests 1–7

This research investigates the spatial characteristics of the arch structure formed by vital particles within the hang-up system created by caved ore and rock. Test 1, with $N_{\mathrm{d}} = 50$, serves as a representative case, with Fig. 6.22a, b illustrating both the bottom view and the vertical section of the drawpoint located at the base of the model. The findings indicate that the particles closest to the drawpoint predominantly consist of non-vital particles, characterized by lower Degrees, specifically identified as red particles. Notably, the arch structure composed of vital particles is not confined to the outermost layer of the hang-up system involving caved ore and rock; rather, it is located within the interior region, approximately 1.5–2.0 times the distance from the outermost layer of non-vital particles. This observation aligns with natural phenomena, particularly seen in the construction of ant tunnels. In these tunnels, the vital arch is strategically positioned above the tunnel surface to enhance overall structural stability. Ants demonstrate an adaptive approach by removing low-force grains from the tunnel surface, as these grains do not contribute meaningfully to the overall integrity of the stress arch and are instead supported by normal forces. Due to the lower normal forces acting on these grains, it requires less external force to overcome the friction among them, thereby facilitating the easier removal of non-vital grains [71]. The implications of the analysis conducted are significant for managing hang-up occurrences in real-world caving mines. A noteworthy conclusion is that it may be more advantageous to focus treatment efforts on caved ore and rock that lies at a distance from the outermost particles, rather than concentrating only on the outermost layer itself. This strategic shift in focus could lead to more effective mitigation of hang-ups, ultimately resulting in a more efficient process for addressing instability issues associated with hang-ups in caving mines.

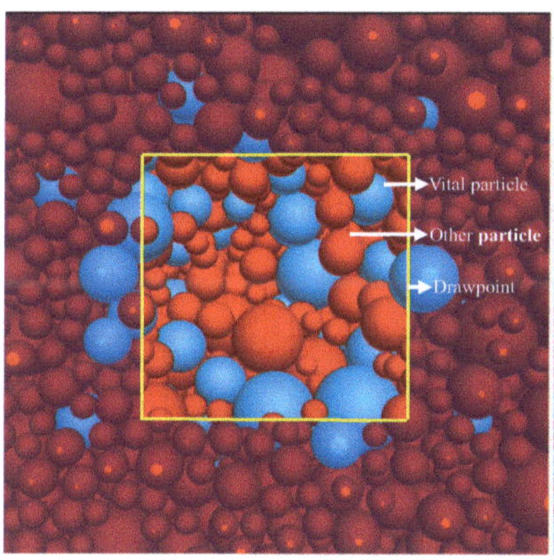

(a) Bottom view

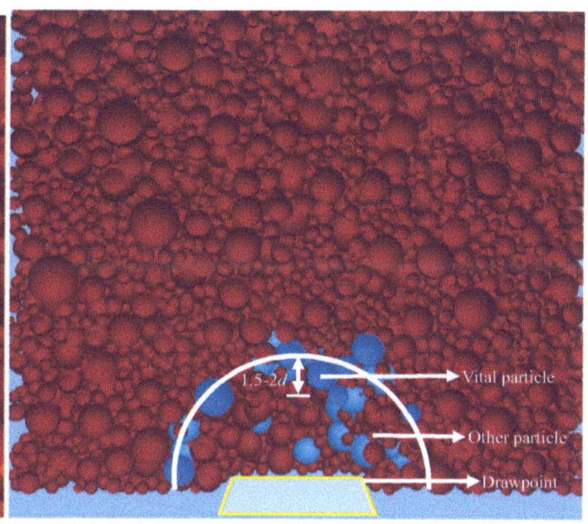

(b) Vertical section

Fig. 6.22 Spatial location characteristics of arch structure in the hang-up system of caved ore and rock in Test 1

6.2.3.4 Clustering and Connectivity of Vital Particles

The Clustering coefficient [66, 72] and the Shortest path [66, 73] are important indicators used to describe the short-range and medium-range topological characteristics of complex networks.

The Clustering coefficient evaluates the interconnectedness of neighboring nodes around a specific node in the network, providing valuable insights into the short-range properties of essential particles. A higher Clustering coefficient for a node indicates stronger connections with its neighboring nodes, resulting in increased local clustering throughout the network. This characteristic is particularly significant for the local hang-up systems of caved ore and rock, as a higher Clustering coefficient for a particle signifies enhanced mechanical stability within its immediate environment. The formula for calculating the Clustering coefficient C_i is presented in Eq. (6.5):

$$C_{\mathrm{i}} = \frac{2L_{\mathrm{i}}}{\mathrm{k}_i(k_{\mathrm{i}} - 1)} \tag{6.5}$$

Here, L_i is the number of connections between neighbors of node i, and k_i is the Degree of node i.

The Clustering coefficient C_i can also be defined as the ratio of the number of closed triangles to the number of connected triples in a complex network [66]. Taking Node 2 in Fig. 6.23 as an example, its neighboring nodes include Nodes 1, 4, and 7. Among these, only Node 1 is connected to Node 4, and Node 4 is connected to Node 7, resulting in a Clustering coefficient of $C_2 = 2/3$. The Clustering coefficients of the remaining nodes are also calculated and displayed in Fig. 6.23.

The mean Clustering coefficient of vital particles within the hang-up system acts as a measure of the intensity of short-range interactions between these particles, providing insight into the degree of hang-up. Figure 6.24 illustrates the connection between the mean Clustering coefficient of vital particles and the number N_{d}, taking into account three influencing factors. It is important to note that when the number of vital particles is low (e.g., $N_{\mathrm{d}} \leq 10$), their spatial distributions are quite dispersed,

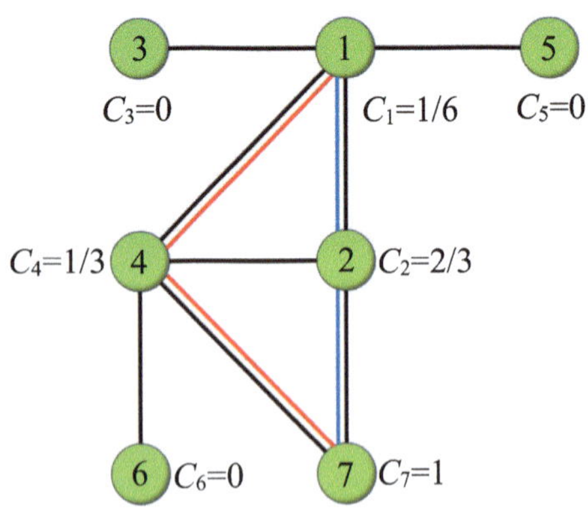

Fig. 6.23 Schematic diagram of clustering coefficient and shortest path

leading to significant variations in the mean Clustering coefficient that lack statistical relevance. Consequently, the range of vital particles N_d represented in Fig. 6.24 extends from 10 to 100.

The average Clustering coefficient exhibited significant fluctuations in the light blue region of Fig. 6.24a when the count of vital particles was low (e.g., $N_d \le 20$). In contrast, a generally upward trend was noted in the light-yellow area of Fig. 6.24a as the number of vital particles increased (e.g., $N_d \ge 20$) along with the rise in the ratio of drawpoint size to the average particle size D/d. Test 1 recorded the highest average clustering coefficient for vital particles, followed by Test 2, and then Test 3. For example, at $N_d = 100$, the average clustering coefficient of vital particles in Test 2 was approximately 10.3% and 27.6% lower than that of Test 1 and Test 3,

Fig. 6.24 Relationship between average clustering coefficient and number of vital particles under the influence of three factors

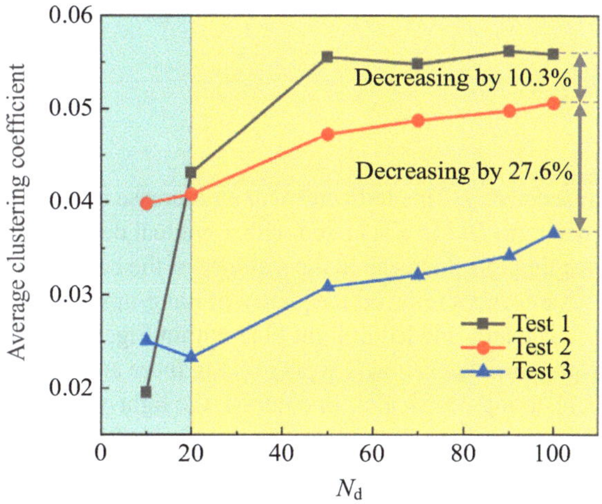

(a) D/d

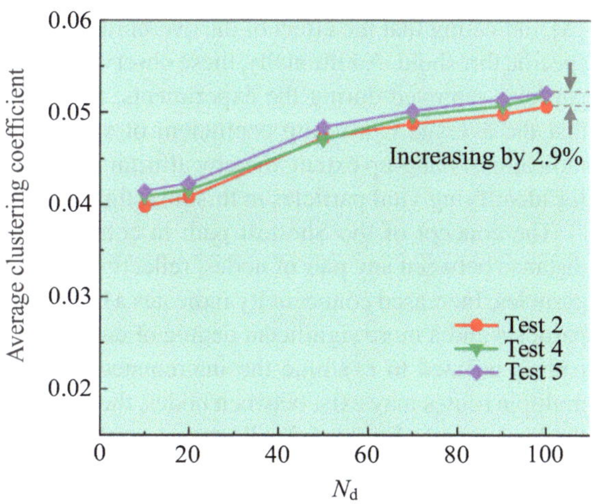

(b) μ

Fig. 6.24 (continued)

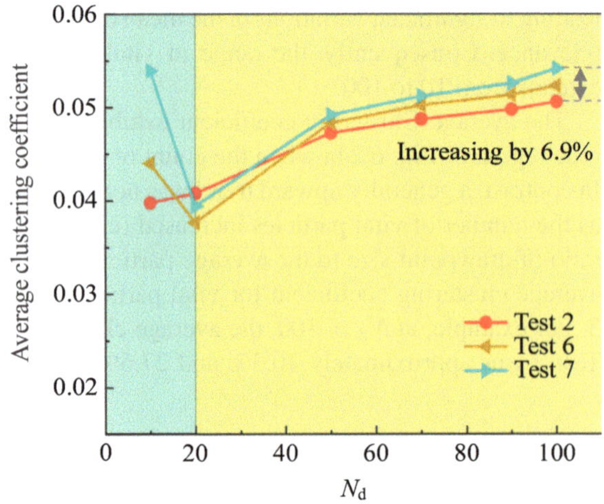

(c) p

respectively. This indicates that an increase in the drawpoint size to average particle size ratio D/d is associated with a gradual decline in the clustering of vital particles, leading to a decrease in the stability of the contact network within the local hang-up system and a reduced frequency of hang-ups.

The area shaded in light blue within Fig. 6.24c illustrates that when the count of vital particles is low (e.g., $N_d \leq 20$), there are significant fluctuations in the average Clustering coefficient. In contrast, the light-yellow area in Fig. 6.24c highlights that as the number of vital particles increases (e.g., $N_d \geq 20$), the three curves typically demonstrate an upward trend with only minor variations. Notably, at $N_d = 100$, the average differences in Clustering coefficient among the three tests approximate 3.5%. These findings align with the previous research conducted by Castro [14, 15], indicating that the effect of the overburden stress p stabilizes once it surpasses a specific threshold. Additionally, these observations correlate with the number of vital particles removed during the experiments, as outlined in Table 6.4. This suggests that the average Clustering coefficient of vital particles can serve as an effective indicator of hang-up extent, thereby affirming the Degree-based approach employed for identifying vital particles in this investigation.

The concept of the Shortest path in complex networks refers to the minimum distance between any pair of nodes, reflecting the interconnectivity among essential particles. Increased connectivity indicates a higher level of interaction between these particles and a more significant degree of entanglement. Consequently, the Shortest path is utilized to examine the interconnectivity among vital particles. Although multiple routes may exist between nodes, the Shortest path is defined as the one with the least length. Figure 6.23 illustrates a schematic representation, while Eq. (6.6) provides the formula for calculating the average Shortest path. Considering Nodes 1 and 7, as depicted in Fig. 6.23, there are two Shortest paths available: 1–4–7 (indicated by the red edge) and 1–2–7 (shown by the blue edge), each having a path length of $d_{17} = 3$.

$$< d >= \frac{2}{N(N-1)} \sum_{i \geq j} d_{ij} \quad (6.6)$$

Here, $< d >$ is the average Shortest path length of the network, N is the total number of nodes, d_{ij} is the Shortest path length between nodes i and j.

The typical shortest distance of vital particles within a localized hang-up system involving caved ore and rock can illuminate the degree of medium-range interactions among these particles, thereby offering insights into the level of hang-up. Figure 6.25 illustrates the correlation between the average shortest path of vital particles and the number Nd, taking into account three different factors. Just like the average Clustering coefficient, when the count of vital particles is low (for instance, $N_{\text{d}} \leq 10$), their spatial arrangements tend to be scattered, leading to an average shortest path that lacks statistical significance. Consequently, the count of vital particles N_{d} in Fig. 6.25 varies from 10 to 100.

As illustrated in the light blue region of Fig. 6.25a, all three curves show a marked upward trend as the number of vital particles N_{d} increases. The same figure indicates that these three curves reveal a significant upward progression correlated with the rise in vital particle count N_{d}. As N_{d} transitions from 50 to 100, the curves display a gradual reduction in steepness, implying a stabilization in the interconnections among the vital particles. Test 3 demonstrates the highest average shortest path of vital particles, followed by Test 2, while Test 1 has the lowest average. For $N_{\text{d}} = 100$, the average lengths of shortest paths in Tests 1 through 3 settle around 7.18, 8.23, and 9.64, respectively. The average lengths of shortest paths in Test 2 are approximately 17.1% and 12.8% longer than those in Tests 1 and 3, respectively. This suggests that an increase in the ratio of drawpoint size to average particle size *D/d* leads to longer shortest paths among vital particles, which ultimately diminishes connectivity and stability within the local hang-up system. Consequently, there is a decrease in the extent of hang-up.

The average shortest path exhibits significant variation when the count of vital particles is low (for instance, $N_{\text{d}} \leq 20$), as demonstrated in the light blue region of Fig. 6.25b. Conversely, in the light-yellow area of Fig. 6.25b, where the count of vital particles is elevated (e.g., $N_{\text{d}} \geq 20$), the three curves display minimal alterations. The average shortest path for vital particles across the three experiments stays relatively stable, fluctuating between 7.20 and 8.33. In Test 4, this average is approximately 4.2% shorter compared to Test 5. Additionally, the effect of the particle friction coefficient on the average shortest path aligns with the average Clustering coefficient of the vital particles, as illustrated in Fig. 6.24b.

The three curves presented in Fig. 6.25c illustrate a gradual upward trend as the quantity of vital particles N_{d} increases. More specifically, within the light-yellow area of Fig. 6.25c, the smoothness of these curves enhances as N_{d} escalates from 50 to 100, signifying a stabilization in the interconnectivity of vital particles. While the average shortest path of vital particles in Test 7 is marginally elevated compared

Fig. 6.25 Relationship between average shortest path and number of vital particles under the influence of three factors

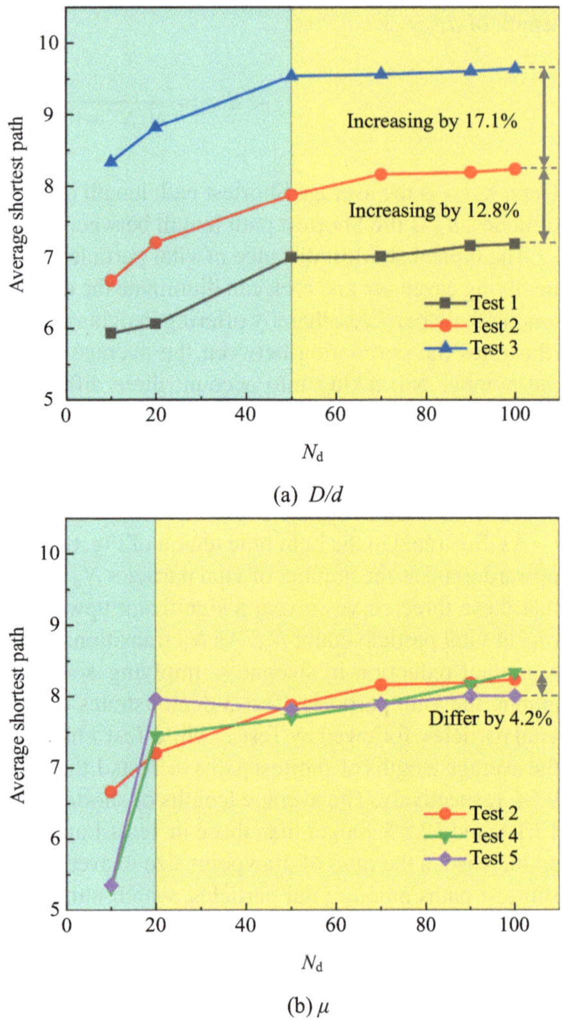

to that in Test 6 and Test 2, the observed differences are not deemed statistically significant. At $N_d = 100$, the average shortest path of the vital particles varies between 7.88 and 8.75 throughout the three tests. There is a 6.9% variance in the average shortest path of vital particles when comparing Test 2 with Test 7. The correlation between overburden stress and the average shortest path is consistent with the average Clustering coefficient of vital particles illustrated in Fig. 6.24c.

The Clustering coefficient, as shown in Figs. 6.24 and 6.25, represents the degree of clustering among neighboring vital particles, while the shortest path reflects the connectivity between any two vital particles. Both metrics complement each other in illustrating how different factors influence the occurrence of hang-ups, particularly in short to medium ranges. Prior research [74, 75] has established that coarse particles

Fig. 6.25 (continued)

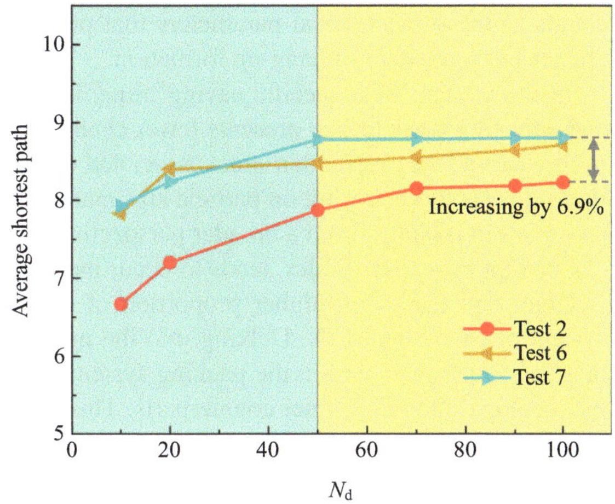

(c) *p*

serve as a vital component in the contact force chain network, facilitating significant bearing capacity and stress transfer functions. Conversely, finer particles enhance this network. This research categorizes larger particles as essential based on their degree. From a meso-mechanical perspective, factors such as the ratio of drawpoint size to average particle size, the friction coefficient of particles, and the stress from overburden can be quantitatively described by the geometry and spatial distribution of these vital particles, as well as the strength of their interactions. Therefore, in caving mines, it is imperative to avoid forming large aggregates of coarse particles during the draw process to minimize hang-up incidents, enhance draw efficiency, and improve emergency management. The clustering and connectivity of vital particles within the network of caved ore and rock provide a quantitative assessment of hang-up severity and serve as reliable predictors of this issue.

6.3 Conclusions

For the numerical study of the influence of multiple parameters on hang-ups, the main conclusions and discussions are as follows:

This research employs relevant structural and mechanical analysis techniques within the domain of computational mechanics for granular materials to examine the characteristics associated with the hang-up of caved rock and ore, considering various influencing factors at both macroscopic and mesoscopic dimensions. Unlike earlier studies that primarily focused on the phenomenological aspects of hang-up characteristics, this research emphasizes the significance of the ratio between drawpoint size and average particle size as a macroscopic variable, the force chain as a meso-mechanical structure, and the quantity and average strength of the force chains as meso-mechanical parameters that play a crucial role in influencing the challenges and extent of hang-up formation.

In the context of a specific caving mine, it has been observed that managing and adjusting particle size presents fewer challenges than controlling the frictional properties and the overburden stress associated with caved ore and rock. This distinction suggests that focusing on particle size management can streamline operational processes in mining. From a broader perspective, the formation of hang-ups, which can disrupt mining activities, tends to occur more frequently when the particle size is larger and there is a higher proportion of coarse particles within the packing system of caved materials. Delving into the mesoscopic level, it becomes evident that coarse particles within the packing system exhibit greater resistance to movement compared to their finer counterparts. This resistance is primarily attributed to the anchoring effect exerted by coarse particles, which enables them to establish a more interconnected and robust network of contact forces. These coarse particles function as force chain attractors, stabilizing the local packing arrangement of caved ore and rock, thereby complicating the movement and flow of these materials. Given these dynamics, it is imperative for researchers and professionals in the caving industry to intensify their investigations into the behavior and characteristics of coarse particles within the particle packing system [74, 76]. Specifically, the static control exerted by these coarse particles and the dynamic evolution mechanisms of force chains that include multiple coarse particles in the local contact force network warrant further study. This is essential, particularly above the drawpoint, where the interactions become critical. Advanced methodologies such as computational mechanics of granular materials [77], statistical physics [78, 79], and complex network theory [80, 81] must be employed to explore these phenomena comprehensively. To fully understand the intricate transformation mechanisms that occur during the transition between the hang-up and flow phases of the local particle system, it is vital to elucidate the differing roles of coarse and fine particles. Moreover, a thorough investigation into the interaction mechanisms involving coarse particles is necessary, alongside an exploration of the geometric, structural, and mechanical properties of what can be termed 'vital caved ore and rock' and 'vital force chains'. Gaining such insights will facilitate more precise control over the size of caved ore and rock and the parameters within the stope, ultimately leading to more effective solutions for managing hang-up issues encountered in caving mines.

This research conducted isolated numerical draw tests to investigate the influence of various macro factors, including particle size, friction properties, and overburden stress, on the hang-up and flowability of caved ore and rock. Among the findings is the development of a 3D phase diagram illustrating hang-ups, which can be utilized in cave mining applications, such as the design of drawpoint layouts and draw strategies. Furthermore, the study evaluated the meso-structural and mechanical properties of the local packing system of caved ore and rock under varying conditions of hang-up and flow. It also analyzed the relationships and sensitivities among macro-influencing factors, meso-structural parameters, and meso-mechanical properties [44]. The key conclusions are as follows:

(1) The ratio of the drawpoint size to the average particle size *D/d* plays a significant role in the frequency of hang-up events, which diminishes as this ratio increases. In contrast, a rise in the particle friction coefficient μ and overburden stress p leads to a higher occurrence of hang-ups. The contribution of these factors to the incidence of hang-ups is quantified as 54.562% for *D/d*, 33.569% for μ, and 11.869% for p, respectively. Notably, the frequency of hang-up events is most influenced by the *D/d* ratio, followed by μ, with p having the least effect. Therefore, it is essential for block caving mines to focus on optimizing drawpoint layouts and assessing the cavability of the rock mass to effectively minimize the frequency of hang-up events.
(2) In the parameter values examined in this study, the effects of the drawpoint size to average particle size ratio *D/d* and overburden stress p on the local structural characteristics of caved ore and the rock situated above the drawpoint are minimal, with the maximum changes in void ratio v, average anisotropy coefficient $< \beta >$, and bond orientational order Q_6 being below 5%. In contrast, a rise in the friction coefficient of the caved ore and rock results in an increase in both the void ratio and the overall structural anisotropy of the local particle packing system of caved ore and rock.
(3) Under varying conditions related to the ratio of drawpoint size to the average particle size *D/d*, the particle friction coefficient μ, and the overburden stress p, there is no significant change in the average coordination number. To comprehend the mechanism behind the formation of hang-ups, it is essential to examine the medium-range mechanical structure of force chains rather than focusing solely on the short-range interactions between particles. As the ratio of drawpoint size to average particle size increases, the number of force chains in the vicinity above the drawpoint rises; however, this number diminishes considerably with higher particle friction coefficients and increased overburden stress. Conversely, while the ratio of drawpoint size to average particle size increases, the average strength of the force chains decreases. In contrast, an increase in the particle friction coefficient and overburden stress enhances the average strength of these chains. Notably, the relationship between the number of force chains and their average strength is inversely correlated.
(4) The proportion of drawpoint size to the mean particle size *D/d* significantly impacts hang-ups by affecting the average strength of force chains σ_p. Additionally, the particle friction coefficient μ predominantly influences hang-ups by altering the collimation coefficient δ of the force chain and the void ratio v. Furthermore, the overburden stress p primarily affects hang-ups by its effect on both the average force chain strength σ_p and the quantity of force chains n. The characteristics and effects of hang-ups are predominantly determined by the local mechanical properties of the caved ore and rock, as well as their evolution, rather than by the structural attributes of the local particle packing system. Essential meso-mechanical factors influencing hang-ups include the average strength of force chains and the number of force chains within the local particle packing system of caved ore and rock.

Additionally, this study investigates the identification and characterization of vital particles within the hang-up system of caved ore and rock. Initially, complex network analysis was employed to determine the optimal identification index for vital particles in the particle packing system of caved ore and rock across various hang-up states. Subsequently, the 'premeditated attack method' was utilized to rank the significance of caved ore and rock, while also examining the geometry, spatial distribution, clustering, and connectivity of these particles under diverse influencing factors [82]. The primary conclusions are as follows:

(1) The contact network of caved ore and rock within the hang-up system follows a random network model, characterized by an average Degree distribution that resembles a Poisson distribution. Approximately 75% of particles exhibit Degrees ranging from 1 to 4, while the remaining 25% possess degrees exceeding 4.
(2) The correlation coefficients among Degree, H-index, and Cycle ratio all exceed 0.72, indicating a strong positive correlation among these three factors. However, no significant correlation exists between Coreness and the aforementioned indicators. In instances of instability within the local hang-up system, Degree is regarded as the preferred index for vital particle identification due to its lower complexity and calculation cost, as well as its identification of the fewest vital particles.
(3) Regarding the characteristics of particle size, the hang-up systems of caved ore and rock are influenced by various factors. The average size of vital particles diminishes in a power function manner as the number increases in descending order. When the hang-up is removed, the average size of the discharged vital particles is over 2.6 times greater than the average particle size of the system. Even at a count of 100 vital particles, their average size can still be about 1.8 times larger than that of the system's average. Spatially, critical particles generally cluster together to create an arch formation, measuring approximately 17–25*d* in height and 27–33*d* in width. Notably, the arch structure formed by vital particles is not confined to the outermost hang-up system of caved ore and rock; it also penetrates into the inner zones of the outermost non-vital particles by a distance of 1.5–2.0*d*.
(4) The analysis of the Clustering coefficient and Shortest path provides significant insights into the factors influencing hang-up occurrences at both short and medium distances. An increase in the number of vital particles enhances clustering and connectivity, which, in turn, contributes to greater overall stability. Conversely, a reduction in the ratio of drawpoint size to the average particle size, coupled with an increase in the friction coefficient of particles and overburden stress, ultimately reinforces the stability of the local hang-up system. This research establishes a theoretical foundation for further investigations based on complex networks by directly informing effective strategies for mitigating hang-ups. Our findings suggest that managing the size distribution of caved ore and rock, particularly through the targeted removal of larger particles, is an exceptionally effective approach to addressing hang-up issues.

Note. Figures 6.1, 6.2, 6.3, 6.4, 6.5, 6.6, 6.7, 6.8, 6.9, 6.10, 6.11 and 6.12 and Tables 6.1, 6.2 in Sect. 6.1 are reprinted from the authors' previously published work [44]:

Sun H, Zhao L, Wei L, et al. Numerical study of the influence of multiple parameters on hang-ups: insight from a structural and mechanical characteristics analysis [J]. Rock Mechanics and Rock Engineering, 2024, 57: 4073–4087.

Figures 6.13, 6.14, 6.15, 6.16, 6.17, 6.18, 6.19, 6.20, 6.21, 6.22, 6.23, 6.24 and 6.25 and Tables 6.3, 6.4 in Sect. 6.2 are reprinted from the authors' previously published work [82]:

Sun H, Zhou S G, Zhao L S, et al. Identifying Vital Particles in Hang-Up Systems of Caved Ore and Rock via Complex Networks [J]. Rock Mechanics and Rock Engineering, 2025.

References

1. Brown ET (2007) Block caving geomechanics, 2nd edn. Julius Kruttschnitt Mineral Research Centre, University of Queensland, Brisbane
2. Chitombo GP (2010) Cave mining: 16 years after Laubscher's 1994 paper 'Cave mining a state of the art'. In: Proceedings of the 2nd international symposium on block and sublevel caving, Perth, Australia, pp 45–61
3. Castro R, Trueman R, Halim A (2007) A study of isolated draw zones in block caving mines by means of a large 3D physical model. Int J Rock Mech Min Sci 44(6):860–870
4. Litwiniszyn J (1956) Application of the equation of stochastic processes to mechanics of loose bodies. Arch Mech 8(4):393–411
5. Jin AB, Sun H, Wu SC et al (2017) Confirmation of the upside-down drop shape theory in gravity flow and development of a new empirical equation to calculate the shape. Int J Rock Mech Min Sci 92:91–98
6. Dorador L (2016) Experimental investigation of the effect of broken ore properties on secondary fragmentation during block caving. The University of British Columbia, Vancouver
7. Sun H, Gao YT, Elmo D et al (2019) A study of gravity flow based on the upside-down drop shape theory and considering rock shape and breakage. Rock Mech Rock Eng 52(3):881–893
8. Sun H, Jin AB, Davide E et al (2020) A numerical based approach to calculate ore dilution rates using rolling resistance model and upside-down drop shape theory. Rock Mech Rock Eng 53(10):4639–4652
9. Castro R, Arancibia L, Gomez R (2022) Quantifying fines migration in block caving through 3D experiments. Int J Rock Mech Min Sci 151:105033
10. Trueman R, Castro R, Halim A (2008) Study of multiple draw-zone interaction in block caving mines by means of a large 3D physical model. Int J Rock Mech Min Sci 45:1044–1051
11. Castro R, Pineda M (2015) The role of gravity flow in the design and planning of large sublevel stopes. J South Afr Inst Min Metall 115(2):113–118
12. Kvapil R (1965) Gravity flow of granular material in hoppers and bins Part 2. Int J Rock Mech Min Sci 2:277–304
13. Hadjigeorgiou J, Lessard JF (2010) Strategies for restoring material flow in ore and waste pass systems. Int J Min Reclam Environ 24:267–282
14. Castro RL, Fuenzalida MA, Lund F (2014) Experimental study of gravity flow under confined conditions. Int J Rock Mech Min Sci 67:164–169
15. Castro RL, Gómez R, Hekmat A (2016) Experimental quantification of hang-up for block caving applications. Int J Rock Mech Min Sci 85:1–9

16. Lozano C, Janda A, Garcimartin A et al (2012) Flow and clogging in a silo with an obstacle above the orifice. Phys Rev E 86:03136
17. Lozano C, Zuriguel I, Garcimartin A (2015) Stability of clogging arches in a silo submitted to vertical vibrations. Phys Rev E 91:062203
18. Gella D, Maza D, Zuriguel I (2017) Linking bottleneck clogging with flow kinematics in granular materials: the role of silo width. Phys Rev Fluids 2:084304
19. Fang YG, Gou LF, Hou MX (2020) Arching effect analysis of granular media based on force chain visualization. Powder Technol 363:621–628
20. Park KH, Baek SH, Jung YH (2020) Investigation of arch structure of granular assembly in the trapdoor test using digital RGB photoelastic analysis. Powder Technol 366:560–570
21. Meng Y, Zhu HJ, Chung YK et al (2018) Effect of coefficient of friction on arch network in shearing process under low confinement. Powder Technol 335:1–10
22. Bao N, Wei J, Chen JF et al (2022) Interaction of soil arching under tapdoor condition: Insights from 2D discrete-element analysis. Int J Geomech 22(6):04022066
23. Hadjigeorgiou J, Lessard JF (2007) Numerical investigations of ore pass hang-up phenomena. Int J Rock Mech Min Sci 44:820–834
24. Vo T, Yang HW, Russell AR (2016) Cohesion and suction induced hang-up in ore passes. Int J Rock Mech Min Sci 87:113–128
25. Hekmatnejad A, Crespin B, Pan PZ et al (2021) A hybrid approach to predict hang-up frequency in real scale block cave mining at El Teniente mine, Chile. Tunn Undergr Space Technol 118:104160
26. Wensrich CM, Katterfeld A (2012) Rolling friction as a technique for modelling particle shape in DEM. Powder Technol 217:409–417
27. Jiang MJ, Shen ZF, Wang J (2015) A novel three-dimensional contact model for granulates incorporating rolling and twisting resistances. Comput Geotech 65:147–163
28. ITASCA CONSULTING GROUP (2016) PFC3D (particle flow code in 3 dimensions) user's guide in PFC2D version 5.0. Itasca Consulting Group Inc., Minneapolis
29. Wei H, Tang XJ, Ge Y et al (2019) Numerical and experimental studies of the effect of iron ore particle shape on repose angle and porosity of a heap. Powder Technol 353:526–534
30. Horabik J, Parafiniuk P, Wiacek J et al (2022) DEM modelling of the influence of initial stress state on the discharge rate of spherical particles from a model silo. Powder Technol 403:117402
31. Nedderman RM (1992) The method of differential slices: statics and kinematics of granular material. Cambridge University, Cambridge
32. Castro R (2006) Study of the mechanisms of gravity flow for block caving. University of Queensland
33. Klymenko OV, Kucherenko S, Shah N (2017) Constrained global sensitivity analysis: sobol' indices for problems in non-rectangular domains. Comput Aided Chem Eng 40:151–156
34. Wilkinson L (2006) Revising the Pareto chart. Am Stat 60(4):332–334
35. Schröder-Turk GE, Mickel W, Kapfer SC et al (2011) Minkowski tensor shape analysis of cellular, granular and porous structures. Adv Mater 23(22–23):2535–2553
36. Rycroft CH (2009) Voro++: a three-dimensional Voronoi cell library in C++. Lawrence Berkeley National Laboratory
37. Zhao SW, Zhao JD, Gou N (2020) Universality of internal structure characteristics in granular media under shear. Phys Rev E 101(1):012906
38. Steinhardt PJ, Nelson DR, Ronchetti M (1983) Bond-orientational order in liquids and glasses. Phys Rev B 28(2):784–805
39. Zhang LG, Nguyen NGH, Lambert S et al (2017) The role of force chains in granular materials: from statics to dynamics. Eur J Environ Civil Eng 21(7–8):874–895
40. Papadopoulos L, Porter MA, Daniels KE et al (2018) Network analysis of particles and grains. Journal of Complex Networks 6(4):485–565
41. Walker DM, Tordesillas A (2010) Topological evolution in dense granular materials: a complex networks perspective. Int J Solids Struct 47(5):624–639
42. Robert PB, Bulbul C (2018) The physics of jamming for granular materials: a review. Rep Prog Phys 82(1):012601

43. Sato A, Tang HW (2020) Analysis of ore pass hang-ups in long vertical ore passes by 3-D DEM. Int J Min Eng Min Process 9(1):1–11
44. Sun H, Zhao L, Wei L et al (2024) Numerical study of the influence of multiple parameters on hang-ups: insight from a structural and mechanical characteristics analysis. Rock Mech Rock Eng 57:4073–4087
45. Gómez R, Sanhueza V, Fustos R et al (2022) A new approach for hang-up prediction in block caving mines based on drawbell geometry and vertical stress. Min Metall Expl 39(1):2393–2400
46. Gómez R, Loyola M, Palma S et al (2023) Experimental study of the inrush of fines events in caving mining. Int J Rock Mech Min Sci 169:105436
47. Jia CQ, Hu J, Chen DY et al (2022) Adaptive event-triggered state estimation for a class of stochastic complex networks subject to coding-decoding schemes and missing measurements. Neurocomputing 494:297–307
48. Yu XY, Chen Z, Liu F et al (2023) How urban metro networks grow: from a complex network perspective. Tunn Undergr Space Technol 131:104841
49. Feng JR, Zhao MK, Yu GH et al (2023) Dynamic risk analysis of accidents chain and system protection strategy based on complex network and node structure importance. Reliab Eng Syst Saf 238:109413
50. Bellingeri M, Bevacqua D, Turchetto M et al (2022) Network structure indexes to forecast epidemic spreading in real-world complex networks. Front Phys 10:1017015
51. Henkes S, Quint DA, Fily Y et al (2016) Rigid cluster decomposition reveals criticality in frictional jamming. Phys Rev Lett 116(2):028301
52. Hou SQ, Wang W, Wang ZY et al (2018) Force chain characteristics and effects of a dense granular flow system in a third body interface during the shear dilatancy process. J Appl Mech Tech Phys 59(1):153–162
53. Yu WX, Wang SD, Miao YH (2023) Characterizing force-chain network in aggregate blend using discrete element method and complex network theory. Constr Build Mater 400:132724
54. Shi LW, Xiao X, Wang X et al (2022) Mesostructural characteristics and evaluation of asphalt mixture contact chain complex networks. Constr Build Mater 340:127753
55. Liu GQ (2021) Asphalt mixture main force chains characteristics and skeleton composition force mechanism. Southeast University
56. Liu XF, You SH, Xie CK (2020) Study on instability of clay particle slope accumulation based on complex network. Appl Math Mech 41(9):931–942
57. Arevalo R, Zuriguel I, Trevijano SA et al (2010) Third order loops of contacts in a granular force network. Int J Bifurcat Chaos 20(3):897–903
58. Csermely P, Korcsmáros T, Kiss HJM et al (2013) Structure and dynamics of molecular networks: a novel paradigm of drug discovery: a comprehensive review. Pharmacol Ther 138(3):333–408
59. Kitsak M, Gallos LK, Havlin S et al (2010) Identification of influential spreaders in complex networks. Nat Phys 6:888–893
60. Pei S, Muchnik L, Andrade JS Jr et al (2014) Searching for superspreaders of information in real-world social media. Sci Rep 4:5547
61. Motter AE, Lai YC (2002) Cascade-based attacks on complex networks. Phys Rev E 66(6):065102
62. Normile D (2013) The metropole, superspreaders and other mysteries. Science 339:1272–1273
63. Kim HJ, Kim JM (2005) Cyclic topology in complex networks. Phys Rev E 72:036109
64. Fan TL, Lü LY, Shi DH et al (2021) Characterizing cycle structure in complex networks. Commun Phys 4:272
65. Smart AG, Ottino JM (2008) Evolving loop structure in gradually tilted two-dimensional granular packings. Phys Rev E 77:041307
66. Barabási AL, Albert R (1999) Emergence of scaling in random networks. Science 286(5439):509–512
67. Chen G, Wang X, Li X (2012) Introduction to complex networks: models, structures and dynamics. Higher Education Press, Beijing

68. Hirsch JE (2005) An index to quantify an individual's scientific research output. Proc Natl Acad Sci USA 102(46):16569–16572
69. D'Souza RM, di Bernardo M, Liu YY (2023) Controlling complex networks with complex nodes. Nature Rev Phys 5:250–262
70. Jebli I, Belouadha FZ, Kabbaj MI et al (2012) Prediction of solarenergy guided by pearson correlation using machine learning. Energy 224:120109
71. Buarque DE Macedo R, Andò E, Joy S et al (2021) Unearthing real-time 3D ant tunneling mechanics. In: Proceedings of the National Academy of Sciences 118(36):e2102267118
72. Newman MEJ (2003) Mixing patterns in networks. Phys Rev E 67(2):026126
73. Newman MEJ (2001) The structure of scientific collaboration networks. Proc Natl Acad Sci 98(2):404–409
74. Wang T, Wautiier A, Zhu JG et al (2023) Stabilizing role of coarse grains in cohesionless overfilled binary mixtures: a DEM investigation. Comput Geotech 162:105625
75. Zhang LR, Nguyen NGH, Lambert S et al (2017) The role of force chains in granular materials: from statics to dynamics. Eur J Environ Civ Eng 21(7–8):874–895
76. Voivret C, Radjaï F, Delenne JY et al (2009) Multiscale force networks in highly polydisperse granular media. Phys Rev Lett 102:178001
77. Yang H, Xu WJ, Sun QC et al (2016) Study on the meso-structure development in direct shear tests of a granular material. Powder Technol 314:129–139
78. Bernal JD, Mason J (1960) Packing of spheres: coordination of randomly packed spheres. Nature 188(4754):910–911
79. Binquan K, Yixin C, Jindong L et al (2017) Granular materials flow like complex fluids. Nature 551:360–363
80. Yin H, Austin RB, Jure L (2017) Higher-order clustering in networks. Phys Rev E 97(5):1–11
81. Bao XG, Hu QT, Ji P et al (2022) Impact of basic network motifs on the collective response to perturbations. Nat Commun 13(1):5301
82. Sun H, Zhou SG, Zhao LS et al (2025) Identifying vital particles in hang-up systems of caved ore and rock via complex networks. Rock Mechanics and Rock Engineering

MIX
Papier aus verantwortungsvollen Quellen
Paper from responsible sources
FSC® C105338

If you have any concerns about our products,
you can contact us on
ProductSafety@springernature.com

In case Publisher is established outside the EU,
the EU authorized representative is:
Springer Nature Customer Service Center GmbH
Europaplatz 3, 69115 Heidelberg, Germany

Printed by Libri Plureos GmbH
in Hamburg, Germany